Geodiversity

Geodiversity
valuing and conserving abiotic nature

Murray Gray
Department of Geography,
Queen Mary, University of London

John Wiley & Sons, Ltd

Other Wiley Editorial Offices

John Wiley & Sons Inc., 111 River Street, Hoboken, NJ 07030, USA

Jossey-Bass, 989 Market Street, San Francisco, CA 94103-1741, USA

Wiley-VCH Verlag GmbH, Boschstr. 12, D-69469 Weinheim, Germany

John Wiley & Sons Australia Ltd, 33 Park Road, Milton, Queensland 4064, Australia

John Wiley & Sons (Asia) Pte Ltd, 2 Clementi Loop #02-01, Jin Xing Distripark, Singapore 129809

John Wiley & Sons Canada Ltd, 22 Worcester Road, Etobicoke, Ontario, Canada M9W 1L1

Wiley also publishes its books in a variety of electronic formats. Some content that appears
in print may not be available in electronic books.

British Library Cataloguing in Publication Data

A catalogue record for this book is available from the British Library

ISBN 0-470-84895-2 (HB)
ISBN 0-470-84896-0 (PB)

Typeset in 10/12pt Times-Roman by Laserwords Private Limited, Chennai, India

Contents

Preface

I began writing this book on 11 September 2001 and completed it as the Iraq War drew to an end in April 2003. The nineteen months spent in writing the book were dominated by the "war on terror", and during that time much was written about globalisation, religious conflicts and cultural diversity. While Fukuyama (2001) sees these events as only blips in the trend towards modernity, Sacks (2002) believes that we should not just be tolerant of difference but should celebrate it. "Only when we realise the danger of wishing that everyone should be the same – the same faith on the one hand, the same McWorld on the other – will we prevent the clash of civilizations..." (Sacks, 2002, p. 209).

This book is about the value of difference, diversity and distinctiveness in the natural world. A few weeks after 9/11, when Osama bin Laden appeared on television to denounce the aerial bombing of Afghanistan, I predicted that specialists in that country's geology would be able to identify his whereabouts from the bedrock lithology and colour shown in the background. And sure enough, it was not long (*The Times*, 19 October 2001) before newspapers were reporting that Dr John Ford Shroder of the University of Nebraska had identified the rocks as being from the Pliocene Shaigalu or Eocene Siahan Shale Formations of the Katawaz Basin in south-east Afghanistan. This allowed search activities to be focused in these areas. This is a very unusual but perfectly clear example of one of the applications of the principles of geological diversity. If Afghanistan had been composed of a single rock type, bin Laden's general whereabouts would have been much more difficult to locate.

On the first anniversary of 9/11, I was attending a conference on the geological foundations of landscape in Dublin Castle, Ireland. And as I stood for a minute's silence, I was not only reminded of Ireland's tragic past and present but also of the way in which geology and landscape transcend administrative boundaries and can bring people together to value their heritage, conserve its integrity and overcome political barriers and national bureaucracies. In this case, the Royal Irish Academy and Geological Surveys of both the Republic of Ireland and Northern Ireland worked together to organise the conference and field trips.

The mountains and tectonics of Afghanistan, the deserts and rivers of Iraq and the bogs and coastline of Ireland, illustrate landscape geodiversity very clearly, but there is also diversity in their economic geology resources. The oil wealth of Iraq is well known and is attributed by some as the underlining reason for the war. But Afghanistan also has huge geological wealth, as yet largely unexploited. This includes an estimated 300 million barrels of oil, about 100 million tonnes of coal and large reserves of copper, gold, iron, chromite and industrial minerals (Stephenson & Penn, 2003). We also know about the Afghan gemstones – the lapis lazuli, ruby and spinel of Badkhshan, the emeralds in the Panjshir valley, the rubies and sapphires of Jegdalek and Gandamak and the pegmatites of Nuriston (Bowersox & Chamberlin, 1995). I hope that the people of Afghanistan will be able to realise the future projected for them by Bowersox & Chamberlin (1995, p. xv) when they said that "At present, the Afghan gemstone wealth is undetermined, mostly undiscovered, and certainly unexploited. We believe the potential of the country is so great as to promise an acceptable standard of living for every man, woman and child...".

This book is aimed at several types of readers. First, I hope it will be of interest to those closely involved in geoconservation whether in universities, nature conservation agencies, geological surveys or other organisations around the world. The book is intended to highlight best practice and I hope there are ideas in it that specialists can use. I apologise if I have omitted brilliant initiatives in geoconservation in one country or the other and hope that I will be told about these (j.m.gray@qmul.ac.uk) for a possible second edition. Secondly, the book is intended to stimulate discussion and thought on geoconservation by those whose primary concern has been wildlife conservation. It is therefore at least partly written with biologists and other non-geologists in mind. Thirdly, I hope it will stimulate further university courses in geodiversity and geoconservation and will become an established university textbook for second- and third-year undergraduate or postgraduate courses.

The book would not have been written without my three decades of experience working in the geosciences and public life during which time I have benefited from the ideas of countless people. I was born and educated in Edinburgh and must thank my late parents for the educational opportunities they gave me. At the University of Edinburgh, I studied both geography and geology and gained greatly from that education. There can be few better places to study the geosciences and conservation than Edinburgh and Scotland, given their diverse geology and geomorphology, their association with founders of the subject like James Hutton, John Playfair and Charles Lyell, and the innovative conservation efforts that have been nurtured there from the famous nineteenth-century American conservationist, John Muir, who was born in Dunbar, only 30 km east of Edinburgh, to the modern-day geoconservation work of Scottish Natural Heritage.

In public life, I have been a member of South Norfolk District Council and Chairman of its planning committee for many years and have served on regional planning panels and environment agency committees for the East of England. I am grateful to these organisations for the invaluable opportunities they have given me to understand how geoconservation can be promoted through local and regional government.

The book was largely written during sabbatical leave from my teaching duties at Queen Mary, University of London in 2001–2002, and I wish to thank the College authorities and my Head of Department, Prof. Roger Lee, for their generous support and encouragement for this project. My geography colleagues not only covered my teaching but also made many valuable suggestions for improving the book. Ed Oliver, cartographer in the Geography Department at Queen Mary, skillfully drew all the diagrams for this book in an unfairly short timescale.

During the sabbatical year, I benefited from discussions with many people including John Gordon (Scottish Natural Heritage), Stewart Campbell (Countryside Council for Wales), Colin Prosser (English Nature), and by e-mail with Kevin Kiernan, Mike Pemberton and Chris Sharples in Tasmania, Archie Landals and David Welch in Canada, Vince Santucci in the United States. I spent May 2002 in the Canadian and American Rockies learning about geoconservation in those two countries and benefited from discussions with Dave Dalman (Banff National Park), Archie Landals (Alberta government), Dan Fagre (USGS, Glacier National Park), Arvid Aase (Fossil Butte National Monument), Hank Heasler and Lee Whittlesey (Yellowstone National Park), Bill Dolan (Waterton Lakes National Park) and numerous others. Several people, too numerous to mention, have helped by providing detailed information or by supplying diagrams or photographs, and I am grateful to all of them.

The following people kindly read and commented on drafts of all or part of the book – Matthew Bennett, Cynthia Burek, Lars Erikstad, John Gordon, Archie Landals, Andrew McMillan, Mike Pemberton, Colin Prosser, Vince Santucci and Chris Sharples – and I am extremely grateful to all of them for their encouragement and valuable suggestions. This is a better book as a result of their comments. Cynthia Burek kindly allowed me to participate in the workshops aimed at creating a Local Geodiversity Action Plan (LGAP) in Cheshire.

I am grateful to many copyright holders for permission to reproduce figures. The sources are acknowledged in the figure captions, and where I received no reply, I have assumed that there is no objection to inclusion of the material in this book.

I must thank John Wiley and Sons for their faith in this project and particularly to Sally Wilkinson, Sue Barclay and Keily Larkins for their efficient handling of the book material. Laserwords Private Limited skillfully turned my inadequate typescript into a set of proofs. Finally, my thanks to Pauline for putting up with my obsession with the geosciences in general and this book in particular over a long period of time.

Murray Gray
Norfolk, April 2003

1

Defining Geodiversity

1.1 A Diverse World

Imagine, if you can, a very uniform planet. A planet composed of a single monomineralic rock such as a pure quartzite. A planet that is a perfect sphere with no topography and where there is no such thing as plate tectonics. Although it has weather, this is very similar everywhere with a solid cloud cover, light rain and no winds, so that there is little variation in surface processes or weathering. Consequently, the soil is also very uniform. The absence of gradients and surface processes means that there is little erosion, transportation or deposition of sediments. This planet has seen few changes in its 4.6 thousand million year history and there is, in any case, no sedimentary record of these changes. To say the least, this is not a diverse or dynamic planet.

It has to be admitted that our imagined planet has certain attractions. In fact, Medieval and Renaissance writers deplored the rough and disorderly shape of the Earth and "infestation by mountains which prevented it from being the perfect sphere that God must surely have intended to create" (Midgley, 2001, p. 7). Furthermore, there are no natural hazards such as earthquakes or avalanches to cause death and destruction. Civil engineering is very simple, given the predictability of the ground conditions. Walking is easy with no gradients to negotiate or rivers to cross. But think of the disadvantages. In a planet made entirely of quartz, there are no metals and therefore no metallic products. And in any case, since there is no coal, oil or natural gas, and no geothermal, wave, tidal or wind power, the energy to produce any goods or electricity is lacking. Every place looks the same, so getting lost is easy and there is no sense of place. Employment and entertainment are limited, given the absence of materials and lack of environmental diversity. The quartzite is too hard and massive to quarry in the absence of mechanical equipment or explosives, so the buildings are primitive, being constructed from soil and the simple vegetation types that exist on our planet, for, in the absence of physical diversity and habitat variation, little biological evolution of advanced plants and animals has been able to take place. This means that we humans would probably not exist on this planet, but if we did we would certainly find this to be a very primitive and boring place.

Thankfully our world is not like this. It is highly diverse in almost all senses – physical, biological and cultural – and although this produces problems for society and even conflicts and war, would we really want a less diverse and interesting home? The diversity of the physical world is huge and humans have put this diversity to good use. We have found that the diversity of physical resources on the planet is valuable to us in an enormous number of ways, even if we often fail to fully appreciate

Geodiversity: valuing and conserving abiotic nature   Murray Gray
© 2004 John Wiley & Sons, Ltd   ISBNs: 0-470-84895-2 (HB); 0-470-84896-0 (PB)

this fact. Diversity also brings with it flexibility of technologies and a greater ability to adapt to change.

Although our medieval ancestors hated the physical chaos of the Earth, our modern aesthetic appreciation of planetary diversity is probably deeply buried in our evolutionary psyche so that we value it more than uniformity. The broad diversity of places, materials, living things, experiences and peoples not only makes the world a more useful and interesting place, but probably also stimulates creativity and progress in a wide range of ways. Diversity therefore brings a range of values, and it is the thesis of this book that things of value ought to be conserved if they are threatened. And, as we shall see, there are many threats to planetary diversity induced by human actions both directly and indirectly.

The term "conservation" is used in preference to "preservation" in this book since the latter implies protection of the *status quo*, whereas nature conservation must allow natural processes to operate and natural change to occur. Unfortunately, human action has often accelerated or tried to stop natural processes, and has thus destroyed much that is valuable in the natural environment. While change through human action is inevitable, we should at least understand the consequences of our actions and hopefully minimise the impacts and losses. Conservation is therefore about the management of change.

I referred in the Preface to a growing respect for diversity and a realisation that there is value in difference. Marsh (1997, p. 2) comments cogently that

"There is a predictable sameness creeping over the face of the North American landscape. Since the early 1970s, highways, shopping centres, residential subdivisions, and most other forms of development have taken on a remarkable similarity from coast to coast. Not only do they look alike, but modern developments also tend to function alike, including the way they relate to the environment, that is, in the way land is cleared and graded, storm water is drained, buildings are situated and landscaping is arranged. We know that this façade of development masks an inherently diverse landscape in North America. If we look a little deeper, it is apparent that landscape diversity is rooted in the varied physiographic and ecological character of the continent and this in turn reflects differences in the way the terrestrial environment functions. Does it not seem reasonable then that development and land use should also reflect these differences if they are to be responsive to the environment?".

Similarly, in the United Kingdom, there has been a trend away from a "one-style-fits-all" approach to house design across the country's housing estates and towards one that respects and reflects local vernacular character, architectural distinctiveness and harmony between buildings, settlements and landscapes (Countryside Commission, 1996, 1998). Local initiatives including farmers' markets, local economic trading schemes and Local Agenda 21 are reconnecting people with their local social, economic and environmental scene. Despite the pressures of globalisation, local diversity is fighting back.

1.2 Biodiversity

Nowhere has this trend towards the value of diversity been more evident than in the field of biology. In recent decades, the growing concern about species decline and extinction, loss of habitats and landscape change led to a realisation of the multi-functioning nature of the biosphere. For example, it acts as a source of fibre, food and medicines, it sustains concentrations of atmospheric gases, it buffers environmental change and it contains millions of species of plants and animals, most of which have unknown value and ecosystem function and deserve respect in their own right. Yet, of the 1.5 to 1.8 million known species, it is estimated that up to a third could be extinct in the next 30 years (Grant, 1995).

Concern for species and habitat loss led to some important international environmental agreements and legislation including the Ramsar Convention on Wetland Conservation (1971), Convention on International Trade in Endangered Species (CITES) (1973) and the Bonn Convention on Conservation of Migratory Species (1979). More recently, the European Union has played an active role in biological conservation, for example, through the Habitats Directive and Birds Directive.

An International Convention on Biological Diversity was first proposed in 1974 and during the 1980s the phrase "biological diversity" started to be shortened to "biodiversity". An important meeting of the US National Forum on Biodiversity took place in Washington, D.C. in 1986 under the auspices of the American National Academy of Sciences and Smithsonian Institution. The conference papers (Wilson, 1988) mark an important milestone in the history of nature conservation and caused the issue to be taken seriously by politicians both inside and beyond America.

International recognition of the need for biosphere conservation led to the UN Convention on Biodiversity agreed at the Rio Earth summit in 1992, ratified in 1994 and signed by over 160 countries, though it did not include the United States. The agreement was far reaching and the main features are listed in Table 1.1. Since then, great attention has been given at international, national, regional and local levels to protecting and enhancing the biological diversity of the planet. These are usually classified into genetic diversity (conserving the gene pool), species diversity (reducing species loss) and ecosystem diversity (maintaining and enhancing habitats and their biological systems). And biodiversity is not just about numbers of species or ecosystems but about the countless interconnections between them. A wealth of strategies and action plans are being implemented to carry forward the aims of the UN Convention. Every signatory country must prepare a national plan for conserving and sustaining biodiversity, has a responsibility for safeguarding key ecosystems and is responsible for monitoring genetic stock. International designations include Ramsar sites (under the Ramsar (Wetland Habitat) Convention), Special Protection Areas (under the European Union Birds Directive) and Special Areas for Conservation (under the European Union Habitats Directive). The International Union for Nature Conservation (IUCN) has helped over 75 countries to prepare and implement national conservation and biodiversity strategies. Jerie *et al.* (2001, p. 329) have referred to this as "the torrent of effort being put into the management of biodiversity". Even some ecologists have spoken of the obsession with loss of species and habitats rather than focusing on the more important issue of functional significance of species in a variety of ecosystems (Dolman, 2000).

In the United Kingdom, the national *Biodiversity Action Plan* (1994) is being supplemented by many regional and local initiatives including Local Biodiversity Action Plans (LBAPs), Species Recovery Programmes (SRPs) and Habitat Action Plans (HAPs). These are being implemented by the national conservation bodies (English Nature, Scottish Natural Heritage, Countryside Council for Wales) in collaboration with local authorities and a wide range of wildlife and conservation organisations (e.g. County Wildlife Trusts, Royal Society for the Protection of Birds, Campaign to Protect Rural England). By 1999, 400 species action plans, 100 species statements for priority

Table 1.1 *Main features of the Convention on Biodiversity (1992) (after Mather & Chapman, 1995)*

- Development of national plans, strategies or programmes for the conservation and sustainable use of biodiversity.
- Inventory and monitoring of biodiversity and of the processes that impact on it.
- Development and strengthening of the current mechanism for conservation of biodiversity both within and outside protected areas, and the development of new mechanisms.
- Restoration of degraded ecosystems and endangered species.
- Preservation and maintenance of indigenous and local systems of biological resource management and equitable sharing of benefits with local communities.
- Assessment of impacts on biodiversity of proposed projects, programmes and policies.
- Recognition of the sovereign right of states over their natural resources.
- Sharing in a fair and equitable way the results of research and development and the benefits arising from commercial and other utilisation of genetic resources.
- Regulation of the release of genetically modified organisms.

species and 40 action plans for priority habitats had been published and were being implemented (Simonson & Thomas, 1999).

More than 20 books have appeared with "biodiversity" in the title (Wilson, 1988; Groombridge, 1992; World Conservation Monitoring Centre, 1992; Reid, 1993; Schulze & Mooney, 1994; Heywood & Watson, 1995; Hawksworth, 1995; Jermy *et al.*, 1996; Gaston, 1996; Reaka-Kudla *et al.*, 1997; Jeffries, 1997; Rushton, 2000). Wilson (1997, p. 1) refers to "biodiversity" as "one of the most commonly used expressions in the biological sciences and has become a household word". "Biodiversity science" and "biodiversity studies" have been born and the origin and maintenance of biodiversity "pose some of the most fundamental problems of the biological sciences" (Wilson, 1997, p. 2).

1.3 Geodiversity

Geological and geomorphological conservation (geoconservation) have a long history. In the first 20 years of the nineteenth century, the quarrying of stone from Salisbury Crags in Edinburgh, Scotland was having such a serious impact on the city landscape that legal action was taken in 1819 to prevent further deterioration (McMillan *et al.*, 1999). Germany established the first geological nature reserve in the world at Siebengebirge as early as 1836. Yellowstone National Park, USA, was established in 1872 largely for its scenic beauty and geological wonders (see Box 5.1). Also in the 1870s, Fritz Muhlberg campaigned to protect giant erratic boulders in Switzerland that were being exploited as kerbstones (Jackli, 1979), and in Scotland the "Boulder Committee" was established, under the direction of David Milne Home, to identify all remarkable erratics and to recommend measures for their conservation (Milne Home, 1872a,b; Gordon, 1994a). Some of the first specific geological sites to be protected were also in Scotland where City Councils acted to enclose Agassiz Rock striations in Edinburgh (1880) and the Fossil Grove Carboniferous lycopod stumps in Glasgow (1887). Other initiatives have followed and many countries now have areas and sites protected at least partially for their geological or landscape interest. But despite some recent international conferences and books (Martini, 1994; O'Halloran *et al.*, 1994; Stevens *et al.*, 1994; Wilson, 1994; Alexandrowicz, 1999; Barenttino *et al.*, 1999; Gordon &

Leys, 2001a), in most countries geoconservation is weakly developed and lags severely behind biological conservation.

Geologists and geomorphologists started using the term "geodiversity" in the 1990s to describe the variety within abiotic nature. The major attention being given to biodiversity and wildlife conservation was simply reinforcing the long-standing imbalance within nature conservation policy and practice between the biotic and abiotic elements of nature. Although geological and geomorphological conservation had been practised for over 100 years, these were usually the "Cinderella" of nature conservation (Gray, 1997a). Many international nature conservation organisations, although using the general term "nature conservation" appeared to see this as synonymous with "wildlife conservation" and focused most or all of their attention on the latter. Milton (2002, p. 115) summarised the situation well in stating that "Diversity in nature is usually taken to mean diversity of living nature . . .".

The situation in Australia can be used to emphasise the point. Jerie *et al.* (2001, p. 331) noted that "the recent history of conservation management has permitted an enormous bias to develop towards assessment and management of biotic elements of natural systems". Similarly, Pemberton (2001a) believed that "nature conservation agencies and governments across the country, and overseas, tend to emphasize the need for the conservation of biodiversity whilst virtually ignoring the geological foundation on which this is built and has evolved". He attributed this to the lack of training of earth scientists in geoconservation theory, policy and practice. He made the interesting observation that "the majority of earth scientists are trained and employed in the extractive industries. To be involved in conservation could be seen to be contrary to the goals of the profession . . ." and he compares this with the biological sciences where conservation is a major graduate employer. "This has generally meant that geoconservation has remained something of an oddity, divorced from mainstream nature conservation, and so it has generally had low priority within land management agencies" (Pemberton, 2001a). Although geoconservation has not yet been accorded great prominence in Australian nature conservation, Kiernan (1996, p. 6) believes that "few professional land managers would, having been made aware that a particular landform was, say, the only example in Australia of its type, seriously argue against the validity of safeguarding it, just as they would wish to safeguard the continued existence of a biological species".

A similar situation has existed in the United Kingdom, where although geoconservation policy and practice have long been actively pursued and developed by several groups and organisations, this has not always been recognised by the wider nature conservation community or public. This situation has not been helped by the fact that UK geoconservation has been lumbered with terms like "Earth heritage conservation" and the even more opaque "natural features" (Gray, 2001). Therefore, some geologists and geomorphologists saw "geodiversity" not only as a very useful new way of thinking about the abiotic world but also as a means of promoting geoconservation and putting it on a par with wildlife conservation (Prosser, 2002a).

In the United States, most nature conservation effort has been directed through the national parks system, and although several units of the system (see Section 5.8) have been established for their geological or geomorphological interest, this is not always recognised. For example, Sellars' (1997) eloquent history of nature conservation in the US National Parks is almost entirely dominated by wildlife issues, reflecting the major concerns of the parks system over the last 130 years. An amazing modern example of this is the fact that at the time of writing, Grand Canyon National Park employs scores of ecologists but not a single professional geologist!

It is difficult to trace the first usage of the term "geodiversity", and indeed it is likely that several earth scientists coined the term independently, as a natural twin to the term "biodiversity". Some of the first uses appear to have been in Tasmania, Australia. Kevin Kiernan (personal communication) was using the terms "landform diversity" and "geomorphic diversity" in the 1980s and drawing analogies with biological concepts by using terms such as "landform species" and "landform communities". The term "geodiversity" was used by Sharples (1993), Kiernan (1994, 1996, 1997a) and Dixon (1995, 1996a, b) in studies of geological and geomorphological conservation in Tasmania, in particular, or Australia, in general. Sharples (1993) used it to cover "the diversity of earth features and systems" while Dixon (1996a), Eberhard (1997), Sharples (2002a) and Australian Heritage Commission (2002) have defined it as

> "the range or diversity of geological (bedrock), geomorphological (landform) and soil features, assemblages, systems and processes".

The term is now very well understood in Tasmanian nature conservation. Sharples (2002a) stresses the importance of distinguishing the terms "geodiversity", "geoconservation" and "geoheritage". He defines them as follows:

- "geodiversity" is the *quality* we are trying to conserve,
- "geoconservation" is the *endeavour* of trying to conserve it, and
- "geoheritage" comprises *concrete examples* of it which may be specifically identified as having conservation significance.

Sharples sees these terms as much preferable to the alternatives of "geological diversity", "geological conservation" and "geological heritage" since these terms are associated with solid rock rather than the range of abiotic forms, materials and processes.

The most important landmark for geoconservation in Australia was adoption of the Australian Natural Heritage Charter in 1996 and subsequent update in 2002 (Australian Heritage Commission, 1996, 2002), which has the term and substance of geodiversity interwoven throughout its Articles (see Section 6.8.5). As a result, geodiversity is now a widely used and understood term in Australian nature conservation.

In the proceedings of an important conference on geoconservation held at Malvern, UK in 1993, the term "geodiversity" was used by Wiedenbein (1994) in relation to geotope conservation in German-speaking countries. In the same volume, Erikstad (1994a), Harley (1994) and Todorov (1994) used the term "geological diversity", but not the shortened form, while Joyce (1994), in a discussion of an appropriate terminology for geological or earth science conservation, did not refer to either "geological diversity" or "geodiversity". Later, Joyce (1997, p. 38) states that at the Malvern conference, "the possible use of the term geodiversity was suggested by some participants ... but failed to receive significant support ...". Joyce (1997, p. 39) is critical of the term "geodiversity" since it "... may be attempting to draw too strong a parallel between sites, landscape features and processes in biology and geology". This issue is discussed in Section 7.2.

In the United Kingdom, Gray (1997a, p. 323) suggested that "perhaps one day we will see .... a Geodiversity Action Plan for the UK to rank alongside its biological counterpart" and this was repeated by Gray (2000). Slightly later, it was used as the title of an article by Stanley (2000) and was adopted by the Royal Society for Nature Conservation (RSNC) in the title of its quarterly earth science newsletter *Geodiversity Update* launched in January 2001 but terminated in 2002 when the RSNC decided to cease all earth science work! In the first edition, Stanley (2001) asked the question, "So what is geodiversity?", to which the answer is given that

> "It is the link between people, landscapes and culture; it is the variety of geological environments, phenomena and processes that make those landscapes, rocks, minerals, fossils and soils which provide the framework for life on Earth".

This is a very wide definition, but Stanley (2002) goes even further in arguing that "biodiversity is part of geodiversity". Prosser (2002a,c), in useful discussions of terminology, accepts the validity of the term "geodiversity". The traditional term used in the United Kingdom for geoconservation is "Earth heritage conservation", but this is longer, duplicates the shorter term and has an ambiguous meaning since many people see it as including biological components of the Earth.

The Nordic Council of Ministers introduced the term "geodiversity" into nature conservation in their countries in 1996. Johansson (2000) has published an excellent book on the geodiversity of the Nordic countries, and an English summary was produced in 2003 (Nordic Council of Ministers, 2003). This sees geodiversity as

> "the complex variation of bedrock, unconsolidated deposits, landforms and processes that form landscapes . . . . . . Geodiversity can be described as the diversity of geological and geomorphological phenomena in a defined area" (Johansson, 2000, p. 13).

Erikstad & Stabbetorp (2001) used the term in relation to natural areas and environmental impact assessment, and it appears now to be in common usage in the Scandinavian countries.

Burnett *et al.* (1998) and Nichols *et al.* (1998) writing in the United States were still using the term "geomorphological heterogeneity", suggesting that "geodiversity" had yet to reach common usage there by 1998, and this has been confirmed by many North American contacts in 2002.

Although there have been some books written on geological/geomorphological conservation, all of which contain much useful material, they have tended to be collections of papers on particular topics or countries (O'Halloran *et al.*, 1994; Stevens *et al.*, 1994). The exception is the book edited by Wilson (1994), which covers traditional approaches to geoconservation in the United Kingdom in considerable detail. However, there is no book yet published that deals with the wider aspects of geodiversity, with an international perspective or covering some of the new approaches taken to earth science conservation or integrated landscape management. This last aspect is discussed in the excellent recent book by Gordon & Leys (2001a), though again it is an edited collection mainly devoted to Scotland. The present book aims to be a coherent, wide-ranging and international discussion of the principles of geodiversity and practices of geoconservation, and is the first such treatment of these subjects.

Having considered other definitions of "geodiversity", it is appropriate to end this section by giving the definition used in this book, which is a modification of that used in Australia (see above):

"Geodiversity: the natural range (diversity) of geological (rocks, minerals, fossils), geomorphological (land form, processes) and soil features. It includes their assemblages, relationships, properties, interpretations and systems".

1.4   Geodiversity as a Resource

A widely used method for assessing the availability of geological resources is the McKelvey box (Fig. 1.1). In this scheme, the outer box is the resource base, which is the total amount of a material that exists on Earth (= total resource). However, much has yet to be discovered (= hypothetical resource) and many geomaterials could not be economically or technically exploited (= conditional reserves). Therefore, there is an interior box representing the reserves, which is that part of the total resource whose exploitation might conceivably be economic and technically feasible in the foreseeable future. However, estimates of reserves and resources will vary with changes in economic conditions and geological knowledge. Shortages in reserves will raise the price of a commodity and therefore increase the pace of geological exploration, which in turn will increase the reserves. Shortage of reserves will also increase price, making it more economic to exploit lower-grade and less-accessible resources. A price rise will also increase interest in recycling of materials or substitution by cheaper materials (see Section 6.5).

We can now apply this resource/reserve concept to geodiversity and geoheritage (Fig. 1.2). The outer box would then refer to all the geodiversity in the world, at least

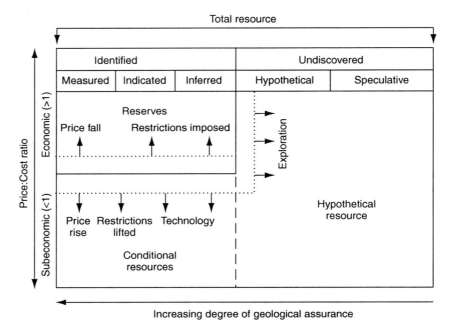

Figure 1.1   *The McKelvey box showing the relationship between resources and reserves. The size of a reserve relative to the resource varies, expanding because of exploration, price rises, improved technology, etc. but contracting with price falls or governmental restrictions Reproduced from Bennett, M.R. & Doyle, P. (1997) Environmental Geology by permission of Wiley.*

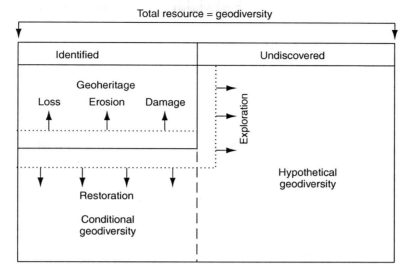

Figure 1.2   *The McKelvey box with geodiversity as the resource. Geoheritage is part of geodiversity and can expand with restoration or contract with erosion, loss or damage*

some of which has yet to be described and mapped, and which has been increasing since the Earth formed (see Section 2.3). But it is not necessary or desirable to conserve all this diversity, given the pragmatic problems of doing so and the needs of society to utilise some of the georesource. We therefore have an inner box related to the geoheritage. These are the elements of geodiversity that are judged to be significant and worthy of conservation because of their values (see Chapter 3). Geoconservation should aim to avoid the further loss of the geoheritage (Chapter 4), while georestoration (Section 6.3) and field research (Section 7.3.4) should seek to increase it.

1.5   Aims and Structure of the Book

The aims of this book are

- to raise awareness of the values of the geodiversity of the planet;
- to point out the threats to this diversity;
- to examine the ways that this diversity can be conserved, managed and restored;
- to outline the need for a more holistic approach to nature conservation and land management; and
- to propose a vision for the future based on current best practice.

I hope the book will stimulate interest in these topics not just amongst geologists, geomorphologists and soil scientists, many of whom are at least aware of the issues, but also amongst the wider academic and non-academic community – biologists, nature conservationists, landscape architects, planners and politicians – for the world, in general, has not paid sufficient attention to these issues.

The chapters of the book follow a natural sequence. In Chapter 2, I shall attempt the extremely difficult task of describing the geodiversity of the planet. The aim of this chapter is not to catalogue every variation in rocks, minerals, sediments, processes, landforms, soils and fossils, but rather to outline the principles and causes of diversity in the abiotic world. As stated above, biodiversity is often classified into genetic, species and ecosystem diversity, a classification that is based primarily on scale. Such a system is not as appropriate for geodiversity, although we can certainly recognise scale differences in the abiotic world. For example, landforms combine to form landscapes that combine to make continents or tectonic plates. Nonetheless, the obvious way to classify abiotic nature is by using a well-tried tripartite subdivision of *material, form* and *process*. The focus of the book is on the terrestrial systems and the solid Earth (which I shall refer to as the "geosphere"), rather than conservation or environmental management of all planetary abiotic systems. Consequently, I shall say little about conservation of atmosphere and oceans. The term "lithosphere" is inappropriate since terrestrial systems include fluvial, coastal and glacial processes and, therefore, parts of the hydrosphere. But the main diversity in earth materials is in lithospheric materials – rocks, minerals, sediments and soils. Form comprises landforms and physical landscapes while numerous processes act on the materials to produce landforms. Anyone who already has a good understanding of the planet's geodiversity can probably skip this chapter.

This is followed by Chapter 3, which discusses the value of this geodiversity in terms of intrinsic, cultural, aesthetic, economic, functional and research/educational values. A summary table at the end of the chapter recognises over 30 types of abiotic value. Chapter 4 outlines the main threats to this valuable geodiversity including mineral extraction, river engineering, fossil collecting, urban expansion, coastal erosion and defence, waste disposal, agricultural practices and afforestation. The impact of these activities will be greater on some physical materials and systems than on others, and the issue of sensitivity to human modification emerges from this chapter as a very important factor.

Chapter 5 then follows logically from the previous two using the simple conservation formula:

$$\text{Value} + \text{Threat} = \text{Conservation need}$$

Chapter 5 covers the traditional methods of conservation, namely by designating protected areas and using legislation and other approaches to conserve and manage them. A survey of national and sub-national systems reveals the diversity of approaches used and the effectiveness of geoconservation work, and attempts to give examples of good practice. This is also a major objective of Chapter 6, which looks at a range of new approaches and initiatives to extend geoconservation beyond protected areas to the sustainable management of the wider planetary georesource.

Chapter 7 revisits some important issues for geodiversity in relation to biodiversity and discusses a basic paradox for geodiversity. On the one hand the subject needs to establish itself as a distinctive, independent and essential field of nature conservation, but on the other there is a growing need and trend towards an integrated approach to nature conservation incorporating geological and biological systems and indeed to integrated land management in general. Finally, Chapter 8 draws some conclusions and tries to establish a vision for the year 2025.

This structure of valuing geodiversity, understanding the threats and conserving and managing the resource, and indeed the whole impetus for writing the book follows a logical pattern reflected in the following quote from African conservationist, Baba Dioum:

"For in the end we will conserve only what we love.
We will love only what we understand.
And we will understand only what we are taught".

I hope this book stimulates a much greater interest in the values of geodiversity and the need to conserve them.

2
Describing Geodiversity

2.1 Introduction

The geodiversity of the Earth is hardly less remarkable than its biodiversity. Any attempt to fully describe this geodiversity would fill several books, and so the challenge here is to summarise it in a few thousand words concentrating on the factors producing the diversity rather than an encyclopaedic catalogue of the whole range of minerals, rocks, sediments, fossils, soils, landforms and processes on the planet. There are several books that describe the geology, geomorphology and pedology of the Earth in more detail, many of them in lavish colour. Readers are referred to those by Summerfield (1991), Scott (1991, 1992), Bradshaw & Weaver (1993), Skinner & Porter (1997), Allen (1997), Davidson et al. (1997), Tarbuck & Lutgens (1999), Press & Siever (2000), Christopherson (2000), Hamblin & Christiansen (2001), Strahler & Strahler (2002) and particularly to the superbly illustrated recent book by Marshak (2001). Anyone who is already familiar with the geodiversity of the planet may wish to skip this chapter.

Geodiversity applies at various scales from the global scale of continents and oceans to elemental scale of atoms and ions. This is no different to the scale issue in biodiversity where bioscientists have to deal with habitat variations on a global scale, but also with genetic diversity and biotechnology at the microbiological scale. This chapter begins with the early history of the Earth and plate tectonics since they are the keys to understanding much of the planetary geodiversity at many scales.

2.2 Origin of the Earth

To understand the current geodiversity of the planet, we need to understand how the planet originated and how it has evolved through its 4,600,000,000 year (4.6 Ga) history. Astrophysicists believe that the Earth and other planets of our solar system developed from a residual disk of gas and dust, called a planetary nebula, left over from the formation of the Sun, perhaps 15 Ga. All 92 naturally occurring elements were already in existence at this time, having been formed by nuclear fusion at very high temperatures from an original mix of hydrogen and helium, the two lightest elements.

There then followed a long series of collisions and combinations in which gaseous atoms coalesced to form molecules, molecules combined to form dust, and dust accreted to become small pieces of rock called planetesimals (Marshak, 2001). Over millions of years, further collisions resulted in larger and larger rock lumps, which in turn attracted

Geodiversity: valuing and conserving abiotic nature   Murray Gray
© 2004 John Wiley & Sons, Ltd   ISBNs: 0-470-84895-2 (HB); 0-470-84896-0 (PB)

smaller pieces, often as meteorites. Thus "the Earth was conceived and grew violently, a chaos of impact and fragmentation and annealing. All was instability" (Fortey, 1997, p. 31). Eventually, a series of protoplanets revolving (in the same direction and, with the exception of Pluto, in the same plane) around the Sun evolved into the planetary system we see today, with smaller lumps (asteroids and meteoroids) still hurtling around space and occasionally colliding with the planets. The Earth's Moon is believed to have been formed when a particularly large collision at 4.5 Ga blasted debris into orbit, which subsequently coalesced (Canup & Asphaug, 2001). It is still pockmarked by impact craters, whereas the Earth's have been largely eradicated by subsequent geological evolution (see below). The near spherical shape of the planets is produced by gravitational pull aided by geothermal heat as a result of planetesimal collision and radioactive decay of elements.

The earth's basic composition was established at this early stage with iron (35%), oxygen (30%), silicon (15%) and magnesium (13%) forming most of the Earth's matter (Table 2.1). The solar wind is believed to have blown most of the light elements (hydrogen, helium, etc.) into the outer parts of our solar system where they make up significant parts of Jupiter, Saturn and the other outer planets. But the young Earth would have been "melted and cauterized by the feverish concatenation of impacts. Elements would have been shuffled and recombined as new minerals in a frantic alchemy of creation" (Fortey, 1997, p. 31).

2.3   Early History of the Earth

Even though at this early stage the Earth had a geodiversity, with meteorites bringing new minerals, which melted on impact and added to the nascent mix, and a surface pockmarked by meteorite craters similar to the Moon, overall the primitive Earth was a fairly homogeneous planet. From this time to the present day, we can assume that the Earth's geodiversity has progressively increased. Furthermore, unlike biodiversity, the Earth's geodiversity has probably had few major setbacks caused by the meteorite

Table 2.1 *Composition (% by weight) of the whole Earth and of the crust. Note that differentiation has created a light crust depleted in iron and rich in oxygen, silicon, aluminium, calcium, potassium and sodium (after Press & Siever, 2000)*

Element	Whole Earth	Crust
Iron	35	6
Oxygen	30	46
Silicon	15	28
Magnesium	13	4
Nickel	2.4	<1
Sulphur	1.9	<1
Aluminium	1.1	8
Calcium	1.1	2.4
Potassium	<1	2.3
Sodium	<1	2.1
Other	<1	<1

impacts or climate change that brought mass extinctions to the biological world (see below). Although these impacts no doubt destroyed some rocks, minerals, landforms or soils, they also created new materials and features that added geodiversity to the planet. The main natural threat to geodiversity is erosion, which removes material, but reforms the material elsewhere, often in a new way. Deposition buries materials previously at the surface, but they are not lost and may subsequently be re-exposed by natural erosion or human excavation. For further discussion of the geological evolution of the Earth see Windley (1995).

At this early stage, the Earth's atmosphere also developed from volcanic gases, as the more volatile compounds found their way to the surface in a process known as out-gassing. "You might say that our atmosphere, and the possibility of life itself, was the consequence of a vast, terrestrial flatulence risen from the bowels of the Earth" (Fortey, 1997, p. 35). At this stage, the Earth's surface would have been largely molten and the atmosphere would have consisted of an unbreathable cocktail of volcanic gases including nitrogen, ammonia, methane, carbon monoxide, carbon dioxide, sulphur dioxide and water vapour. This atmosphere was held in place by the Earth's gravitational field. The large amount of water vapour subsequently condensed to form clouds and rain, which then filled the oceans.

The continental drift hypothesis proposed by the German scientist Alfred Wegener in the early decades of the last century was replaced in the 1960s by the theory of plate tectonics. Much of Wegener's evidence for the matching of continental margins, rocks and fossils was confirmed, but rather than continental blocks drifting across the Earth's surface like supertankers, the plate tectonics theory proposes that the whole crust is broken into pieces like an eggshell, and that these pieces are continuously in motion. In fact, it is now believed that a surface crust began to form on the cooling Earth over four thousand million years ago (4 Ga) and that this crust quickly broke into plates, which have been on the move ever since. However, the position of the margins has changed through time. To begin with, it is believed that the plates were small, thin and rapidly moving, since at this stage the Earth had cooled only sufficiently for a thin surface crust to form. It also follows that the intense heat below the surface meant that volcanic activity and volcanic gas emission would have been intense, particularly along plate margins. The resultant volcanic arcs were relatively light and therefore were differentiated from the denser crustal materials. Here we began to see the origin of the continents, a process that was aided by erosion, sedimentation and uplift of thick sediment sequences and granitic intrusions as in modern mountain building (see below). These blocks of lighter material became the continental crust, segregated from the denser oceanic crust below (see Box 2.1).

Collisions and volcanic suturing created progressively larger plates and larger continents, and as the Earth continued to cool, these so-called protocontinents became thicker and stronger. By 2.7 Ga, the core areas (cratons) of our current continents had formed, comprising rocks dating back to more than 4 Ga discovered in Canada, Greenland, China and Wyoming, USA. The 4 Ga date is therefore taken as distinguishing the very early phase of Earth history when the earth was too hot for significant solid rock areas to stabilise (Hadean, Fig. 2.2), from the Archean period of protocontinent formation. It should be noted, however, that zircon grains dating back to 4.4 to 4.2 Ga have been discovered in rocks in Australia and elsewhere, indicating that some minerals were crystallising from the molten rock at this very early date (Marshak, 2001).

Box 2.1   *Internal Structure of the Earth*

The Earth's internal layers, comprising crust, mantle and core also formed at an early stage in the life of the planet. Since iron is more dense than the other common minerals, it quickly migrated in a molten state to the earth's interior, with the result that the layers increase in density from the surface to the centre (Table 2.1, Fig. 2.1). The present-day structure has been deciphered from various types of evidence including research on planetary density and the progress through the Earth of earthquake waves. Three layers are recognised each of which is divided into two.

- *Core*   The inner core is believed to be a solid iron alloy with a density of $c.13\,g/cm^3$ and a temperature of over $4,300\,°C$. The outer core is a liquid iron alloy with a slightly lower density and temperature.
- *Mantle*   The mantle is also divided into two sections separated by a transition zone. The lower mantle has a density of 4 to $5.5\,g/cm^3$ while the density of the upper mantle is closer to $3.5\,g/cm^3$. Most of the mantle is solid rock, though at a depth of 100 to 200 km below the ocean floors, there is a low-velocity zone where a mixture of molten rock and solid material exists. But even though the mantle is solid, the high temperatures mean that it is able to deform internally at very slow rates (a few centimetres/year) and these movements occur as convection currents (see below).
- *Crust*   The crust is also divided into two, a lower oceanic crust with a density of about $3.0\,g/cm^3$ and an upper continental crust at $c.2.7\,g/cm^3$, separated by the Mohorovicic Discontinuity, often referred to simply as the Moho. Crustal rocks are dominated by silicon and oxygen, most commonly combined as silica (SiO_2). In turn, silica is combined with other elements (iron, magnesium, aluminium, calcium, potassium and sodium) to form the so-called silicate minerals and silicate rocks that dominate the Earth's crust (see below).

By 3.8 Ga, condensing water vapour and rainfall had caused rivers to run over the lifeless continents and a global ocean had formed. This is deduced from the presence of water-rounded mineral grains from this time indicating that the processes of erosion, transportation and deposition were established early in the age of the continents. It is also likely that the ocean rapidly became salty as surface and groundwater dissolved salts from the rocks and transported the material to the sea. Archean rocks also contain the first record of life, with simple cells, bacteria and algae appearing around 3.8 Ga. It seems likely that the free oxygen produced by these bacteria precipitated the iron out from the seas and rivers, to form vast sedimentary deposits of banded irons, such as those in western Australia. These "represent a turning point in the history of life on Earth; they show for the first time the influence of life on the structure of the planet itself .... It is a unique two way interaction" (Manning, 2001, p. 22). Thus by the end of the Archean, 2.5 Ga ago, the Earth had begun to resemble the planet as we know it today with atmosphere, oceans, dry land and the early evolution of life forms.

The Proterozoic (from the Greek meaning "first life") is clearly misnamed, but this is often the case in geology as new discoveries make older names inappropriate.

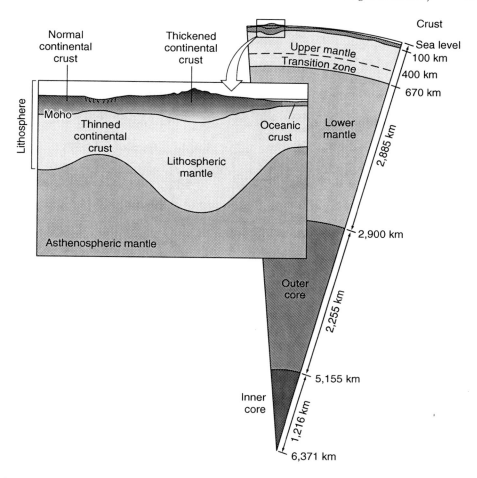

Figure 2.1 *Internal structure of the Earth (After Marshak, S. (2001) Earth: Portrait of a Planet. W.W. Norton & Co, New York, by permission of W.W. Norton & Co.)*

Nonetheless, there was a significant evolution of multicellular life during the c.2 Ga of Proterozoic time. In particular, the evolution of photosynthetic organisms changed the atmosphere from an oxygen-poor volcanic gas mix to something much more like the oxygen-rich air we breathe today, though it was not until the Phanerozoic that today's levels of 21% oxygen were reached. It was the production of this oxygen that allowed the development of energy-producing metabolism and atmospheric ozone to screen ultraviolet radiation. These two developments eventually allowed the great diversification of life to occur as living organisms had the energy and protection to be able to conquer the land.

Also during the Proterozoic eon, the continuing cooling of the Earth brought a slowing of plate tectonic processes including the accretion of volcanic provinces. By 1.8 Ga, most of the large continental cratons that exist today had formed (see Fig. 2.3),

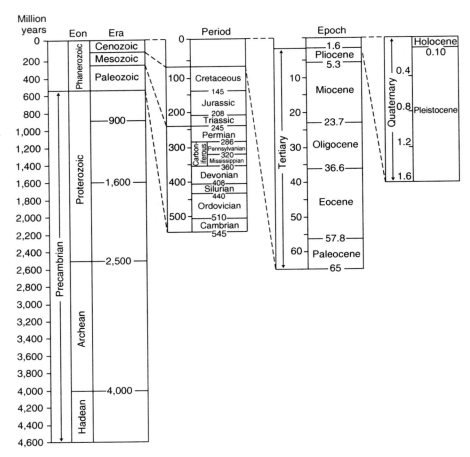

Figure 2.2 *The history of the Earth (Modified after Marshak, S. (2001) Earth: Portrait of a Planet. W.W. Norton & Co, New York, by permission of W.W. Norton & Co.)*

though parts of the cratons have been buried by later deposition. For example, Fig. 2.4 shows that the North American craton is a collage of different plate fragments, volcanic belts and accretionary provinces fused together between 1 and 2 Ga ago. The evidence of collision between Archean blocks is particularly well exposed along the Trans-Hudson belt in central Canada. In the United States, most of the Precambrian craton complex is buried below Phanerozoic rocks, which form a so-called cratonic or continental platform. The Phanerozoic mountain belts and coastal plain are also the result of subsequent events (see below).

2.4   Plate Tectonics

By the beginning of the Phanerozoic, many of the continents had begun to resemble their present-day outlines, including North America, South America, Africa, Antarctica and Australia. The Palaeozoic Era saw many further changes to the continental masses as they drifted about, but by the end of the Palaeozoic, they had come together in the

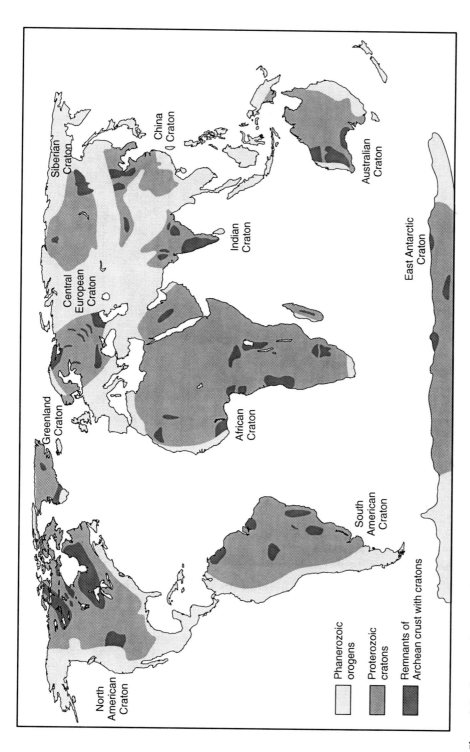

Figure 2.3 *Proterozoic cratons incorporating remnants of Archaean crust. Note that many parts of these cratons are buried by later sediments (After Marshak, S. (2001) Earth: Portrait of a Planet. W.W. Norton & Co, New York, by permission of W.W. Norton & Co.)*

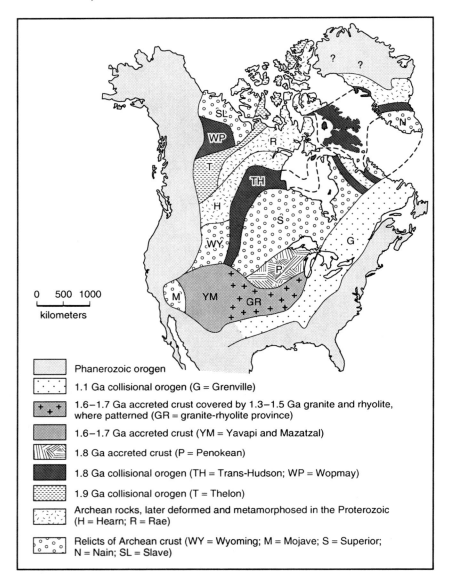

Figure 2.4   *The North American craton as a collage of different fragments assembled by collision and accretion in Precambrian time (After Marshak, S. (2001) Earth: Portrait of a Planet. W.W. Norton & Co, New York, by permission of W.W. Norton & Co.)*

supercontinent of Pangaea (Fig. 2.5). It is with the break-up of Pangaea over the last 200 Ma that the modern era of plate tectonics begins.

Some basic principles of plate tectonics were mentioned above, particularly in relation to the movement of plates. This is driven by convection currents in the Earth's mantle and aided by the penetration of water (Boulton, 2001). The internal areas of plates remain relatively rigid and intact, but there is a high level of geological activity along the plate margins. In fact, three types of plate margin can be recognised:

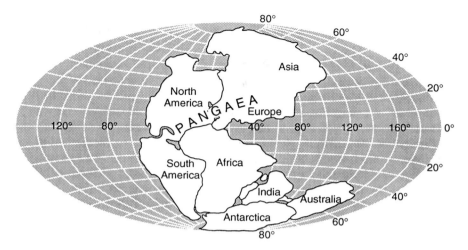

Figure 2.5 *Some 200 million years ago, all the Earth's present continents were joined together in a huge supercontinent called Pangaea ("all lands") (Reproduced from Press, F. & Siever, R. (2000) Understanding Earth. 3rd ed. W.H. Freeman, New York, by permission of W.H. Freeman and Company/Worth Publishers)*

- *Divergent margins*, where the convection currents pull two plates apart by a process known as *sea-floor spreading*. Molten rock (magma) wells up in the fissures that open along the divergent plate margin producing an oceanic ridge and creating new oceanic crust. Examples include the mid-Atlantic Ridge, the higher parts of which protrude above sea level, most notably in Iceland. America and Europe are moving apart at about 20 mm per year. Existing plates may start to rift apart, as exemplified by the East African Rift Valley.
- *Convergent boundaries*, where the convection currents move two plates together. Three sub-types of convergent margin are possible – ocean/ocean, ocean/continent and continent/continent. At the ocean/ocean margin, one plate slides below the other in a process known as subduction. The surface expression of this process is a deep ocean-trench, such as the Mariana Trench in the western Pacific, the lowest point on the Earth's surface. Subducted plates are carried down into the mantle where they melt and produce gaseous magma, which rises to the surface producing volcanic island arcs. Examples include the Aleutian Islands of Alaska, the Indonesian archipelago and the West Indies. In the case of ocean/continent margins, the oceanic crust is subducted below the continent, and major mountain building may result from multiple intrusions of igneous rock, eruptions of lava and accretion of sediments scraped off the subducting plate. Examples include the Andes. Where two continents move together, they are too buoyant to be subducted and instead a "head-on" collision may occur in which continental crust fractures and is thrust upwards into mountain belts reinforced by igneous intrusions. Examples include the Himalayas, where India is moving into the Asian continental plate, and the Alps, where the African and European plates collided.
- *Transform boundaries*, where one plate slides laterally past another along a fault line, without the creation or loss of plate material. Transform boundaries or transform faults link sections of oceanic ridges but also occur in other locations. For example,

much of the famous San Andreas Fault in North America is a transform fault where the North American and Pacific plates slide past each other.

From the above descriptions, it should be clear that crustal movement produces a diverse range of processes, materials and morphologies depending on the type of plate margins. As a simplification, divergent margins are typified by shallow-seated earthquakes and relatively quiet, fissure-type eruptions of basalt, resulting in low-angle, shield volcanoes. Convergent margins, on the other hand, are typified by deep-seated, destructive earthquakes, explosive volcanic eruptions of more viscous lavas and pyroclastic emissions resulting in steep-sided volcanic cones. They are also typi-fied by violent mountain building, a process known as *orogenesis*. Summerfield (1991, 2000) describes the development of major mountain ranges including the Andes, the Himalayas, the Tibetan Plateau and the southern Alps of New Zealand.

2.5   Landscapes of Plate Interiors

Within this overall pattern there are, of course, exceptions and complications. For example, the Hawaiian Islands are volcanically active, but lie in the middle of the Pacific Plate, far from a plate boundary. The explanation is that they lie above a so-called "hot-spot", where a plume of magma rises from deep in the mantle and is able to penetrate the crust and erupt on the surface. Since these hot spots remain in the same place as the crustal plates drift over them, it is not surprising to find that the Hawaiian Islands form a linear chain of volcanic islands with the oldest being furthest from the current position of the hot spot. Another well-known hot spot occurs at Yellowstone in Wyoming, where no lava has erupted for 600,000 years, but where hot water and steam heated by an underground magma chamber reach the surface as hot springs and geysers (see Figs. 5.2 and 5.3). Another hot spot lying away from a plate margin occurs below the western Spanish Canary Islands, but some hot spots lie on plate boundaries as indicated by the presence of more intense volcanic activity there than elsewhere on the same margin, for example, Iceland.

Another complication involves the accretion of microplate terranes onto larger con-tinental plates. For example, North America has received several of these through its history. Most of Florida is probably a fragment of Africa! The Appalachians comprise several microplate remnants of ancient Europe, Africa and oceanic islands welded onto an older eastern seaboard. And most of the western seaboard and western part of the North American Cordillera comprise accreted terranes. As a result, the phys-iography of North America has what Marsh (1997) describes as a "startling variety", celebrated as part of the national heritage in both Canada and the USA (Fig. 2.6 and Table 2.2). Summerfield (1991) describes the landform and structure of other plate interiors and continents.

It should be clear from the above discussion that a knowledge of Earth history and plate tectonics leads to an understanding of the distribution and nature of the major structural features of the Earth's surface – continents and oceans, mountain ranges and ocean trenches, volcanic island arcs and rift valleys, ocean ridges, continental shields and transform faults. All these help to give the planet a variety of form and structure – a global geodiversity – and form the backdrop for examining smaller-scale geodiversity.

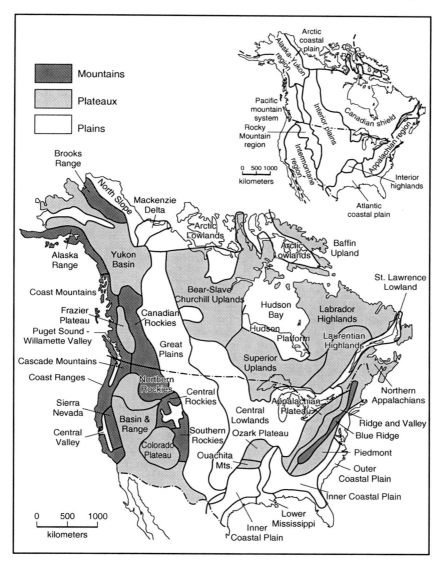

Figure 2.6   *The physiographic diversity of North America (Reproduced from Marsh, W.M. (1997) Landscape Planning: Environmental Applications. 3rd ed., Wiley, New York)*

2.6   Earth Materials

In this section, the variety of abiotic materials comprising the Earth's crust will be outlined, beginning with minerals, the building blocks of sediments and rocks and therefore of the planet itself.

2.6.1   Minerals

Geological minerals are defined as naturally occurring, crystalline, solid, inorganic substances with a specific chemical composition and an internal structure characterised

Table 2.2 *Physiographic regions and provinces of the USA and Canada (after Marsh, 1997)*

Regions	Provinces
Canadian Shield	Baffin Upland
	Bear-Slave-Churchill Uplands
	Superior Uplands
	Laurentian Highlands
	Labrador Highlands
	Hudson Platform
Appalachian Mountains	Blue Ridge
	Piedmont
	Ridge & Valley
	Appalachian Plateau
	Northern Appalachians
Interior Highlands	Ozark Plateau
	Ouachita Mountains
Atlantic Coastal Plain	Outer Coastal Plain
	Inner Coastal Plain
	Lower Mississippi
Interior Plains	Central Lowlands
	Great Plains
	St Lawrence Lowlands
Rocky Mountain Region	Canadian Rockies
	Northern Rockies
	Central Rockies
	Southern Rockies
Intermontane Region	Colorado Plateau
	Columbia Plateau
	Basin & Range
Pacific Mountain System	Alaska Range
	Coast Mountains
	Frazier Plateau
	Cascade Mountains
	Coast Ranges
	Sierra Nevada
	Central Lowlands
	Puget Sound – Williamette Valley
Alaska-Yukon Region	Brooks Range
	Yukon Basin
Arctic Coastal Plain	North Slope
	Mackenzie Delta
	Arctic Lowlands

by an orderly arrangement of atoms, ions or molecules in a lattice. This orderly arrangement occurs during the process of crystallisation, when the constituent atoms, ions or molecules come together in the appropriate chemical proportions and alignments related to chemical bonding and electron sharing. During crystallisation, original microscopic crystals grow larger with well-formed crystal faces if they are free to grow (see below).

Minerals may crystallise in a variety of ways. When magma cools, minerals will crystallise when the temperature drops below their melting point, which varies from mineral to mineral. Crystallisation also occurs by evaporation of a solution. When water evaporates, any minerals dissolved in the water will be left behind. As the solution becomes saturated, the dissolved mineral starts precipitating in crystalline form. For example, salt crystals form when sea water evaporates and this is still used as a method of producing crystalline sea salt in many countries. Finally, new minerals may form when minerals are heated. As the atoms and ions become more mobile, they may rearrange themselves to become new minerals with different crystal structures, a process known as *solid-state diffusion* (Press & Siever, 2000).

With 92 elements in the Periodic Table, it should not be surprising to find that there are 3000 to 5000 known minerals, though many are quite rare and only about 1% of them commonly occur in rocks. Some of the more common are listed in Table 2.3. Minerals have diversity in several respects, and these properties are not only vital in identifying minerals, they also determine the uses to which the minerals can be put (see Chapter 3). Anyone requiring more information on mineralogy may wish to consult Putnis (1992), Batty & Pring (1997), Perkins (2002) and Klein (2002). In outline, the variations include the following:

- *Chemical composition.* We have already noted that the most common earth material is silica (SiO_2), which, in its crystalline form, occurs as the mineral quartz, but others have very complex formulae. For example, biotite, a black mica, has the formula $K(Mg,Fe)_3(AlSi_3O_{10})(OH)_2$. The proportions of iron and magnesium in biotite can vary, so that even within a single mineral, we find diversity. Biotite mica is a silicate, the most common mineral group. Silicates are built from the basic silicate tetrahedral ion $(SiO_4)^{4-}$ in which four oxygen ions surround, and share electrons with, a silicon ion. These can then be combined into rings, chains, double chains,

Table 2.3   Some common minerals and their characteristics (after Marshak, 2001)

Amphibole	Dark-coloured, non-metallic mineral
Calcite ($CaCO_3$)	White, pink or clear with milky lustre; fizzes in acid
Chlorite	Dark green; composed of thin flakes
Clay	Group of minerals occurring as very thin flakes
Corundum (Al_2O_3)	Very hard – used as an abrasive
Diamond (C)	Clear and glassy; hardest mineral known
Dolomite ($CaMg[CO_3]_2$)	Similar to calcite, but less reactive to acid
Feldspar	Group of common, light-coloured, hard minerals
Orthoclase ($KalSi_3O_8$)	Pinkish, common feldspar also known as K-feldspar
Plagioclase	Common feldspar containing Na or Ca; white, grey or blue
Galena (PbS)	Dark grey; metallic cubic crystals
Garnet	Dark brown to maroon, glassy, roundish grains
Graphite	Very soft, grey and metallic
Gypsum	Clear and glassy, very soft
Halite (NaCl)	Clear or grey, cubic grains
Haematite	Either earthy coloured or shiny bluish grey and metallic
Kyanite	Bluish, blade-shaped
Magnetite (Fe_3O_4)	Dark grey to black, metallic and magnetic
Mica	Group of minerals occurring in thin flakes or sheets
Biotite	A black mica
Muscovite	A clear to light brown mica
Olivine ($[Mg,Fe]_2SiO_4$)	Olive green, glassy; occurs in clusters
Pyrite (FeS_2)	Golden bronze; metallic; cubic form
Pyroxene	Dark-coloured; non-metallic mineral
Quartz (SiO_2)	Glassy and hard; typically white, grey or clear
Serpentine	Group of minerals forming thin, thread-like needles
Talc	Very soft mineral

sheets and frameworks to form several common minerals including feldspar (Deer *et al.*, 2001), mica, amphibole and pyroxene. Apart from the silicates, other important mineral groups include carbonates such as calcium carbonate ($CaCO_3$), oxides such as the common iron oxide haematite (Fe_2O_3), sulphides such as pyrite (FeS_2), sulphates such as gypsum ($CaSO_4.2\,H_2O$), halides such as fluorite (CaF_2) and native metals such as copper (Cu).

- *Crystal size.* Large, well-formed crystals occur in conditions where they are able to grow slowly and in an unrestricted environment. For example, this may be in open spaces in rocks, such as cavities and fractures. Garcia-Guinea & Calafarra (2001) describe giant crystals of gypsum discovered in an underground cavity in South East Spain. However, if space is restricted or growth is rapid, then crystal growth is limited and they may coalesce to form a solid crystalline mass composed of mineral grains. Natural glass such as obsidian is formed from molten material that solidifies so quickly that there is no internal atomic order and crystals have not had time to form.

- *Crystal form and habit.* Minerals are defined as crystalline solids, which means that the atoms, ions and molecules form a regular pattern known as a *crystal lattice*. Galena (PbS), for example, consists of lead and sulphur atoms packed together in a regular array. Given the freedom to grow (see above), this configuration results in galena crystals with a cubic form. Common salt crystals (NaCl) also have a cubic form. However, crystals can come in a very wide diversity of shapes. For example, they may be plate-like as in the micas, trapezoid as in calcite, pyramidal as in diamond, and so on. Some substances with exactly the same chemical composition have more than one crystal structure and therefore form more than one mineral. For example, carbon exists in two mineral forms as diamond and graphite. Diamond, which has a closely packed tetrahedral crystal structure and a density of 3.5 g/cm^3 forms at very high temperatures and pressures, whereas graphite, which has hexagonally arranged atoms in weakly bonded sheets and a density of only 2.1 g/cm^3, is formed at lower temperatures and pressures. Crystal habit refers to the general shape or character of a crystal. For example, asbestos has a fibrous habit, a characteristic that is now known to cause lung disease.

- *Hardness.* Mineral hardness varies from very hard, like diamond, to very soft, like talc. In 1822, the Austrian mineralogist, Friedrich Mohs, devised a scale of mineral hardness based on the ability of a mineral to scratch others (Table 2.4). Hardness is related to the strength of the chemical bonding. Most silicate minerals lie in the range 5 to 7, except for sheet silicates like mica that lie in the range 2 to 3.

- *Cleavage.* This refers to the presence of splitting planes within mineral crystals. Muscovite mica, for example, splits into thin sheets less than a millimetre thick very easily, whereas quartz and garnet lack cleavage planes. The explanation again lies in crystal structure. Mica is a sheet silicate, whereas quartz and garnet are bonded strongly in all directions. Some minerals have several cleavage directions. Calcite and dolomite, for example, each have three.

- *Fracture.* This refers to the nature of break surfaces other than cleavage planes. The different fracture styles include conchoidal, fibrous and splintery.

- *Lustre.* This refers to the light reflecting qualities of minerals. Terms such as metallic, vitreous, resinous, greasy, pearly, silky and adamantine are used to describe the lustrous properties of different minerals.

Table 2.4 *Mohs' scale of hardness*

10	Diamond
9	Corundum
8	Topaz
7	Quartz
6	K-feldspar
5	Apatite
4	Fluorite
3	Calcite
2	Gypsum
1	Talc

- *Colour and streak.* Colour and streak of minerals may be characteristic of minerals or may be the result of impurities. Pure magnesium olivine is white, but with iron impurities is green. This is unusual since iron normally gives a red or brown colouration to minerals. Quartz may be clear, white, purple, grey or rose. Gemstone colour is discussed in Section 3.5. Streak refers to the colour produced by scraping a mineral against a hard, unpolished white surface, and this again varies.
- *Specific gravity.* This varies greatly in minerals. The difference between the two carbon minerals, diamond and graphite, was referred to above. Specific gravity depends on the atomic weight of the constituent elements and how closely the atoms are packed together. Lead, for example, has a specific gravity of $11 \, g/cm^3$ because it has a high atomic weight, whereas for quartz the value is $2.65 \, g/cm^3$.
- *Chemical properties.* Many minerals have distinctive chemical properties, some of which are valuable in their economic use (see Section 3.5).

Minerals, therefore, display a huge diversity of physical and chemical characteristics that generally result from their chemical composition and atomic structure. These properties give minerals their practical values and in themselves contribute to the diversity of rocks, as we shall now describe.

2.6.2 Rocks and sediments

Rocks are a natural aggregation of minerals and fall into three groups – igneous, metamorphic and sedimentary – each of which has its own incredible diversity. Igneous rocks form by the solidification of molten rock, metamorphic rocks have been transformed under intense heat and pressure from pre-existing rocks, and sedimentary rocks formed by the consolidation and cementation of sediment deposited at the Earth's surface. Together, they provide us with a fairly comprehensive history of the planet, they give us an insight into its internal processes, and they even suggest explanations for the origin of the solar system. Rocks therefore deserve to be better understood and valued by society for their amazing diversity, for the clues they can give us about past environments, and for their practical uses (see Section 3.5).

The theory of plate tectonics allows us to see the three great rock groups as part of the rock cycle originally proposed by Hutton (1795). Igneous rocks formed during plate collisions together with sediments that have accumulated in the vicinity are uplifted at convergent margins during mountain-building episodes (orogenies). The heat and pressure exerted during these episodes also create metamorphic rocks. Once uplifted, mountains become susceptible to weathering and erosional processes, and the debris produced is transported to the oceans by rivers, glaciers and mass movement (see below). Here, the sediments accumulate in thick sequences that are hardened and cemented into sedimentary rocks. At the base of these sequences, the temperature and pressure may be sufficient to form further metamorphic rocks. Finally, these materials may be transported to a convergent plate margin where the cycle recommences as part of a perpetual recycling process (Press & Siever, 2000). Of course, this is a generalised representation of a much greater natural complexity and diversity.

Igneous rocks

Apart from proposing the basis for the rock cycle, James Hutton was also responsible for first identifying a solidified igneous rock at Salisbury Crags in Edinburgh, Scotland, where an underlying block of sedimentary rock had been baked and partially incorporated into the rock above. Hutton argued that the latter must have been molten when emplaced in order to bake the sediment and surround the partially detached block. After more than 200 years of further study, geologists now have a very detailed knowledge of igneous rocks and their formation, and for detailed information, readers are referred to Hall (1996) and Best & Christiansen (2001). Not surprisingly, there is a huge diversity of igneous rocks but their classification is based principally on texture and chemical and mineral composition (Fig. 2.7).

- *Texture*, in this context, refers to the size of the mineral grains. "Coarse-grained" refers to a rock with large crystals, while "fine-grained" is applied to a rock with

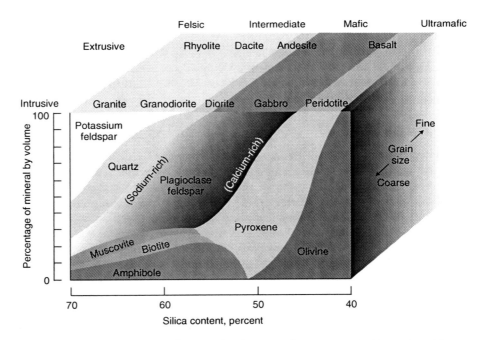

Figure 2.7 *Basic classification showing the diversity of igneous rocks (Reproduced from Press, F. & Siever, R. (2000) Understanding Earth. 3rd ed. W.H. Freeman, New York, by permission of W.H. Freeman and Company/Worth Publishers)*

small crystals. It has already been noted that rapid cooling of magma produces a non-crystalline natural glass (obsidian). On the other hand, slow cooling in restricted environments results in the growth of large crystals. The relevance of this to igneous rocks is that magma that is intruded deep in the Earth's crust cools only very slowly and becomes a coarse-grained intrusive igneous rock, whereas magma that erupts at the surface as lava (extrusive igneous rock), is subject to rapid cooling and therefore develops a fine-grained texture. Even here, however, there may be differences in crystal size between the interior of the lava flow, which cooled more slowly than the margins. Special names are given to other types of texture. For example, the term "porphyritic" refers to an igneous rock in which large crystals (phenocrysts) sit in a finer-grained matrix. A final type of igneous rock comprises materials that have been ejected during violent igneous eruptions when everything from fine-grained ash to large bombs may accumulate around a volcano. These are subsequently hardened into so-called pyroclastic rocks, such as pumice, scoria and tuff.

- *Chemical and mineral composition* is the other important characteristic in igneous rock classification (Fig. 2.7). Chemical composition is usually based on silica content, which normally varies between 40 and 70%. Rocks at the upper end of this range, like rhyolite and granite, are termed "silicic", or sometimes "acidic", whereas those at the lower end of this scale, like basalt and gabbro, contain higher percentages of magnesium (Mg) and iron (Fe) and are therefore given the name "mafic" or sometimes "basic". In terms of mineralogy, silicic rocks contain quartz, potassium- and calcium-rich feldspars and lower percentages of mica and amphibole. Mafic rocks, on the other hand, are dominated by sodium-rich plagioclase, olivine and pyroxene. Mineral composition also produces differences in colour and specific gravity. Silicic rocks are lighter in both specific gravity and colour, usually with light- or mid-grey colourations or pink where orthoclase feldspar dominates. Mafic rocks, on the other hand, are denser and darker, with black, dark grey and green predominating. The above descriptions are a gross oversimplification of the variations in rock composition and texture that occur in nature and readers should refer to the specialist texts for a full appreciation of the diversity of igneous rocks.

The explanation of the variation in the chemical composition of igneous rocks lies in the chemistry of the magma. At divergent margins, where the magma rises from the mantle and erupts at the surface without passing through the continental crust, the lava is usually mafic in composition, contains less gas and is less viscous. This leads to predominantly basaltic lavas, less-violent eruptions and low-angle lava flows and volcanoes. Current examples include Iceland and Hawaii, while past examples include the flood basalts of Washington and Oregon, USA and at the Deccan Plateau in India. On the other hand, at convergent margins, the magma rises through continental crust and as it does so it may melt large quantities of silica-rich sedimentary rocks. The molten mix is therefore much more silicic in composition, contains more gas and is more viscous. This results in violent eruptions, rhyolitic and andesitic lavas and pyroclastic rocks, and steep-sided cones like Fujiyama in Japan. Igneous processes and morphologies are considered further later in this chapter.

Before leaving igneous rocks, it is also important to consider briefly the possible outcomes of fractional crystallisation, a process proposed by Canadian geologist N.L. Bowen early last century. This involves the separation and removal of successive fractions of crystals as magma cools. It occurs because the temperature at which minerals crystallise from a melt varies. For example, olivine has a very high melting point of about 1800 °C. As magma cools, olivine will crystallise at this temperature, and as the olivine crystals form, they may begin to settle to the base of the magma chamber. Further cooling may result in new minerals crystallising with the result that a mineralogically differentiated body is produced. While some igneous intrusions do exhibit this character, including the Palisades cliff on the west bank of the Hudson River, facing New York, others do not, and it is now recognised by geologists that Bowen's proposals were oversimplified. Geological reality produces a much more diverse range of outcomes.

Sediments and sedimentary rocks

Sediment is derived from the weathering, erosion, transportation and deposition of rock fragments (clastic sediments), or from the precipitation from water of chemical and biochemical minerals (chemical and biochemical sediments). The former are much more abundant than the latter and many variables combine to produce a large diversity in clastic sediments. Readers requiring full information on the classification of sedimentary rocks are referred to Greensmith (1989), Prothero & Schwab (1996), Selley (1996), Reading (1996), and Tucker (1996, 2001).

- *Particle size distribution and sorting.* Sediments vary in particle size from massive boulders, through cobbles, pebbles, sand and silt to clay, and several classification systems of particle sizes exist. Not surprisingly, this variability has much to do with the processes of weathering, erosion, transportation and deposition that will be discussed in more detail below. Clays are often the result of chemical weathering of minerals and because of their fine-grained nature, are easily transported by rivers to lakes or oceans where they settle-out in the low-energy, deep-water environments. Boulders, on the other hand, reflect high-energy erosion, transportation and sedimentation, perhaps associated with catastrophic floods. Size is also controlled by bedrock characteristics. Closely jointed rocks such as slates are easily eroded into small rock fragments compared to a massively jointed granite where only large blocks can be detached. But whatever the original size of the entrained particles, they are reduced in size the further the distance of transport due to attrition. Particle sorting refers to the range of particle sizes in a sediment. A sediment with a wide range of sizes from clays to boulders as in many glacial sediments is described as *poorly sorted*, whereas one that is dominated by a single size range, for example, a beach sand, is termed *well sorted*. Some rock units, termed *graded units*, show a change in particle size vertically within the unit, usually becoming finer upwards, due, for example, to larger grains settling out of water faster.
- *Particle composition.* Sediments may be composed of a variety of minerals, depending on the mineralogy of the source rock and subsequent processes. Clays are usually composed of the minerals produced by chemical weathering such as kaolinite, montmorillonite and illite. We all know that the most attractive beaches are sand coloured, and are generally composed of quartz grains stained orange by iron impurities. Similarly for most desert sands. But sands derived from mafic igneous rocks are black, as in Iceland and Hawaii. Coarse clastic sediment particles are usually composed of more than one mineral grain, and it is not unusual to see rock particles brought together and cemented as a new rock, as in the case of conglomerates and breccias.

Sedimentary rock cements also vary in composition, some being silica-based and others being calcitic.

- *Particle shape.* Particle shape is described in various ways. Terms such as blades, rods, discs and spheres are used to describe the three-dimensional shape of coarse clastic particles. Sphericity is a measure of how equal the three axes are, whereas roundness refers to two-dimensional rounding of the edges. Hence a cube has a high sphericity but a low roundness. Mineral grains in an igneous rock are highly irregular where they become interlocked during crystallisation. On the other hand, after erosion and transportation in water, they may become highly rounded as they come into contact with other particles and the river bed. Detailed shape may reveal particle history. Frosted grains result from wind transportation, whereas glacial transport produces characteristic fractures on mineral grains.

Although there are other sedimentary variables such as colour, fossil content, compaction, and so on, these three characteristics – particle size, shape and composition – combine to produce the three main factors used in classifying sedimentary rocks, and Table 2.5 shows some of the main types.

Chemical and biochemical rocks also exhibit significant geodiversity. The main chemical sediments are evaporites, including rock salt and gypsum. Rather like igneous crystallisation, there may also be fractional differentiation of evaporite minerals with gypsum, halite, anhydrite and, finally, salts of potassium and magnesium crystallising as evaporation progressively concentrates the salt solution. All of these minerals are

Table 2.5 *A common classification of sedimentary rocks (after Greensmith, 1989; Press & Siever, 2000)*

Sediments	Particle size or minerals	Rocks
Boulder		
	256 mm	Conglomerate
Cobble		Breccia
	64 mm	Agglomerate
Gravel		
- - - - - - - - - - - - - -	2 mm -	
Sand		Sandstone, coarse tuff
- - - - - - - - - - - - - -	0.062 mm -	
Silt		Siltstone, fine tuff
- - - - - - - - - - - - - -	0.0039 mm -	
Clay		Mudstone (blocky fracture)
		Shale (bedding fracture)
Carbonate sand & mud	Calcite	Limestone
	Dolomite	Dolostone
Iron oxide sediment	Haematite	Iron formation
	Limonite	
	Siderite	
Evaporite sediment	Gypsum	Evaporite
	Anhydrite	
	Halite	
Siliceous sediment	Opal	Chert
	Chalcedony	
	Quartz	
Carbonaceous sediment	Coal, etc.	Organic
Phosphatic sediment	Apatite	Phosphorite

commercially valuable, as we shall see in Chapter 3. Examples of significant evaporite deposits include the Permian rocks of the Zechstein Sea that stretched from Russia to England, but the evaporite sequences are particularly thick in eastern Germany and the Netherlands. Important lake evaporites occur at the Great Salt Lake in Utah, USA and in the East African Rift Valley. Phosphorite, ironstone and coal are other examples of chemical and biochemical rocks, while oil and gas are geologically formed organic fluids. Chert is a chemically or biochemically precipitated form of silica and is similar to flint, layers of which are abundant in the English and French chalk rocks.

Chalk is in fact one of a series of biochemical sedimentary rocks, other examples being limestone and dolomite. Chalk and limestone are calcium carbonate rocks made up of lithified carbonate shells and tests of micro-organisms such as foraminifera. Coral reefs are a particular form of calcium carbonate and give rise to reef limestones. Dolomite is a chemically altered form of limestone by the addition of magnesium ions shortly after deposition. Chemically precipitated calcium carbonate also occurs, including cave stalactites, stalagmites and other features collectively termed *speleothems*.

A variety of sedimentary structures may also occur within individual sedimentary rock units. These include horizontal bedding and lamination, cross-bedding, dune bedding, ripple marks, sole structures, channel scours and mud cracks. Deformational structures commonly seen within rock sequences, and which themselves display great diversity, include slump structures, load casts, dewatering structures, organic structures, joint patterns, faults and folds (Fig. 2.8).

Metamorphic rocks

Metamorphic rocks result from solid-state changes to other rocks induced by high temperatures and pressures. The changes may result in changes in texture, mineralogy or chemical composition of a rock, and sometimes involves all three so that the original nature of the rock is unrecognisable. In these instances, geologists refer to the changes as high-grade metamorphism, whereas lesser changes brought about by lower temperatures and pressures are referred to as low-grade metamorphism.

Figure 2.8   *A fold in Palaeozoic sedimentary strata, Shell Canyon, Wyoming, USA*

Temperature in the crust rises by about 30 °C/km depth. Thus, at a depth of 10 km, the temperature will be $c.300$ °C and at 30 km depth it will be $c.900$ °C. Most metamorphic rocks form within this depth and temperature range, with high-grade rocks forming in the deeper and hotter parts of the crust. Heat can bring about profound changes to the mineralogy and texture of rocks by breaking chemical bonds, altering the crystal structure of minerals and initiating re-crystallisation of new or altered minerals.

The high confining pressures deep in the crust are capable of producing new minerals with denser crystal structures. Furthermore, the directed pressures frequently experienced at convergent plate margins, cause minerals to change size, shape and orientation. The combined effect of heat and pressure is critical in producing the foliated appearance of many metamorphic rocks since these influences cause minerals to segregate into planes and grow perpendicular to the applied stress. Since heat also makes rocks more pliable, these mineral bands often display severe folding and deformation structures.

Other metamorphic processes include chemical and physical changes to the rocks surrounding an igneous intrusion. First, contact metamorphism bakes the rocks along the contact zone, with rocks such as hornfels being formed close to the contact and other mineralogical changes occurring farther away within a so-called metamorphic aureole. Secondly, it is normal for magma intrusion to be accompanied by hydrothermal solutions that permeate into the intruded rocks and react with them, changing their chemical and mineral compositions and sometimes completely replacing one mineral with another without changing the rock's texture, a process known as *metasomatism*. Dissolved minerals may be carried away to re-precipitate as veins in rock fissures. Quartz veins are very common where silica-rich hydrothermal fluids have permeated through a rock, and many valuable metal ores such as copper, lead and zinc are also formed in this way (Fig. 3.11). In the case of divergent, submarine, plate margins, heated sea water plays an important role in altering the chemical composition of the ridge basalts.

Like other rock types, metamorphic rocks exhibit a great diversity of types, whose characteristics are summarised below. For more detailed descriptions of metamorphic rocks, readers are referred to Fry (1992) and Shelley (1993).

- *Cleavage.* This refers to the finely spaced, planar partings in rocks such as slates and phyllites, and this fracture cleavage should not be confused with mineral cleavage discussed above. It results from low-grade metamorphism of fine-grained rocks such as shales, where the pressure causes a realignment of the mineral grains perpendicular to the applied pressure. This produces parallel cleavage planes unrelated to the original bedding, which may still be visible. Slates are generally dark grey or black, but may be coloured red or purple where iron oxide minerals are present or green where chlorite is present.
- *Schistosity.* This is the name given to a coarser foliation produced not only by the lineation of platy minerals such as biotite and muscovite micas but also including quartz, feldspars and other minerals. It gets its name from schist, one of the most common metamorphic rock types, which frequently exhibits wavy splitting planes. Schists are frequently named from the dominant minerals present, and thus we have mica schists, chlorite schists, quartz schist, garnet schists and a diverse range of others.
- *Banding.* This is a still coarser type of foliation common in high-grade metamorphic environments where the micas and chlorite are lost and quartz, feldspars and

Figure 2.9   *Banded gneiss, Helsinki, Finland*

mafic minerals dominate. These rocks frequently display spectacular, deformed, light (quartz and feldspar) and dark (mafic minerals) segregated bands of minerals, and are termed *banded gneisses* (Fig. 2.9).

- *Non-foliated rocks.* Non-foliated metamorphic rocks include quartzites (metamorphosed quartz-rich sandstones) and marbles (derived from limestones and dolomites). Some, like the famous Italian Carrara marble, are pure white, but most contain irregularly coloured bands and streaks, giving the marbles an endless diversity. Other non-foliated metamorphic rocks include argillites (produced by low-grade metamorphism of shaly sedimentary rocks) and greenstones (metamorphosed mafic-rich volcanic rocks, often produced by sea water reactions at divergent oceanic plate margins and receiving their colouration from chlorite).

- *Porphyroblasts.* These are large crystals that have grown faster than the finer-grained matrix in which they are set. The crystals may vary between a few millimetres and several centimetres in diameter, and examples include cubic iron pyrites crystals in slate and garnet crystals in schist.
- *Shear textures.* These form when two rock surfaces slide past each other. Rocks known as mylonites form at the shear surfaces as rocks are crushed and sheared under high pressures deep in the crust.

Where metamorphism has affected a large area, for example, in the orogenic belts of the Alps and Appalachians, systematic changes in the metamorphic grade may be seen, indicating a transition from high to low grade. A sequence of index minerals from chlorite, biotite, garnet, staurolite, kyanite to sillimanite is produced by increased metamorphism in a uniform shale, though there are many local variations caused by compositional variations in the parent rocks (see Fig. 2.10a). Metamorphism of basalt, for example, produces a zeolite, greenschist, amphibolite, pyroxene granulite series, but at very high pressures blueschist and eclogite form instead (see Fig. 2.10b). Migmatites form at very high temperatures and pressures and are transitional to igneous rocks. They are typified by contortions and veins of melted rock.

It should be clear from the above discussion that a great diversity of metamorphic rocks can be formed dependent mainly on temperature, pressure and compositions of the parent rocks.

Sequences and structures

In discovering systematic variations in metamorphic grades, we are beginning to see how rock units relate to each other. In the case of metamorphic grades, we are dealing with spatial variations in rocks brought about at more or less the same time. However, it is more common for geologists to study temporal rock and sediment sequences in order to study the changing geological environments. An infinite diversity of such sequences exists near the surface of the planet, though in many places it is possible to trace the same rock strata over long distances. In other places, systematic changes occur laterally because of environmental changes, such as a reef facies adjacent to a beach facies. Breaks in the sequence are referred to as *unconformities* (see Fig. 5.22), and time-transgressive changes are sometimes observed, for example, where a relative sea-level rise covers different areas at different times. Geological structures are described in Davis & Reynolds (1996).

Vertical sequences show diversity in other variables, for example, palaeomagnetism, and, on a spatial level, geophysical variations in seismic refraction or resistivity are used to detect underground anomalies and patterns for both pure and applied research.

2.6.3 Fossils

The diversity of the fossil record has been evident since before Carl Gustav Linneaus (1707–1778) established a classification system and Charles Darwin (1809–1882) explained it as the result of continued evolution of species from a common source. The generally accepted view is that life began around 3.8 Ga years ago by biochemical processes and has diversified from simple prokaryotes and eukaryotes to the present level of 1.5 to 1.8 million formally named and described species, but possibly over 30 million species in total (Lovejoy, 1997). Fossils commonly exist as hard parts (bones, shells,

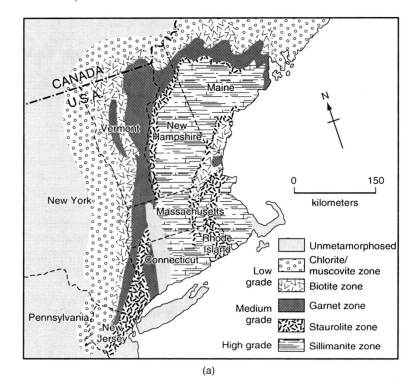

(a)

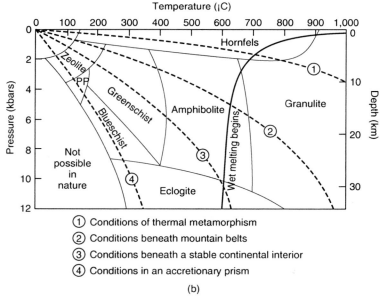

① Conditions of thermal metamorphism
② Conditions beneath mountain belts
③ Conditions beneath a stable continental interior
④ Conditions in an accretionary prism

(b)

Figure 2.10 *Metamorphic diversity: (a) metamorphic zones in New England, USA and (b) common metamorphic facies at varying temperature and pressure conditions (After Marshak, S. (2001) Earth: Portrait of a Planet. W.W. Norton & Co, New York, by permission of W.W. Norton & Co.)*

(a)

(b)

Figure 2.11   *Fossil diversity: (a) crinoids, (b) corals, Donegal Bay, Ireland*

teeth, fish scales, woody tissue) all of which may be mineralised (Fig. 2.11), but soft parts may also be preserved in particular circumstances (e.g. the Cambrian Burgess Shale Fauna, Canada, and frozen woolly mammoths from the Siberian permafrost). In addition, trace fossils include tracks and trails, burrows and borings, root penetration structures, faecal pellets and coprolites, regurgitation pellets and vomit, gastroliths and teeth marks (Bromley, 1996). More details about the evolution and diversity of life on Earth can be found in Clarkson (1993), Doyle (1996), Benton & Harper (1997), Kemp (1999) and Fortey (2002).

Since estimating the number of currently living species is so difficult, it should not be surprising that it is impossible to say how many species have existed during the history of the Earth. However, some authors have attempted to plot the number of fossil families against time. Figure 2.12 shows a compilation by Benton & Harper (1997) and also shows their selection of the ten major events in the history of life.

In addition, the history of life on Earth is punctuated by five major mass extinctions and several smaller events during which a significant proportion of living forms disappeared within a brief period in geological terms. For example, it is estimated that 50% of families and 80 to 95% of species disappeared during the Permo-Triassic event (Benton & Harper, 1997). Alvarez *et al.* (1980) proposed that the Cretaceous-Tertiary extinction was caused by a 10-km-diameter asteroid impacting with the Earth and creating a huge dust cloud that blocked the sun's radiation and prevented photosynthesis. This led initially to the loss of plants, followed inevitably by herbivores and finally carnivores including the dinosaurs. The diversity of the fossil record is therefore not one of simple exponential increase.

The biological recovery time from a mass extinction varies depending on the size of the event. According to Benton & Harper (1997), biotic diversity may have taken 100 million years to re-establish itself from the major Permo-Triassic mass extinction, but was probably of the order of some 10 million years for the Cretaceous–Tertiary event. Figure 2.13 shows the huge expansion of the placental mammals around the Palaeocene-Eocene boundary some 10 million years after the Cretaceous-Tertiary event and illustrates just one fragment of the diversity of life.

2.7   Processes and Landforms

This section aims to provide a brief insight into the diversity of processes and landforms at the surface of the Earth, but does not attempt a full description of all processes and landforms. Readers who require more information are referred to some excellent general geomorphological texts such as Summerfield (1991), Allen (1997), Bloom (1998), Huggett, (2003) or to the more specialised books referred to below.

2.7.1   Igneous processes and forms

We have already discussed some aspects of igneous processes and landforms in describing plate tectonic processes and igneous rocks above. Here we shall concentrate on the diversity of types of eruption, and morphologies of volcanoes and extrusive and intrusive products. For more details, see Francis (1993), Decker & Decker (1998) and Ritchie & Gates (2001).

Table 2.6 is a commonly used classification of volcanic eruptions, which relates the nature of effusive activity to magma type, explosiveness and volcano morphology (see also Fig. 2.14). It will be appreciated that this is a simplified system and that the complexity in nature is greater than shown. As already explained, explosiveness is related to magma viscosity and gas content. A basaltic magma with low viscosity and gas content leads to unexplosive eruptions, whereas an acidic, viscous magma typical of a rhyolite leads to gas retention and explosive eruptions as the gas pressure is released at the surface. An example of the latter was the Mount St. Helens eruption in Washington, USA on 18th May, 1980.

Lava types also vary significantly. Low viscosity lava usually forms a ropy surface, given the name "pahoehoe", where cooling of the surface forms of crust that is dragged into rope-like forms by further flow. Viscous lava, on the other land, forms a blocky,

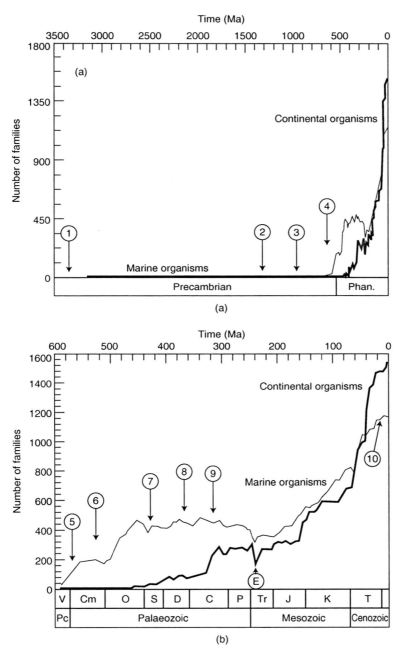

Figure 2.12 *Diversity of fossils through the history of Earth showing the 10 major events that led to increases in diversity and the major Permo-Triassic extinction (E), (a) for the whole of the last 4000 Ma, (b) for the Phanerozoic only. 1, origin of life; 2, eukaryotes and the origin of sex; 3, mulitcellularity; 4, skeletons; 5, predation; 6, biological reefs; 7, terrestrialisation; 8, trees & forests; 10, consciousness (Modified after Benton, M. & Harper, D. (1997) Basic Palaeontology. Pearson Education, Harlow. © Addison Wesley Longman Limited 1997, reprinted by permission of Pearson Education Limited)*

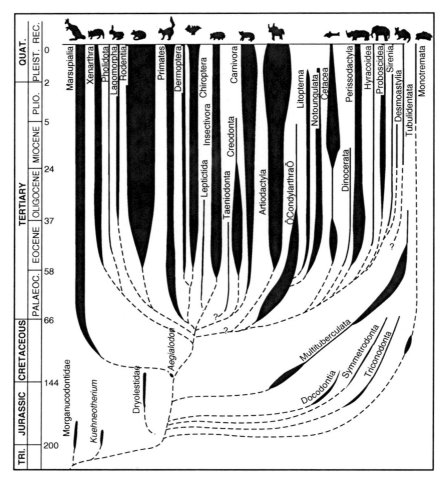

Figure 2.13 *Radiation and diversity in placental mammals after the Cretaceous – Tertiary extinction event (After Benton, M. & Harper, D. (1997) Basic Palaeontology. Pearson Education, Harlow. © Addison Wesley Longman Limited 1997, reprinted by permission of Pearson Education Limited)*

angular lava termed "aa". Among the diversity of other features associated with vulcanicity are lava tunnels or tubes, which form where the lava drains out from below a solidified crust (see Fig. 6.29) and pillow lavas formed by rapid cooling of lava underwater (see Fig. 5.28). Igneous intrusions include batholiths, stocks, laccoliths, lopoliths, sills, dykes, cone sheets and ring dykes and many of these have impressive surface topographic expressions (e.g. Fig. 2.15).

2.7.2   Slope processes and forms

Slopes are an integral part of all but the flattest of landscapes, and they evolve in a diversity of ways and variety of rates. Much depends on the strength and stability of the slope concerned (Selby, 1993). The downslope movement of slope material under the force of gravity is referred to as mass movement. It occurs where the shear stress

Table 2.6   *Diversity of volcanic eruptions (after MacDonald, 1972; Summerfield, 1991)*

Type of eruption	Type of magma	Nature of effusive activity	Nature of explosive activity	Structures formed around vent
Icelandic	Basic, low viscosity	Thick, extensive flows from fissures	Very weak	Very broad lava cones; lava plains with construction of cones along fissures in terminal phase
Hawaiian	Basic, low viscosity	Normally thin, extensive flows from central vents	Very weak	Very broad lava domes and shields
Strombolian	Moderate viscosity; mixed basic & acid	Flows absent, or thick and moderately extensive	Weak to violent	Cinder cones and lava flows
Vulcanian	Acid, viscous	Flows frequently absent; thick if present	Moderate	Ash cones, explosion craters
Vesuvian	Acid, viscous	Flows frequently absent; thick if present	Moderate to violent	Ash cones, explosion craters
Plinian	Acid, viscous	Flows may be absent, variable in thickness when present	Very violent	Widespread pumice and lapilli; generally no cone construction
Peléan	Acid, viscous	Domes and/or short, very thick flows; may be absent	Moderate plus nuées ardentes	Domes; cones of ash and pumice
Krakatauan	Acid, viscous	Absent	Cataclysmic	Large, explosion caldera

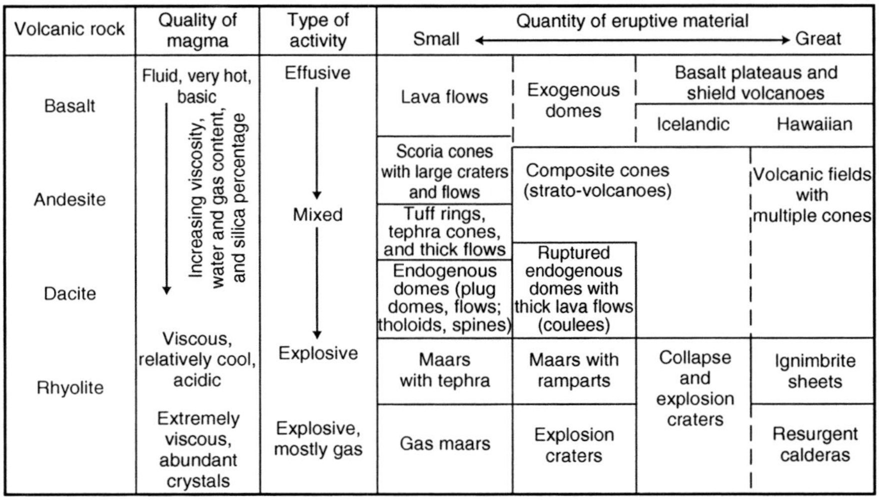

Volcanic rock	Quality of magma	Type of activity	Quantity of eruptive material Small ← → Great			
Basalt	Fluid, very hot, basic	Effusive	Lava flows	Exogenous domes	Basalt plateaus and shield volcanoes	
					Icelandic	Hawaiian
Andesite	*Increasing viscosity, water and gas content, and silica percentage*	Mixed	Scoria cones with large craters and flows	Composite cones (strato-volcanoes)		Volcanic fields with multiple cones
			Tuff rings, tephra cones, and thick flows			
Dacite			Endogenous domes (plug domes, flows; tholoids, spines)	Ruptured endogenous domes with thick lava flows (coulees)		
Rhyolite	Viscous, relatively cool, acidic	Explosive	Maars with tephra	Maars with ramparts	Collapse and explosion craters	Ignimbrite sheets
	Extremely viscous, abundant crystals	Explosive, mostly gas	Gas maars	Explosion craters		Resurgent calderas

Figure 2.14   *Diversity of volcanoes and related landforms (Reprinted from Vulkane und Ihre Tätigkeit, 3. Aufl. Rittmann (1981), with permission from Spektrum Akademischer Verlag)*

Figure 2.15   *Dyke cutting across bedded ash layers, La Palma, Spanish Canary Islands*

exceeds the shear strength of the slope, which is described as actively unstable. Since the operation of many factors potentially leading to instability may vary through time, it should be clear that some slopes are described as conditionally stable, depending on the cumulative magnitude of the factors.

The diverse types of mass movement processes have been classified in various ways but generally six types are recognised – creep, flow, slide, heave, fall and subsidence, each of which has a number of sub-groups. Table 2.7 is an attempt to relate various mass movement types to materials, moisture content, type of movement and rate of movement. However, as Summerfield (1991, p. 168) notes, classifications of this type are valuable in indicating the range of mechanisms and forms of motion, but "it must be appreciated that most movements in reality involve a combination of processes. Debris avalanches, for example, may begin as slides consisting of large masses of rock but then rapidly break up to form flows as the material is pulverised in transit".

Mass movement processes also result in a range of landforms. For example, Fig. 2.16 shows some flow-type morphologies and these do not include major avalanches

Table 2.7 Diversity and characteristics of major mass movement types (after Varnes, 1978; Summerfield, 1991)

Primary mechanism	Mass movement type	Materials in motion	Moisture content	Type of strain and nature of movement	Rate of movement
Creep	Rock creep	Rock (esp. readily deformable types such as shales and clays)	Low	Slow plastic deformation of rock or soil producing a diversity of forms including	Very slow to extremely slow
	Continuous creep	Soil	Low	Cambering, valley bulging and outcrop bedding curvature	
	Dry flow	Sand or silt	Very low	Funnelled flow down steep slopes of non-cohesive sediments	Rapid to extremely rapid
	Solifluction	Soil	High	Widespread flow of saturated soil over low to moderate angle slopes	Very slow to extremely slow
Flow	Gelifluction	Soil	High	Widespread flow of seasonally saturated soil over permanently frozen subsoil	Very slow to extremely slow
	Mud flow	>80% clay size	Extremely high	Confined elongated flow	Slow
	Slow earthflow	>80% clay size	Low	Confined elongated flow	Slow
	Rapid earthflow	Soil containing sensitive clays	Very high	Rapid collapse and lateral spreading of soil following disturbance, often by initial slide	Very rapid
	Debris flow	Mixture of fine and coarse debris (20–80% coarser than sand-sized)	High	Flow usually focused into pre-existing drainage lines	Very rapid
	Debris (rock) avalanche	Rock debris, sometimes with ice and snow	Low	Catastrophic low friction movement of up to several kilometres, usually started by a major rock fall and capable of overriding significant topographic features	Extremely rapid
	Snow avalanche	Snow and ice (plus debris)	Low	Catastrophic low friction movement started by fall or slide	Extremely rapid
	Slush avalanche	Water-saturated snow	Extremely high	Flow along existing drainage lines	Very rapid

(continued overleaf)

Table 2.7 (*continued*)

Primary mechanism	Mass movement type	Materials in motion	Moisture content	Type of strain and nature of movement	Rate of movement
Translational slide	Rock slide	Unfractured rock mass	Low	Shallow slide approx. parallel to ground surface of coherent rock mass along single fracture	Very slow to extremely rapid
	Rock block slide	Fractured rock	Low	Slide approx. parallel to ground surface of fractured rock	Moderate
	Debris/earth slide	Rock debris or soil	Low to moderate	Shallow slide of deformed masses of soil	Very slow to rapid
	Debris/earth block slide	Rock debris or soil	Low to moderate	Shallow slide of largely undeformed masses of soil	Slow
Rotational slide	Rock slump	Rock	Low	Rotational movement along concave failure plane	Extremely slow to moderate
	Debris/earth slump	Rock debris or soil	Moderate	Rotational movement along concave failure plane	Slow
Heave	Soil creep	Soil	Low	Widespread incremental downslope movement of soil or rock particles	Extremely slow
	Talus creep	Rock debris	Low		Slow
Fall	Rock fall	Detached rock joint blocks	Low	Fall of individual blocks from vertical faces	Extremely rapid
	Debris/earth fall (topple)	Detached cohesive units of soil	Low	Toppling of cohesive units of soil from near vertical faces such as river banks	Very rapid
Subsidence	Cavity collapse	Rock or soil	Low	Collapse of rock or soil into underground cavities such as limestone caves	Very rapid
	Settlement	Soil	Low	Lowering of surface owing to ground compaction or shrinkage on withdrawal of water	Slow

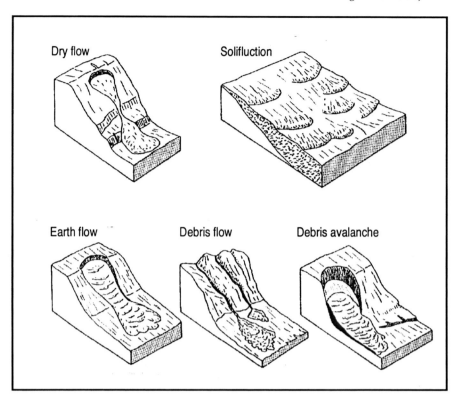

Figure 2.16 *Some flow-type morphologies (after Varnes, 1978). Reproduced with permission of the Transportation Research Board. In Special Report 176: Landslides: Analysis and Control, Transporation Research Board, National Research Council, Washington, D.C., 1978*

like those in 1962 and 1970 at Huascarán Norte in the Peruvian Andes. Slides occur when movement takes place as blocks without internal deformation. Some are translational slides taking place on planar surfaces, but rotational slides are common in homogeneous clays. Falls involve the downward movement of detached rock or soil masses. An example is the Frank Slide in south-west Alberta Canada that occurred in 1903 (Kerr, 1990), while more recently a large section at the summit of Mt Cook in New Zealand collapsed. On a more local scale, gradual fragment detachment from cliffs causes the accumulation of talus slopes and cones at the base of cliffs. Apart from mass movement processes, slopes are also affected by water. For example, rain falling on bare soil or loose sediment dislodges particles by rainsplash erosion while flowing water may produce surface rills or gullies.

Much has been written regarding slope form and slope evolution. Figure 2.17 shows the nine possible three-dimensional slope forms and special slope forms include inselbergs, mesas and buttes.

2.7.3   River environments

In the previous section, we noted that water has an impact on slopes, and here we extend this to outline the diversity of fluvial processes and landforms. For more details,

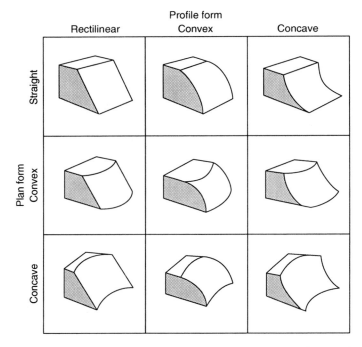

Figure 2.17 *The nine possible shapes of three-dimensional hillslope forms (After Parsons, A.J. (1988) Hillslope Form. Routledge, London, by permission of Thomson Publishing Services)*

see Knighton (1998), Bridge (2002) and Robert (2003). Rivers also transport sediment and Bloom (1998), amongst others, believes that "water flowing down to the sea and immediately beneath the land surface is the dominant agent of landscape alteration". Garrells & Mackenzie (1971) calculated that over the planet as a whole, rivers are responsible for 85 to 90% of sediment transport to the sea, glaciers for about 7% with wind, volcanoes and other processes responsible for the remainder. Of course, the impact of rivers varies greatly over the planet, largely as a result of variations in precipitation. Bloom (1998) believes that perhaps a third of the land surface has no run-off to the oceans since rainfall is very low or when it comes, it evaporates before reaching the sea, but "even arid regions with drainage into closed intermontane basins have landscapes of branching stream valleys" (Bloom, 1998, p. 198).

Of course, not all precipitation runs off into rivers, and in most areas the large majority of water infiltrates the ground on short or long timescales, but infiltration capacity will vary depending on factors such as the duration and intensity of rainfall, water content of the soil, soil texture and mineralogy, type and density of vegetation, slope angles and other factors. While much surface water may infiltrate and eventually reach the groundwater, throughflow or interflow refers to water moving through the soil profile via macropores. This may lead to soil erosion, but more important in this regard is surface runoff whether spread over the surface as sheetflow or concentrated into rills and gullies.

Eventually, the water drains into streams or rivers, which also display a diversity of character (e.g. Fig. 2.18), but all operate as three-dimensional networks within drainage

(a)

(b)

Figure 2.18 Examples of river diversity: (a) rock channel river, Killin, Scotland and (b) braided outwash river, Exit Glacier, Alaska

basins (also known as *catchment areas* or *watersheds*). Various styles of stream networks occur in nature, partly controlled by bedrock structure, tectonic processes and geology (Bloom, 1998; Schumm *et al.*, 2000), and these include trellis, parallel, radial and dendritic systems (Howard, 1967; see Fig. 2.19). Some river channels are occupied by permanent streams but others flow intermittently and still others are ephemeral. Individual river channels vary according to the following:

- *Long profile*   normally concave between headwaters and the sea in graded profiles, but can be can be linear or sometimes convex. At a smaller scale, gradient may vary between and within reaches, and may be interrupted by pools and riffles, waterfalls, rapids, cataracts and lakes.
- *Planform*   may vary between single fairly straight channels, through strongly meandering rivers to the highly complex braided channel networks. In turn, these river forms are related to gradient, sediment load, flow velocity and other factors. Meandering rivers are typified by neck cut-offs, oxbow lakes and point bar deposits on the inside of meander curves.
- *Channel cross section* in meandering rivers is asymmetrical with deeper sections on the outside of meanders, but in straight streams more or less symmetrical cross sections occur. In the latter case, river banks may be vegetated and stable, whereas in other cases they are actively eroding.

Other fluvial landforms include incised meanders where downcutting into bedrock has occurred, natural arches, flood plains with levees, alluvial fans and river terraces. Special features occur in limestone areas where dry valleys, sinkholes and underground caves with a range of speleothems may be present (see also Weathering below for discussion of karst landforms).

The sediment load of rivers is also diverse, and includes a dissolved load, a suspended load of fine particles and a bed load of coarser debris, which rolls, slides or saltates along the channel bed. Stream discharge varies through time as does the sediment load and transport regime. There are also longer timescale changes in rivers induced by climate changes so that some fluvial landforms and deposits are relict features.

Various systems have been used to describe and classify rivers including Geomorphic Characterisation (Rosgen, 1996), River Habitats Surveys (Raven *et al.*, 1997), Stream Reconnaissance Surveys (Thorne, 1998), the River Styles approach (Hardie & Lucas, 2001; Thomson *et al.*, 2001) and Geomorphic Complexity (Bartley & Rutherford, 2001). "Classifications are nonetheless arbitrary divisions of continua and cannot be expected to provide a complete understanding of river functioning" (Soulsby & Boon, 2001, p. 100).

2.7.4   Coastal environments

At a global scale, the world's coastlines can be classified according to their plate tectonic setting and Fig. 2.20 shows Davies' (1980) attempt to present such as classification. Passive margin coasts occur on the trailing edge of the continents and are typified by tectonic stability and wide continental shelves. Examples occur along the eastern coast of both North and South America. On the other hand, convergent margin coasts occur near to subduction zones or continental collisions and are typified by coastal instability and narrow shelves. Examples occur on the western coast of South America and Mediterranean. Island arcs are also common as exemplified in the

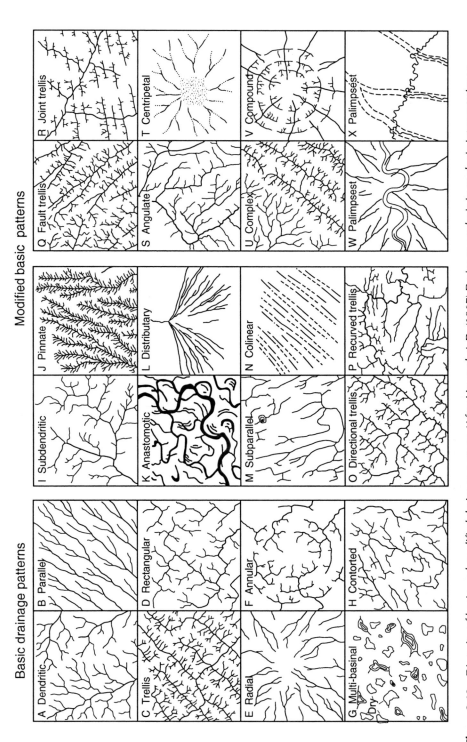

Figure 2.19 *Diversity of basic and modified drainage systems (After Howard, A.D. (1967) Drainage analysis in geologic interpretation: a summation. American Association of Petroleum Geologists Bulletin, **51**, 2246–2259. AAPG © 1967. Reprinted with permission of the American Association of Petroleum Geologists (AAPG) whose permission is required for further use)*

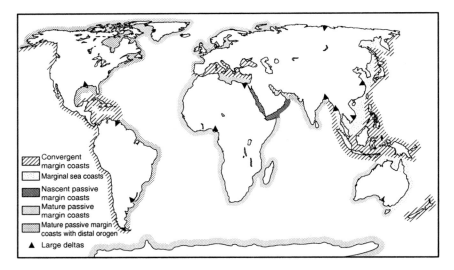

Figure 2.20   *Tectonic setting of the world's coastlines and major deltas (After Davies, J.L. (1980) Geographical Variation in Coastal Development. 2nd ed. Longman, Harlow; Summerfield, M.A. (1991) Global Geomorphology. Longman, Harlow © M.A. Summerfield, 1991, reprinted by permission of Pearson Education Limited)*

western Pacific and continental margins protected by these are termed *marginal coasts* and occur, for example, on coastal China, Vietnam and eastern Australia. Summerfield (1991, p. 315) has noted that large river deltas such as the Mississippi, Niger and Huang Ho are, not surprisingly, more or less confined to passive margin coasts and marginal coasts "since it is only these that provide the outlets for the world's great drainage systems".

Davies (1980) notes several other aspects of global-scale coastal diversity. For example, major wave environments vary mainly according to the nature of the atmospheric circulation, with storm wave environments in the westerly zones and monsoonal influences in some tropical locations. High latitude coasts are often low-energy coasts because of the protection of sea ice. Between these locations, swell environments predominate. Tidal regime and range are also highly variable.

At a smaller scale, Kiernan (1997a) attempted to assess the factors contributing to coastal geodiversity:

- *Bedrock variables* – lithology, structure, orogenic and tectonic movements, bedrock preparation, sediment characteristics.
- *Topographic variables* – submarine topography, hinterland topography, shoreline topography.
- *Oceanographic variables* – waves, tsunamis, tides, storm surges, currents, sea-level change.
- *Coastal process variables* – shore weathering, sediment transport.
- *Temporal variables* – duration of processes, rate of sea-level change, number of change cycles.

Kiernan (1997a) went on to classify Tasmania's coastal landforms and to specifically identify their geodiversity.

We can recognise both erosional and depositional processes and landforms (Bird, 2000; Haslett, 2000; Davis & Fitzgerald, 2002; Masselink & Hughes, 2003). Coastal erosion commonly results in cliffs and shore platforms whose detailed morphology will largely be determined by geological structure and lithology. Platforms on dipping sedimentary rocks form series of mini-escarpments compared with the flat platforms that form on more homogeneous rocks. The variety of lithologies, joint patterns, folds and dip angles mean that cliffs and platforms, as well as related features such as caves, undercuts, natural arches and sea stacks, display a very large morphological diversity. Wave erosion is not the only important process either. Sub-aerial processes and mass movements operate on the cliffs, solutional processes operate on carbonate rocks and a range of biochemical processes operate on the shore platforms. Bloom (1998) notes that "Animals from at least 12 phyla, and numerous kinds of plants and microbes, will graze, browse, burrow, or bore into rocks .... Many use chemical secretions to dissolve rocks, especially limestone. Many of the animals have abrasive appendages or teeth by which they remove surface layers of rocks or bore into them". The net effect may be rapid bioerosion and a range of micromorphological features.

Biological processes are also responsible for the development of coral reefs and atolls, which grow into morphological features and are represented in the geological record. Barrier reefs, such as the Great Barrier Reef in Australia, are distinguished from fringing reefs, which are attached to a coast and extend seaward, and island reefs of which Davies (1980) identified various types. Detailed morphology depends partly on the reef-building coral genera present, of which there are over 50.

Depositional coastal processes and morphologies also vary. In some places, sandy beaches occur but these vary in particle size, gradients, morphology and planform. Bars, berms, beach cusps, sand waves, ripples and stream channels are examples of the morphological forms found on sandy beaches. They may or may not be backed by storm ridges and coastal dunes of which there are several types (Davies, 1980). In places offshore barriers separated from the mainland by lagoons form, while spits, baymouth bars, cuspate forelands and tombolos are other well-known landforms. On lower energy coasts, mudflats and saltmarshes with dendritic tidal creeks and ponds occur, sometimes backed by cheniers (beach ridges of sand and shell debris). Passive tropical coasts are often dominated by mangrove swamps. Delta morphology is also diverse (see Fig. 2.21).

Since sea level changes, relict coastal features occur both above and below present sea level and also at present sea level, which may have reoccupied a much older coastline. Most of the coastal landforms described above occur in relict form.

2.7.5 Glacial environments

Glaciers currently cover about 10% of the Earth's land surface, but during full glacial periods the percentage rises to around 30%, mainly by expansion of North American and European ice sheets. Thus, in considering the diversity within glacial environments, we must consider both the glacier ice itself (as a solid component of the lithosphere) and the areas of former glaciation where the geomorphological impacts still dominate the landscape. Fuller details of the diversity of glacial environments and landsystems can be found in Bennett & Glasser (1996), Benn & Evans (1998), Martini *et al.* (2001) and Evans (2003).

Kiernan (1996) has specifically described the geodiversity of glacial landforms, describing both glacial landform "species" and glacial landform "communities". He

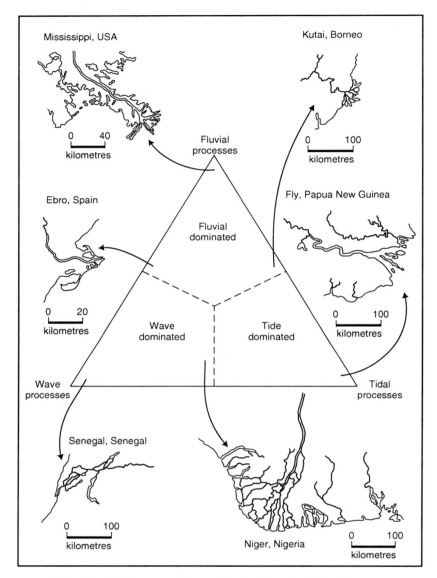

Figure 2.21 *Diversity of delta morphology related to fluvial, tidal and wave influences (After Galloway, W.E. (1975) Process framework for describing the morphologic and stratigraphic evolution of the deltaic depositional systems. In Broussard, M.L. (ed) Deltas: Models for Exploration. Houston Geological Society, Houston, 87–98; Summerfield, M.A. (1991) Global Geomorphology. Longman, Harlow. © J.L. Davies 1973, 1980, reprinted by permission of Pearson Education Limited)*

sees the major controls on glacial landscape evolution and therefore on geodiversity as being the following:

- *Glacier variables* – ice temperature, glacier morphology, glacier constraint, glacier gradient; glacier movement and velocity, ice thickness, glacial processes.

- *Bedrock variables* – lithology, structure, orogenic and tectonic history, bedrock preparation, glacial sediment.
- *Topographic variables* – preglacial topography, contemporaneous topography, post-glacial topography.
- *Temporal variables* – duration of glaciation, number of glacial stages.

Ice masses can be classified in various ways but the one most relevant to this book is the so-called morphological classification, which considers size, shape and environmental location (Table 2.8). Only two ice sheets exist at present (Greenland and Antarctica) but there are thousands of cirque and valley glaciers. Surface features may include a variety of crevasse patterns, ogives, supraglacial stream channels and ponds, moulins, lateral moraines, medial moraines and broader debris spreads. Glaciers terminating in the sea or a lake may calve to form impressive icebergs arrays (Fig. 2.22) or large rafts.

Glacial erosion produces a very large range of features at several scales (Fig. 2.23). At the continental scale, areas such as the Canadian Shield are scoured to produce a complex interplay of rock and freshwater lakes. Mountain areas, on the other hand, are carved into horns, aretes, cirques and glacial troughs (or fjords if flooded by the sea). Smaller features such as crag-and-tails and roches moutonnees occur, while at the smallest scale, striae, polished surfaces, and a range of friction cracks is observed.

Glacial depositional processes produce a range of till types, which in turn vary in sedimentological properties. Glacial depositional landforms include till sheets and a range of moraines and drumlinoid forms. Drumlins themselves are highly variable with a range of parameters being used to describe their size, shape and distribution (Rose & Letzer, 1977). Erratics are glacially transported boulders foreign to the area and like striae, drumlins and so on can be used to reconstruct directions of glacier movement (Fig. 3.23).

Glaciofluvial action also produces a range of landforms and sediments. Erosional landforms include "p-forms", potholes and meltwater channels, and, at the largest scale the channelled scablands of Washington State, USA produced by catastrophic floods as glacier lake ice dams collapsed.

Glaciofluvial sediments deposited in contact with the ice may form kames and kettle holes, eskers and kame terraces, while beyond the ice braided meltwater streams (Fig. 2.18b) deposit outwash fans, trains or plains (also known as sandar) that may become terraced by fluvial downcutting to produce outwash terraces. On reaching the sea or lake, the traction load is usually deposited as glaciofluvial deltas, while the finer

Table 2.8 *Diversity of types of ice mass (modified after Martini et al., 2001)*

Ice sheets	Continental ice sheets	Shield-like domes > 25,000 km²
	Lowland ice caps	Smaller ice domes in lowland areas
	Plateau glaciers	Flat plateau ice fields with ice cascades
	Highland ice caps	Mountain ice fields with many nunataks
Valley Glaciers	Ice streams	Fast moving ribbons within ice sheets
	Reticular glaciers	Transitional from ice sheets and glaciers
	Outlet glaciers	Glaciers descending from ice sheets/ice caps
	Alpine glaciers	Glaciers originating in cirques
	Cirque glaciers	Glaciers confined to cirques
	Cliff or niche glaciers	Small glaciers on steep slopes or cliffs
Lowland Glaciers	Piedmont glaciers	Low-angle ice-lobes laterally unconstrained
	Expanded foot glaciers	Smaller lobes where valley glaciers fan out
	Ice shelves	Floating sections of glaciers or ice sheets

Figure 2.22 *Iceberg calving from outlet glaciers from the Greenland ice sheet*

material settles as rhythmite sediments called varves. Lakes are frequently formed around the margins of glaciers and ice sheets including around the southern margin of the North American ice sheet, for example, Lake Agassiz west of Lake Superior. These lakes were frequently large enough to have shoreline features such as beaches and spits, but other lakes have erosional shorelines as seen at the Parallel Roads of Glen Roy in Scotland.

2.7.6 Periglacial environments

Like glacial environments, periglacial activity also exists in both active and fossil forms. The periglacial zone has been defined in various ways. French (1996), for example,

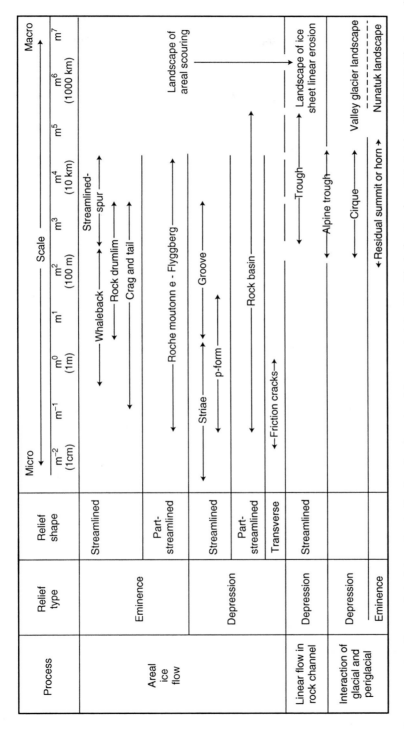

Figure 2.23 *Landforms of glacial erosion indicating the diversity of forms and scales (Modified after Sugden, D.E. & John, B.S. (1976) Glaciers and Landscape: A Geomorphological Approach. Edward Arnold, London. Reproduced by permission of Hodder Arnold)*

defines it as the area where frost action processes predominate, but Bloom (1998) believes it should be restricted to those non-glacial areas of high latitude or altitude that are underlain by permafrost.

Permafrost is ground that remains frozen for at least two years. Its thickness increases with latitude from sporadic patches, through a discontinuous zone to a continuous zone where depths of over 1000 m are possible. Unfrozen areas in, on, below or between permafrost are termed *talik*. Above the permafrost is the active layer that freezes and thaws annually, but this also varies in thickness, reaching a maximum of 2 to 3 m in the discontinuous permafrost zone. Permafrost ice may exist in several forms including pore ice, ice layers and lenses, vein ice, intrusive ice and needle ice.

Periglacial processes include frost weathering responsible for the disintegration of rock by ice expansion upon freezing of water in pores, joints, and so on. The effects will vary depending on the nature of the rock. For example, porous rocks such as chalk will be reduced to a paste, whereas a well-cleaved rock like slate will split into thin rock fragments. On horizontal ground this results in blockfields or felsenmeer, but on steep slopes the material collects as talus slopes or protalus ramparts in cases where the material accumulates below a snowpatch. Rock glaciers are tongue-shaped masses of frozen debris (Giardino *et al.*, 1987). Involutions result from sediment distortion when coherent beds freeze.

Frost heaving and thrusting are responsible for vertical and horizontal movement of coarse material and the formation of many types of patterned ground, such as sorted polygons and stone stripes. Other types, such as frost wedge polygons are the result of frost cracking in which the ground fractures because of contraction at low temperatures. Washburn (1979) related patterned ground morphologies to formative processes. Solifluction lobes form where thawed material moves slowly downslope, and they may be either stone-banked or turf-banked.

Pingos are ice-cored hills up to 70 m high and 700 m in diameter, which are particularly common on the Mackenzie delta area of the Canadian arctic. These are classified as closed system types since they form by sealing of talik below shallow lakes and subsequent freezing and upward expansion of the water/ice. Open system types form where spring water rises to the surface and freezes and they often occur in groups. A variety of other frost mounds have been described, including palsas, which are smaller mounds of peat or lens-ice typical of the discontinuous zone.

Melting of the permafrost creates a variety of features collectively known as *thermokarst*. The features include thaw lakes, thaw mounds where ice wedge polygons thaw, and alas depressions and lakes where large-scale collapse of the tundra surface occurs (Czudek & Demek, 1970). In river banks and coastlines, ice lenses may melt when exposed to the atmospheric temperatures and water action in summer. Lateral slope retreat is often the result of this thermal erosion. Fluvial regimes in periglacial areas show very dramatic annual and diurnal cycles. On the coast, shore platform formation may be rapid because of frost shattering of the cliffs.

The presence of many of these features have been recognised in fossil form south of the former North American and European ice sheets, as well as in other parts of the world. In China, eastern Europe, parts of the USA and other areas, large thicknesses of wind blown silt (loess) have accumulated under periglacial conditions when there was abundant fine debris, little vegetation and strong wind action. However, the particle size of loess varies depending on the local sources of dust.

2.7.7   Arid environments

"The processes of landscape development in dry regions differ only in frequency and intensity, rather than in kind, from those in humid regions. . .. The largest single identifiable climatic region on earth is the dry region, where either seasonal or annual precipitation is insufficient to maintain vegetational cover and permit perennial streams to flow" (Bloom, 1998, p. 277). Nonetheless, there have been various attempts to identify the dry region, some of which distinguish hyperarid, arid and semi-arid areas. Together they currently cover over 25% of the earth's land surface principally in two belts coinciding with the subtropical anticyclonic belts around 15 to 30° N and S.

In these areas, fluvial processes may be limited, but when rainfall does occur, the normally dry river channels (wadis) are filled with sediment-laden water (Graf, 1988; Bull & Kirkby, 2002). Bullard & Livingstone (2002) describe the interactions between fluvial and aeolian systems in dryland environments. In mountainous desert areas, alluvial fans coalesce to form bajadas, which grade into playas or salt lakes.

Aeolian erosion produces coarse debris lag deposits termed *desert pavements* often with faceted stones (ventifacts). Yardangs are wind abraded hills and ridges with steeper stoss slopes, for example, at Qaidam in China. Deflation hollows and larger depressions such as the Qattara in Egypt also occur.

Aeolian sediment transport occurs by suspension of fine dust, saltation or surface creep. The sediment often accumulates in ripples, dunes or megadunes. Several types of dune have been recognised, but the basic classification is into free dunes, whose form is related to the wind regime, and impeded dunes, whose morphology is influenced by vegetation, topographic barriers, localised sediment supplies or other factors. The main types of free dunes are related to wind regime (Fig. 2.24 and Table 2.9). At a larger scale, sand seas or ergs occur in certain parts of the world such as the Rub'al Khali erg in Saudi Arabia, which covers 560,000 km². Summerfield (1991) shows the distribution of ergs both at the present day and relict, illustrating the fact that aeolian processes were more extensive in the past as demonstrated by stabilised dunes, for example, in many parts of Africa. Further details on aeolian environments, sediments and landforms can be found in Cooke *et al.* (1982, 1993), Abrahams & Parsons (1994), Lancaster (1995), Livingstone & Warren (1996) and Goudie *et al.* (1999).

2.7.8   Weathering environments

This section will outline the diversity of weathering, particularly mechanical weathering, solution and the landforms produced. For further details see Jennings (1985), Summerfield (1991), Bland & Rolls (1998) or Taylor & Eggleton (2001).

Erosion of rock over long periods of time releases the confining pressures on the rocks below. This promotes the development of pressure release joints that are parallel to the surface (Bloom, 1998). If the surface has a steep gradient, slope failure of rock slabs may occur, as frequently observed on the sides of glacial troughs in, for example, Yosemite Valley, USA. At a smaller scale, thermal expansion and contraction of rocks aided by the presence of moisture and salts may lead to the spalling of thin sheets of rock in a process known as exfoliation or onion weathering. The end product tends to be a rounding of the weathering mass, sometimes producing exfoliation domes or rounded inselbergs known as *bornhardts*. Tafoni and honeycomb weathering pits are also believed to be related to salt weathering since they commonly, though not exclusively, occur in the coastal spray zone.

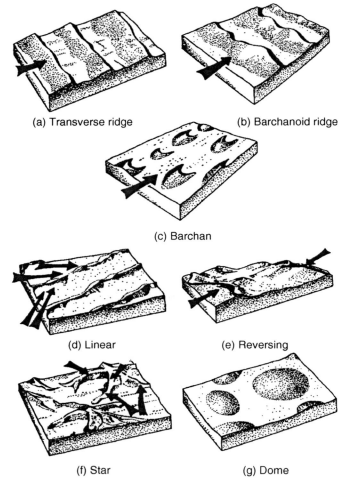

(a) Transverse ridge   (b) Barchanoid ridge

(c) Barchan

(d) Linear   (e) Reversing

(f) Star   (g) Dome

Figure 2.24 *Diversity of major types of free dunes (After McKee, E.D. (ed) (1979) A study of global sand seas. US Geological Survey Professional Paper, 1052, by permission of USGS)*

Table 2.9 *Diversity in basic dune types (after Summerfield, 1991)*

Number & geometry of slip faces	Inferred primary wind regime	Dune type	Morphology
One; unidirectional	Unidirectional	Transverse ridge	Asymmetric ridge
		Barchanoid ridge	Row of continuous crescents
		Barchan	Crescentic form
Two; opposing	Bidirectional at 180°	Reversing	Asymmetric ridge
Two; opposing	Bidirectional oblique	Linear	Symmetric ridge
Three or more; multidirectional	Multidirectional	Star	Central peak with three or more arms
None		Dome	Circular or elliptical mound

Karst landscapes are predominantly the result of limestone solution (Ford & Williams, 1989). The features of karst terrain range from a huge diversity of small-scale solutional forms called karren in Germany or lapies in France (Table 2.10), to major landforms such as solutional hollows known collectively as *dolines*. Sub-types include collapse dolines, solution dolines and subsidence dolines. Uvalas are larger hollows typically resulting from coalescing of doline complexes, while poljes are major depressions sometimes covering over 100 km² of landscape. Cockpit or cone karst describes the landscape resulting from major limestone solution and collapse with only residual rounded hills remaining. Donovan (2002) describes an example from Jamaica, and similar hills are called mogotes in Cuba and pepino hills in Puerto Rico. Tower karst is a spectacular form of this landscape, seen, for example, near Guilin in China, where the residual hills are steep-sided and often over 100 m high. In some places in the humid tropics, thin pinnacles of limestone produce a remarkable pinnacle karst landscape. Caves and chambers are very common in karst areas and redeposition of calcium carbonate in these caves is common. The resulting speleothems take a diversity of forms including stalactites, stalagmites, columns, helictites, curtains, tufa rims, terraces and delicate mineral flowers growing in cave pools.

Finally, duricrusts are hard-surface or near-surface layers resulting from the accumulation of one or more materials. Several types of these crusts exist, including ferricretes (iron), alcretes (aluminium), silcretes (silica), calcretes (calcium carbonate) and gypcretes (gypsum). They are typically 1 to 10 m thick and may form capping layers on residual hills or survive as boulder lags called gibber plains, seen for example, in central Australia.

2.7.9   Soils

Soil is the product of rock and sediment weathering at the Earth's surface, and "is highly variable from place to place on Earth. In fact, the soil is a collection of individually different soil bodies" (Brady & Weil, 2002, p. 11). Several properties of soil are subject to variability:

- *Colour*. This varies considerably in soils from red to yellows to blacks. Colour is usually described by reference to the internationally recognised Munsell colour chart. Three components of colour are used: hue, chroma (intensity or brightness, grey $= 0$) and value (lightness or darkness, black $= 0$).
- *Particle size distribution*. This was discussed with reference to sediments (see above) and the same systems are used to describe particle sizes of soils. The percentages of clay, silt and sand present can be used to provide descriptive soil names.
- *Structure*. This refers to the cohesive arrangement of soil particles into groupings called *aggregates* or *peds*, and these have a great influence on properties such as water movement, aeration and porosity. The four principle structural types are spheroidal, platy, prismlike and blocklike.
- *Density*. Soil density is usually measured as the bulk density of dry soil. Clays and clay loams usually have lower bulk densities than sandy soils because the former usually form aggregates in which pores exist both between and within the granules. Bulk density is easily affected by a range of human activities (see Chapter 4).
- *Pore spaces*. Pore sizes in soil vary, with macropores, mesopores, micropores, ultramicropores and cryptopores being recognised in descending size order (Brady & Weil, 2002).

Table 2.10 *Diversity of solutional microforms developed in limestone (after Jennings, 1985; Summerfield, 1991)*

	Form	Typical dimensions	Comments
Forms developed on bare limestone by areal wetting	Rainpit	<30 mm diameter <20 mm deep	Formed by rain falling on bare, gentle rock slopes. Coalescence gives irregular carious appearance
	Solution ripples	20–30 mm high can be >100 mm long	Wave-like form transverse to dowslope water movement
	Solution flutes (rillenkarren)	20–40 mm across 10–20 mm deep	V-shaped or semi-circular channels formed by concentrated flow down steep slopes
	Solution bevels	0.2–1.0 m long 30–50 mm high	Flat, smooth elements found below flutes. Flow over them occurs as a thin sheet
	Solution runnels (rinnenkarren)	400–500 mm across 300–400 mm deep 10–20 m long	Larger channels formed by increased water flow. May have meandering form
Forms developed on bare limestone concentration of run-off	Grykes (kluftkarren)	500 mm across; up to several metres deep	Formed by the solutional widening of near-vertical joints or bedding
	Clints (flackkarren)	Up to several metres across	Tabular blocks separated by the development of grikes
	Solution spikes (spitzkarren)	Up to several metres	Sharply pointed projections between grikes
Forms developed on partly covered limestone	Solution pans	10–500 mm deep 0.03–3.0 m wide	Dish-shaped depressions usually floored by thin soil, vegetation or algae
	Undercut solution runnels (hohlkarren)	400–500 mm across 300–400 mm deep 10–20 m long	Like runnels but become larger with depth. Recession at depth probably associated with accumulation of humus or soil that keeps sides and base constantly wet
	Solution notches (korrosionkehlen)	1 m high and wide 10 m long	Produced by active solution where soil abuts against projecting rock, giving rise to curved incuts
Forms developed on covered limestone	Rounded solution runnels (rundkarren)	400–500 mm across 300–400 mm deep 10–20 m long	Runnels developed beneath a soil cover that becomes smoothed by more active corrosion associated with acid soil waters
	Solution pipes	1 m across 2–5 m deep	Funnels becoming narrower with depth. Found on soft limestones such as chalk as well as on stronger and less permeable types

- *Other properties.* Several other properties of soils are used, sometimes in particular circumstances. For example, tillage and crusting properties vary and are relevant to agriculture, while a number of mechanical properties of soils are used by engineers.
- *Horizonation.* Soil profiles are characterized by horizons with five major types being recognised (O, A, E, B and C) plus transition zones. In addition subordinate descriptions are indicated by lower case letters (see Brady & Weil, 2002, Fig. 2.35 and Table 2.6)

The huge diversity of soils is a function of five main factors (Jenny, 1941), all of which are closely interrelated.

- *Parent material.* The nature of the surface rocks and sediments has a profound influence on soil characteristics and properties. For example, weathering of a quartz-rich sandstone will invariably produce a sandy-textured soil with strong vertical drainage properties leading to translocation of fine soil particles and plant nutrients. We have already seen that there is a rich geodiversity of rocks and sediments, and therefore it should not be surprising that this factor alone creates an immense variety of soils.
- *Climate.* This has a major influence on soils because it determines the nature and intensity of weathering processes. In turn, this influences several soil characteristics and the rate of soil formation. Water is essential for all chemical weathering processes, and in areas of high rainfall, water is able to percolate deep into the rocks to sustain these weathering processes, particularly where temperatures are also high. Water also leaches soluble and suspended materials and thus leads to soil horizonation. On the other hand, in arid areas, the lack of water limits soil formation and may lead to the build up of soluble salts in the surface layers. In turn, these lead to restricted plant growth and limited organic matter in the soil. Thin soils also occur in cold areas of the world and these can be contrasted with the deeply weathered profiles, sometimes reaching over 50 m depth, of the humid tropics.
- *Biota.* Plants and animals have a strong influence on several soil processes including accumulation of organic matter, biochemical weathering, nutrient cycling, aggregate stability, soil mixing and rates of soil erosion. Since vegetation varies greatly across the Earth's surface, there is a resultant variation in soils. There are also differences in biogeochemical processes between different tree types. The acidic needle litter from coniferous trees decomposes very slowly and recycles only small amounts of calcium, magnesium and potassium. The result is an acidic soil with a thick organic horizon (O-horizon). In contrast, the leaves of deciduous trees are more readily broken down, releasing large amounts of Ca, Mg and K that are then recycled by the trees. The result is a less acidic soil and a thinner forest floor with litter mixed into the A-horizon (Brady & Weil, 2002). Soil organisms (earthworms, termites, etc.) are extremely diverse and they play a diverse set of roles within the soil (Brady & Weil, 2002, Table 11.1).
- *Topography.* Elevation, slope angle, aspect and landscape setting all play a role in soil development, particularly through their impact on other variables. For example, steep slopes encourage water run-off and greater erosion, and therefore tend to have relatively thin soils. On the other hand, thick soils may accumulate at the base of

Table 2.11 *Diversity of World Soils according to the UN Food & Agriculture Organisation (FAO)/UNESCO Soil Map of the World (after Brady & Weil, 2002)*

Acrisols	Low base status soils with argillic horizons
Andosols	Soils formed in volcanic ash that have dark surfaces
Arenosols	Soils formed from sand
Cambisols	Soils with slight colour, structure or consistency change due to weathering
Chernozems	Soils with black surface and high humus under prairie vegetation
Cryosols	Soils of cold climates with permafrost
Ferralsols	Highly weathered soils with sesquioxide-rich clays
Fluvisols	Water-deposited soils with little alteration
Gleysols	Soils with mottled or reduced horizons due to wetness
Greyzems	Soils with dark surface, bleached E horizon and Textural B horizon
Histosols	Organic soils
Kastanozems	Soils with chestnut surface colour under steppe vegetation
Lithosols	Shallow soils over hard rock
Luvisols	Medium to high base status soils with argillic horizon
Nitosols	Soils with low cation exchange capacity clay in argillic horizons
Planosols	Soils with abrupt A-B horizon contact
Phaenozems	Soils with dark surface, more leached than Kastanozems or Chernozems
Podozols	Soils with light-coloured eluvial horizon and subsoil accumulation of iron, aluminium and humus
Podzoluvisols	Soils with leached horizons tonguing into argillic B horizons
Rankers	Thin soils over siliceous material
Regosols	Thin soils over unconsolidated material
Rendzinas	Shallow soils over limestone
Solonchaks	Soils with soluble salt accumulation
Solonetz	Soils with high sodium content
Vertisols	Self-mulching, inverting soils, rich in smectite clay
Xerosols	Dry soils of semi-arid regions
Yermosols	Desert soils

slopes. In low points in the landscape, the soils may be waterlogged so that aeration is limited and so-called gleyed soils result. A group of soils developed in a systematic way across a topographically varied landscape, is referred to as a *catena*.

- *Time*. Rock weathering and soil development takes time, so that we would expect that rocks and sediments recently exposed (e.g. those in front of retreating glaciers) would have very thin profiles compared to those where deep weathering has been ongoing for millions of years (e.g. much of the humid tropics). Rates of soil formation clearly depend on other factors, such as climate, but are generally very slow, so that soil can be regarded as a non-renewable resource (see Chapters 3 and 6).

There is an almost infinite diversity of soils across the surface of the Earth, but in order to make sense of this variation and provide a common language to describe soils, systems of soil classification have been devised. However, different countries have devised their own schemes, and there is as yet no fully internationally applied scheme, though Table 2.11 gives the UN classification used on the World Soils Map (Brady & Weil, 2002). For further details, readers are referred to White (1997), Gerrard (2000), Ashman & Puri (2002) and Brady & Weil (2002). The latter give detailed descriptions

of the US soil types. As well as the main soil orders, the US system recognises subor-
ders (many of which are indicative of the moisture regimes), great groups, sub-groups,
families and, finally, series. There are 19,000 soil series in the United States (Brady
& Weil, 2002) and even these do not fully describe the soil variability of the country.

2.8   Conclusions

In this chapter, I have tried to outline the huge diversity of materials, landforms and
processes of the abiotic world. The aim has not been to describe the full geodiversity
of the planet – there are enough books already available that do this – but simply to
give a flavour of the factors that have produced this amazing diversity. The chapter has
hardly done justice to the real diversity of nature, and readers with limited knowledge
of geology and geomorphology are encouraged to delve more deeply into the extensive
reading available on any of these aspects, or simply to travel with eyes open within
the physical environment that is all around, even in our cities. It is to the value of this
geodiversity that I now turn.

3

Valuing Geodiversity

3.1 Introduction

Why is it important to conserve and manage the geodiversity of the planet? This chapter aims to answer this question by discussing the many values of geodiversity and the reasons for treating the physical basis of our environment with care and respect.

Several authors have tried to outline the value of nature or the rationale for nature conservation, in general (Huxley, 1947; Nature Conservancy Council, 1984; de Groot, 1992; Daily, 1997; Constanza *et al.*, 1997; English Nature, 2002) or earth science conservation, in particular (Nature Conservancy Council, 1990; Wilson, 1994; Kiernan, 1996; Doyle & Bennett, 1998; Page, 1998; Sharples, 2002a). Wilson (1994) recognised two main types of value in the earth's physical resources. Firstly, the economic value in exploiting the physical resources of the planet, and secondly the cultural or heritage value in protecting the aesthetic and research resource of the physical environment. This twofold division is a useful one, but other writers have expanded the classification into four groups (Bennett & Doyle, 1997; Doyle & Bennett, 1998):

- intrinsic value;
- cultural and aesthetic value;
- economic value;
- research and educational value.

However, most of these schemes omit an important role of the physical environment, viz. its functional value for both physical and ecological processes. Thus, an expanded version of the above subdivision is used in this chapter and a summary table is provided at the end. It is also true to say that there is a difference between the value of a resource and the value of the diversity of a resource, and it is the latter that I shall try to focus on in this chapter.

3.2 Intrinsic or Existence Value

Intrinsic value refers to the ethical belief that some things (in this case the geodiversity of nature) are of value simply for what they are rather than what they can be used for by humans (utilitarian value). This is the most difficult value to describe since it involves ethical and philosophical dimensions of the relationships between society and nature. These have been discussed by a wide range of authors, and Beckerman

Geodiversity: valuing and conserving abiotic nature   Murray Gray
© 2004 John Wiley & Sons, Ltd   ISBNs: 0-470-84895-2 (HB); 0-470-84896-0 (PB)

& Pasek (2001, p. 129) refer to the issue as "one of the most recalcitrant problems of environmental ethics". Some have argued that there is no such thing as intrinsic value since the value of nature depends on whichever ethical or belief system that we adopt. Other philosophers argue that nature is not a social construction but has a value in itself and this is not dependent on any uses of nature that humans might adopt (Norton, 1988).

One view is that the resources of the planet should be freely available for human exploitation and there should be no curbs or restrictions on the use of these physical or biological resources. This "technocentric" or "anthropocentric" view of the human place in the environment is one that demonstrates a "lack of concern for anything non-human: 'nature' is seen as an 'external' environment with no worth or value, except its ability to be manipulated or exploited by society" (Phillips & Mighall, 2000, p. 14). Under this view, the environment is regarded as a commodity like any other for which there is a market. This view has been adopted by different societies at different times. It reflects the neoclassical economic tradition of the last 130 years, but it has an even longer history. For example, during the so-called *scientific revolution* of the sixteenth and seventeenth centuries, philosophers such as René Descartes in France or Francis Bacon in England promoted the idea of people, as rational, thinking beings, separate from nature, which they could analyse, control and dominate (Pepper, 1984).

Political systems, too, have been associated with technocentric attitudes. For example, both capitalism and communism have been blamed for promoting economic growth at the expense of the environment (Barry, 1999; Phillips & Mighall, 2000). The techno-centric philosophy therefore, in its most extreme form, is not one that would readily recognise the intrinsic or existence value of physical or biological nature or the need to conserve and manage it.

At the other end of the nature/society philosophical spectrum is "ecocentrism", which sees value in non-exploitative relationships between society and nature. This view has been shared by many movements, cultures and religions over the years though there are many shades of ecocentrism. There are clearly links with the tra-ditional Judaeo–Christian notion of human responsibility for the stewardship of the planet (Cobb, 1988), though there has been intense debate about the interpretation of human–environment relationships in Christian thought (Attfield, 1999; Barry, 1999; Phillips & Mighall, 2000). For example, White (1967) used the Book of Genesis to argue that Christian thought has encouraged technocentric views:

"and God said unto them 'Be fruitful, and multiply, and replenish the earth, and subdue it: and have dominion over the fish of the sea, and over the birds of the air, and over every living thing that moves upon the earth" (Genesis: 1, 28),

though others such as Passmore (1980) and Hillel (1991) have pointed to the fact that the second chapter of Genesis gives a contradictory view:

"God took man and put him in the Garden of Eden to work it and take care of it" (Genesis: 2, 15).

In this view "Man is not set above nature.... his power is constrained by duty and responsibility... man is a custodian, entrusted with the stewardship of God's garden, and he can enjoy it only on the condition that he discharges his duty faithfully" (Hillel, 1991, p. 13).

The Christian "Great Chain of Being" has had considerable influence on Christian thought and attitudes, perhaps even today in technocentric thought. It is made up of a hierarchical set of relationships in which God is at the top and inanimate nature is at the bottom. According to this view, plants draw their nutrients from the earth, animals feed on plants and these serve human needs. Humans, in turn, serve God. However, not all Christians have accepted this hierarchy. St Francis of Assisi, for example, "famously preached to the animals and developed a Christian pantheism in which the natural environment partook of the divinity and grace of God, and was not simply a set of spiritually meaningless or empty resources to be used" (Barry, 1999, p. 42).

Similarly, many eastern religions espouse the principle of empathy with, and respect for, the natural world. Buddhism displays a marked respect for the natural environment and Islam has its own particular set of rules, taken from the Koran, about the proper way of thinking about, and relating to, the environment (Khalid & O'Brien, 1992; Barry, 1999). For example, there is provision in Islamic law for "himas", tracts of land set aside to remain undeveloped in perpetuity, of which thousands remain to this day (Dutton, 1992). Aboriginal thought is also ecocentric and more inclined to emphasise the continuity rather than separation between the human and the non-human worlds (see, for example, Box 3.3). Ehrlich (1988, p. 22) believes that, given the current state of the planet, a "quasi-religious transformation leading to the appreciation of diversity for its own sake. ... may be required to save other organisms and ourselves".

Arguably, the most extreme form of ecocentrism is "Gaianism", first promoted by James Lovelock (1979, 1995) and named after the Greek Earth goddess. This philosophy is not only holistic in seeing the essential integration between the biological and geological elements of the planet but also sees nature as the primary object of concern. According to Lovelock, it is the health of the planet that matters, not that of some individual species of organism, including humans.

O'Riordan (1981) recognises shades of opinion between the extremes of technocentrism and ecocentrism. For example, one category of technocentrics are "accommodators" or "environmental managers" who recognise that there are serious environmental problems that threaten societies striving for economic growth, but favour the application of scientific, technological or management techniques to reduce the impact of development. The ultimate accommodation, of course, is "sustainable development", which argues that environmental resources should continue to be used but in a more thoughtful way.

It is not difficult to perceive that a central tenet of ecocentrism is the intrinsic or existence value of nature in its own right or for its own sake. As Attfield (1999, p. 27) puts it "it is difficult to credit that nothing but our own species matters" or that "absolutely everything exists for the sake of humanity, and it alone". This is a well-understood principle in relation to wildlife, in which there is a strong belief in many societies of animal rights. Alternative strains of thought include "sentientism", which accords moral recognition to all creatures with feelings and only to such creatures, and "biocentrism", which recognises the moral standing of all living things (Attfield, 1999). It is less clear that society holds the same view of geodiversity or would acknowledge a "geocentrist" ethic. As the author has stated (Gray 1998a, p. 273), "'Save the Dolphin' is always likely to have greater appeal to the public than 'Save the Drumlin'". Indeed, since sentientists do not even extend recognition to all creatures, they are hardly likely to attach much value to abiotic nature. But there is no reason, in principle, for separating animate and inanimate nature in this respect and it would certainly be contrary to the Gaia philosophy to do so. Sharples (1993, p. 7) believes that "it is simply 'biocentric chauvinism' to hold that only living things have intrinsic value whilst non-living things do not".

Another line of argument for intrinsic or existence value relates to natural and human timescales. Bronowski (1973, p. 91), with usual eloquence, described how "The hidden forces within the earth have buckled the strata, and lifted and shifted the land masses. And on the surface, the erosion of snow and rain and storm, of stream and ocean, of sun and wind, have carved out a natural architecture". This architecture has taken thousands of millions of years to evolve, yet can be destroyed or altered within days. Given the potential of this asymmetric cycle of creation and destruction, it is arguable that if we understand the lengths of time and complexity of the processes involved, we may conclude that the end result has some intrinsic value.

A further approach is promoted by Goodwin (1992) who suggests that value of natural processes and landscapes comes about precisely because they are not the work of human hands. According to Goodwin, a "natural" landscape is more valuable than a "humanised" landscape, in the same way as a "fake" or reproduction is never as valuable as the original. In response to Gray's (1998a, p. 313) assertion that unlike the natural vegetation of Britain, "the natural landforms remain generally intact.... substantially unaltered since the major changes of the Pleistocene and modification of the Holocene", Adams (1998, p. 168) argues that "a 'natural' landform is valuable because it is unchanged, or minimally changed by human action". Leopold (1949) was one of the first to argue the case for wilderness, natural landscapes and natural sedimentary systems to be protected in their own right, and this argument plays an important role in the justification for conservation of Norwegian rivers (Daly et al., 1994). On the other hand, a technocentric position might be that "artificial" is superior to "natural" (Barry, 1999). This was certainly true of the landscape gardening movement but is still alive. Reed (1998, pp. 12–13), for example, argues that inert waste "is being put to good use making flat, uninteresting golf courses more environmentally attractive"!

Less extreme forms of technocentrism, as exemplified by environmental management, also recognise the existence or intrinsic value of nature, though in this case they are more open to accommodating or managing the resource. Most western societies are of this type and the global conservation movement generally adopts this type of pragmatic approach. As Sharples (2002a, para 2.3.2) puts it:

"Recognition of intrinsic value .... means that while humanity may have a right to exploit natural resources to fulfill our own legitimate needs and purposes, it should not be done in such a way that the diversity of natural geological, geomorphic and soil features and processes (geodiversity) is unnecessarily reduced by the unnatural elimination of entire classes of things, or in such a way that representative systems of natural processes are no longer able to unfold and evolve in their own ways".

Belshaw (2001) makes the distinction between the "intrinsic" value of an object, that is, that a rock, landform or soil should exist for its own sake, and an "existence" value, meaning a non-instrumental value to humans, that is, that humans value the existence of a rock, landform or soil irrespective of any other value to those humans, including an aesthetic value. Beckerman & Pasek (2001, p. 130) agree that "values cannot exist without a valuer", and they accept (p. 131) that a "subjective approach to valuation still allows the valuer to attribute intrinsic value to something".

This leaves the question of whether diversity, as opposed to uniformity, has intrinsic value. Cuomo (1998, p. 132) argued that:

> "to claim that something is ethically valuable merely because it is unlike something else is incoherent – to be ethically valuable something must itself have a certain quality or status, even if that quality or status is contextually determined. Also, claiming that difference itself renders something morally valuable fails to give attention to the content and origins of the thing itself".

As we saw in Section 1.1, opinions have varied in past centuries about the value of landscape diversity (Midgely, 2001) and the conclusion must be that diversity has no inherent value in itself but can have a subjective intrinsic value, that is, an existence value.

Other arguments must also be considered in relation to both biodiversity and geodiversity. Attfield (1999, p. 135) rightly reminds us that "Arguments for preserving whatever is natural assume (implausibly) that whatever is natural is desirable". Yet there is much in the biological world that is certainly harmful to humans. For example, should we protect and enhance the smallpox virus or the anthrax bacterium, when medical science has spent centuries attempting to eradicate these and other diseases? Should we value spiders or rats as much as pandas and dolphins? Do we need to preserve all of the 30 million or so species estimated to exist on earth, even if we ever manage to identify them all? Since carnivorous species hunt their prey, might they not themselves endanger these species? And since extinction itself is a natural process, should we attempt to prevent it from occurring (Passmore, 1980; Attfield, 1999)? These questions raise important ethical and philosophical issues about whether humans should have the right to decide the fate of species, and if so, how priorities should be identified and resources allocated in conserving biodiversity.

Similar questions can be applied to geodiversity. For example, is geological and geomorphological diversity always beneficial? As was indicated in Section 1.1, the answer must be "no". For example, to the civil engineer the endless variety of rocks, sediments, slopes and drainage courses makes life exceedingly difficult and raises the cost of building projects through the need for site investigations, material testing and geomorphological mapping (Cooke & Doornkamp, 1990; Griffiths, 2001). Projects would be much simpler and cheaper to complete if there was greater uniformity and predictability of rocks, sediments, landforms and processes.

Secondly, hazards such as earthquakes, tsunamis, volcanic eruptions, floods, avalanches and landslides kill thousands of people every year and damage property to the tune of millions of pounds (Alexander, 1993; Murck *et al.*, 1997; Decker & Decker, 1998; Bell, 1999; Smith, 2000; McGuire *et al.*, 2002). Should we conserve such damaging processes or try to eradicate them? By and large, human society, perfectly understandably, tries to prevent these hazards and disasters by various means including sensible planning, predicting events, evacuating populations, engineering solutions, and so on. But we also take a morbid interest in the spectacular dynamics of earth processes, the power involved and the threats they pose, as long as we personally are not affected. It would be unfortunate if all potentially hazardous processes were eradicated from the planet since they are part of its natural evolution.

Several other questions need to be asked about the aims and principles of geodiversity. For example, since erosion is a natural process, should we be concerned if it removes an element of geodiversity? Do we need to preserve all the world's geodiversity even if we could identify it? If not, how do we decide what is sufficiently significant to conserve? How should priorities be identified and what resources should be allocated to conserving geodiversity relative to biodiversity? If we accept Cuomo's (1998) premise that diversity can only have a subjective intrinsic value, then it also allows us to support Sharples (2002a) proposal (see Section 1.3) for a distinction to be made between "geodiversity" as a value-free quality, and "geoheritage" as those elements of geodiversity that are seen as significant according to particular subjective values.

For further discussion of these issues, see Fox (1990), Nash (1990), Cuomo (1998), Attfield (1999), Belshaw (2001) and Beckerman & Pasek (2001).

3.3   Cultural Value

The cultural value of geodiversity is certainly related to the last category, but has a more practical element. By cultural value we mean the value placed by society on some aspect of the physical environment by reason of its social or community significance. It is not difficult to find examples of these attachments in both past and present societies, and because the physical environment is valued in this way, it is appropriate to conserve the landscapes and features involved.

3.3.1   Folklore (geomythology)

Primitive societies often explained the origin of rock formations or landforms in terms of supernatural forces and attached due significance to them, and many still bear the names of these associations. Thus, we have famous explanations of columnar jointing such as the Giant's Causeway in northern Ireland, supposed to be the stepping stones of giants, and the Devil's Tower in Wyoming, USA, where native American legend has it that a giant grizzly bear trying to reach a group of people on its summit left its claw marks on the sides of the tower (Fig. 4.12).

In fact, the Devil is very common in the names of physical features around the world. *The Times World Atlas* identifies mountains or hills (Devil Mountain, Devil's Peak, Devil's Paw, Devilsbit, Devil's Mother, Devil's Riding School, Devil Post Pile, Devil River Peak, Devil's Elbow), hollows (Devil's Beef Tub, Devil's Gorge, Devil's Hole, Devil's Kitchen), passes (Devils Gate), lakes (Devil's Lake) and deserts (Devil's Playground) with associations with the Devil. In Serbia, the Devil's Town is a series of over 200 earth pillars, reputed to be petrified humans or beings from another planet (Vasileva, 1997). The Devil's Marbles are huge rounded residual boulders in the Northern Territories, Australia, though Aboriginal belief is that they are eggs of the Rainbow Serpent (see also Box 3.3). The Devil may have even been held responsible for geomorphological sounds. The *Guardian* (27 August 2001) reported that a cavern in Derbyshire, England, has changed its name back to its Anglo-Saxon one of the Devil's Arse, because of the noise made when water that has built up in the cavern drains away. As a result, visitor numbers have increased by 30%!

Other examples of mythical explanations for landforms can be related. Several examples occur in Scotland (see Box 3.1) but many countries have similar folklore. For example, all the major landforms in Hawaii are attributed to the actions of Pelé, the goddess of volcanoes. She excavated the craters and created eruptions when she

Box 3.1   Scottish Landform Legends

The Dog Stone, a raised, undercut sea stack near Oban, Scotland, is said to be where the Irish Giant Fingal tied up his dog whilst he went hunting in the Hebrides (Fig. 3.1). Not far away, in Glen Roy, there is a series of near-horizontal ice-dammed lake shorelines called the "Parallel Roads of Glen Roy" because they are said to be hillside hunting roads used by Fingal. Also near Oban, a complex shaped kettle hole on the Loch Etive kame terraces (Gray, 1975) is said to be where a giant witch's cow lay down, while a nearby perfectly circular kettle hole represents her cheese mould. At Morebattle in the Scottish Borders, the well-sorted sands of the local kames are reputed to have been sifted by nuns as a penance. There are also tales of what we now know to be glacial erratics reputedly being heaved around the country by giants, witches or apocryphally strong men. By the shores of Loch Etive lies "Rob Roy's Putting Stone". Two other boulders on either side of the Kyle of Durness are reputedly the result of a local witches' feud in which the matter was settled by hurling boulders across the Kyle with the furthest throw being agreed as the victor. Loch Bran in Wester Ross is reputed to be the result of digging by Bran, the legendary hound of Ossianic legend (Robertson, 1995).

Figure 3.1   *The Dog Stone, Oban, Scotland, named from the legend that it was where the giant Finglal tied up his dog Bran*

was angry, while the smooth, glassy lapilli are her tears (Marshak, 2001). Elsewhere in the United States, there is a Chippewa Indian legend to explain the Manitou Islands in Lake Michigan, Wisconsin. A mother bear and her two cubs were driven into the lake by a raging forest fire. They swam and swam but the cubs became tired and lagged behind. The mother bear eventually reached the opposite shore and climbed to the top of a sand dune to await her offspring, but they had drowned. Today the cubs are the Manitou Islands, while the sand dune is called Sleeping Bear Dune.

The small Norwegian island of Torghatten is in the shape of a hat, and in the centre of the island a cave passes all the way through the island or the crown of the hat. Legend has it that this cave was produced when an arrow was shot through the hat that then fell to the ground and turned to stone.

In New Zealand, geologists believe the famous Moeraki Stones to be giant septarian nodules eroded from Palaeocene claystones, but Maori legend has it that they are bread rolls dropped from a giant explorer's food basket. Further south, a Maori god, Tu-trakiwhanoa, is said to have been responsible for carving Doubtful Sound, the largest of New Zealand's fjords, as he tried to create a route from the sea to the interior.

Fossils are also the subject of folklore and legends. In Alberta, Canada, large dinosaur bones were thought to come from giants or the "great grandfather buffalo". The Sioux Indians of South Dakota make reference to the bones of "thunder beasts" found in the White River Badlands. Ammonite fossils are reputed to have once been living serpents or horns of an ancient Egyptian god. Amber is reputedly the hardened tears of the daughters of the sun god. *Gryphaea* has the nickname "Devil's toenail" and belemnites are "thunderbolts".

An appropriate name from such folklore explanations of landforms and fossils must surely be "geomythology"!

3.3.2   Archaeological and historical value

Our early ancestors had a very close relationship with their physical surroundings, and geology and landscape must have played an absolutely crucial role in their lives. Ascherson (2002) describes the Scottish landscape as "a poor woman with little flesh between her skin and bones", and unsurprisingly, that landscape bears the marks of early human occupation in its ancient monuments, standing stones, rock art and burial cairns. And as Jacob Bronowski put it in *The Ascent of Man* (1973. p. 40), when man first made rudimentary stone tools by using a simple blow to put an edge on a pebble, "He had made the fundamental invention, the purposeful act which prepares and stores a pebble for later use. By that lunge of skill and foresight, a symbolic act of discovery of the future, he had released the brake which the environment imposes on all other creatures".

During the Palaeolithic the increasing sophistication of stone implements is evident. From initial, crude chopping tools, through well-crafted hand axes to delicate arrow heads, the history of the human use of stone for hunting, butchery and warfare is clear, as is the ability of humans to seek out the best rock types for these purposes such as flint, obsidian, quartzite or other hard rock displaying a conchoidal fracture. Here is an example of geodiversity being sought and exploited in innovative ways. Lynch (1990) describes a 0.7-m thick, naturally baked, Ordovician shale at Mynydd Rhiw in Wales, which was used for axe and chisel manufacture. The Alibates flint quarries in Texas, USA, date back at least 12,000 years and tools and spear points made from the flint are found in many places in the Great Plains and south-west USA, indicating either trading or extensive nomadic hunting. In Britain, the Council for British Archaeology has established that many of the artefacts were traded over quite long distances. According to Evans (1997, p. 3), "by 3000 B.C. large underground flint mines were in operation at Grime's Graves in Norfolk, UK, and it is clear that the miners had noted that particular horizons in the chalk host rock carried the best and most numerous flints".

Later, when agriculture became important and milling equipment was developed, millstones or "querns" were manufactured with grooves to allow the flour to escape. Clay and other materials began to be used in pottery manufacture, while stone and subsequently metals were used to produce art and coinage. Pigments such as natural ochres (iron oxides), umbers (iron-rich clays), cinnabar (red mercury sulphide), wad (black manganese oxides) and galena (grey lead oxides) were extracted. "When tin and copper minerals were (probably accidentally) smelted together with charcoal about 4600 years ago, the Bronze Age came into being. Not until another 2000 years or so had elapsed did iron smelting become widespread" (Evans, 1997, p. 4). By these discoveries, rock began to achieve an economic value (see below) and its diversity was crucial.

Early humans used the diversity of the natural environment to their advantage in other ways. They sought out natural caves or excavated them in suitable rock types. In western North America, they selected cliff sites as "buffalo jumps" where herds would be driven to their death over suitable precipices. But they also discovered the power of construction using rock materials. They soon began to build crude stone shelters and houses, defensive structures, stone monuments or burial sites, and they began decorating the rock surfaces with carvings (Fig. 3.2), paintings and even writing as in the Rosetta Stone. And as they did so, they utilised the natural diversity of the geomaterials that they found all around them (Fig. 3.3). Important examples include the beautiful temples and massive pyramids of the Nile Valley in Egypt, the huge stone monuments at Borrobodur and Prambanan near Yogjakarta on Java, Indonesia, the giant Buddhas at Bamiyan in Afghanistan (destroyed by the Taliban), the Great Wall of China, the granite city and agricultural terraces of Machu Picchu in Peru, the carved rock city of Petra in Jordan and the cliff-houses of the Mesa Verde, Colorado, Gila Cliff, New Mexico and Canyon de Chelly, Arizona, USA (Box 3.2).

Box 3.2   Canyon de Chelly, USA

The Canyon de Chelly is located in a Navajo Reservation in North-East Arizona and was established as a National Monument in 1931. The vertical red sandstone walls of this and adjacent Canyon del Muerto rise to heights of 300 m and contain several large caves both at the cliff base and within the walls. These caves were occupied as cliff dwellings between the fourth and fourteenth centuries, with the later ones involving construction of houses within the caves sometimes rising to five or six storeys and made up of hundreds of rooms. Notable examples are the White House (occupied from about 1060 to 1275 and the Mummy Cave dating from 1253. The walls of the canyon and caves are decorated with fine examples of rock carvings and paintings including antelope, fish and abstract patterns.

"There is a great intellectual step forward when man splits . . . . a piece of stone, and lays bare the print that nature had put there before he split it. The Pueblo people found that step in the red sandstone cliffs. . .. The tabular strata were there for cutting; and the blocks were laid in courses along the same bedding planes in which they had lain in the cliffs of the Canyon de Chelly" (Bronowski, 1973, p. 95).

The use of stone as part of ritual acts is exemplified in many parts of the world. For example, in Scotland, ancient carved footprints have been discovered in Orkney and Shetland, and examples also occur in Ireland and France. Breeze & Munro (1997, p. 12) believe that "The act of a king or chief standing on a special stone to be invested can be seen as symbolic of a relationship with an object of great antiquity – rock ...

Figure 3.2   *Stone used to portray images and stories. Relief scene and hieroglyphics depicting the crowns of the kingdoms of the Upper and Lower Nile, carved on sandstone temple walls, Edfu, Egypt*

and also with the land from which his people earned their food". But Breeze and Munro go further in noting associations between rock and investiture: "It is but a short step to remove a piece of such rock and make it into an object on which a king sat to be invested". And they quote examples from Kingston upon Thames, England, several sites in Ireland, Zollfeld in Austria, Uppsala in Sweden and Aachen in Germany. But the most notorious example must be the "Stone of Destiny", a block of Perthshire sandstone on which the Scottish Kings had been inaugurated for centuries until it was removed from the abbey at Scone by King Edward I of England in 1296. He took it to Westminster Abbey in London, where for 700 years it sat within the Coronation Chair where the kings and queens of England, and subsequently of the United Kingdom, were crowned. The stone was returned to Scotland in 1996 and today it is on display in Edinburgh Castle.

(a)

(b)

Figure 3.3 *Diversity of historic standing stone monuments: (a) tall angular slabs, Callanish, Scotland and (b) rounded boulders, near Evora, Portugal*

Edinburgh Castle is, of course, an impressive defensive structure, built as it is atop a volcanic neck, and there is also no doubting the role that the physical environment often played in the siting of settlements for defensive, resource or other reasons (Fig. 3.4). Other examples include the hill top towns of Tuscany and Umbria in Italy, the "Gold Rush" towns of the American west or Canadian Yukon, and the clustering of Egyptian settlements in the Nile Valley. Bloom (1998, p. 161) believes that "Residual karst landforms have a curious relevance to the political and social histories of the regions in which they are found. Bandits, partisans, guerilla troups and fugitives who are native to a karst region are able to live safely in the many caves that penetrate karst towers...".

Bennett and Doyle (1997, p. 168) point out that the outcome of many famous battles was influenced by the local landscape: "Waterloo (1815) was fought on a clay plain in Belgium, while the Somme (1916) took place in the dissected chalk upland of northern France. Both of these are commemorated on the ground and are intimately associated with the local landscape". The outcome of the Battle of the Boyne (1690), so important in Irish history, was strongly influenced by the topography of the area (Stout, 2002). The Falklands War (1982) was fought between the United Kingdom and Argentina over the sovereignty of a set of islands (Falklands/Malvinas) in the South Atlantic, though like many conflicts, there may have been a background issue of rights to mineral resources in the South Atlantic. Similarly, the Gulf War and the Iraq War are alleged to have been at least partly about oil resources (see Section 6.5.2). It has also been predicted that future wars may well be fought over water. For further examples of geologically induced conflicts and terrain affecting the outcome of warfare, see Woodcock (1994, Fig. 3.8), Rose & Nathanail (2000) and Doyle & Bennett (2002).

Figure 3.4   *The hilltop defensive position of Marvao Castle, Portugal near the Spanish border*

3.3.3   Spiritual value

Many human societies place spiritual or religious value on the physical environment. "Adam" is derived from the Hebrew "adama" meaning earth or soil. Thus "Adam's name encapsulates man's origin and destiny: his existence and livelihood derive from the soil, to which he is tethered throughout his life and to which he is fated to return at the end of his days" Hillel (1991, p. 14). And since "Eve" is derived from the Hebrew "hava" meaning life, together Adam and Eve signify soil and life.

This is reflected in the views of Jomo Kenyatta, former president of Kenya who is reported as describing the value of land to the Gikuyu people in 1938 as follows: "It supplies them with the material needs of life through which spiritual and mental contentment is achieved. Communion with ancestral spirits is perpetuated through contact with the soil in which ancestors of the tribe lie buried... it is the soil that feeds the child through lifetime; and again after death it is the soil that nurses the spirits of the dead for eternity. Thus the earth is the most sacred thing above all that dwell in or on it" (Mackenzie, 1998, p. 24).

North American Indian tribes each have their own stories about the origin of the world. The Blackfoot, for example, believe that the landscape was made by Old Man or Naapi. Consequently, they regard the Earth as sacred and some even see ploughing as wounding or violating the land. Many natural sites are regarded as sacred where people can communicate with spirits, both good and bad. "....What is important for traditional Indian religious believers is .... a location made holy by the Great Creator, by ancient and enduring myth, by repeated rituals such as sun dances, or by the presence of spirits who dwell in deep canyons, on mountaintops, or in hidden caves" (Gulliford, 2000, p. 69). These sacred places remain integral to tribal histories, religions and identities and should not be disturbed. Gulliford (2000) classifies sacred places into several types, many of which are related to physical features, for example, vision quest sites, group ceremonial sites or burial sites, and Fig. 3.5 shows the locations of selected sacred sites in western United States. Vision quest sites, such as the Sweet Grass Hills and Chief Mountain in Montana, are usually isolated and remote peaks where individuals remained without food or water until the arrival of a spiritual bird or animal to give them guidance.

There is a North American Indian story related to pipestone, a red, durable but easily carved rock type found particularly in Minnesota where the quarry is a national monument. The origin of pipestone is said, in Sioux accounts, to date from an ancient time when the Great Spirit, in the form of a large bird, broke off a piece of red stone, formed it into a pipe and smoked it. He told the gathered tribes that this red stone was their flesh, that they were made from it, that they must all smoke to him through it and that they must use it for nothing but pipes. The ground was sacred to all tribes and no weapons must be used on it. Consequently, the quarrying of the pipestone has always been carried out with respect for the earth and for what it yields. The Sioux traditionally leave an offering of food and tobacco beside a group of boulders known as the Three Maidens in return for the gift of the pipestone.

Along with the vision quest mountains referred to above, many other mountains are regarded as sacred, holy or are visited for religious reasons. These include Mount Kailas in Tibet, which is holy to almost a billion Buddhists and Hindus. Mount Fuji (Japan), T'ai Shan (China), Mount Sinai (Egypt), Ausangate (Peru), Mount Athose (Greece) and Croagh Patrick (Ireland) are other examples (M. Price, 2002).

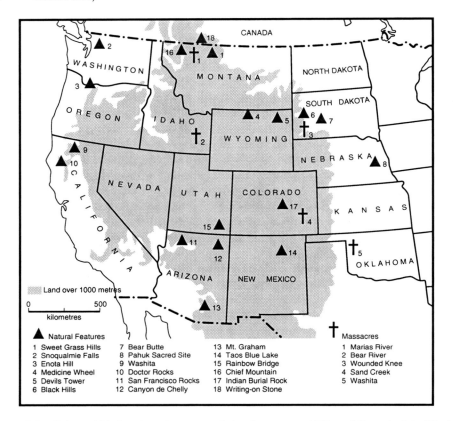

Figure 3.5   *Selected Native American sacred sites in western USA and Canada (Modified after Andrew Gulliford, Sacred Objects and Sacred Places: Preserving Tribal Traditions. (Boulder: University Press of Colorado, 2002, by permission of Andrew Gulliford))*

But perhaps the best example of geospiritual associations is in the Uluru area of Australia. Here, the rock and landforms are held as sacred by the local aboriginal people because of their spiritual beliefs and understandings (see Box 3.3).

Box 3.3   Uluru, Australia

Uluru (formerly Ayers Rock) in Australia still has spiritual significance for the local aboriginal people (Anangu). This is hardly surprising given the dramatic appearance of the hill rising over 300 m from the flat desert floor and the near vertical bedding in the arkosic sandstone bedrock (Fig. 3.6).

The Aborigines believe that there is a hollow below the ground and an energy source known as *Tjukurpa*, a word that, in fact, has several meanings including the creation period. The Anangu believe that the world was once a featureless place. However, during the creation period, the area around the rock was inhabited by dozens of ancestral beings whose activities produced the physical features of the area, such as rocks, sand hills or caves. This gives each of these features, including Uluru itself, eternal spiritual significance as the living presence of Tjukurpa.

Continued on page 79

__ *Continued from page 78* __

Figure 3.6   *Uluru, Australia*

For example, in the creation period, Tatji, the small red lizard, came to Uluru. He threw his kali, a curved throwing stick similar to a boomerang, at the rock but it became stuck in a joint in the rock. Tatji used his hands to try to scoop it out, but in his efforts to retrieve his kali, he left a series of bowl-shaped hollows in the rock. Unable to recover his kali, he finally died in a nearby cave and his implements and bodily remains survive as large boulders on the cave floor. Elsewhere, a fractured slab of sandstone is believed to represent large joints of meat from an emu killed and butchered by two lizard men, Mita and Lungkata. In several caves at Uluru, rock paintings believed to date back many thousands of years but regularly refreshed, relate the many stories of the Tjukurpa.

The spiritual significance of the rock, caves and other features give the physical features of the area a particular value to aboriginal society that is not generally appreciated by tourists to the area. They certainly value the spectacular geomorphology of the rock, its ever-changing colour and the opportunity to climb to the summit, but they are largely ignorant of the sacred lands that lie below or the insult they cause to aboriginal beliefs by doing so (not to mention the footpath erosion).

At about 30 km distance are the Olgas Hills, which are sacred to aboriginal males and contain rocks and pinnacles shaped by wind and water, about which there are many legends. Several are related to fertility, with pinnacles and caves representing male and female genitalia respectively.

3.3.4   Sense of place

Many other present-day societies also feel a strong bond with their physical surroundings and value these ties for cultural, as well as economic, reasons. Agricultural communities are dependent on soil quality and have long valued the material on which their living depends, hence phrases such as "son of the soil" or "mother earth". Trudgill

(2001) sees the cultural links with "soil", "land" and "earth" as representing yield and fertility, provision and abundance, or ownership, patriotism and nationality. The latter is exemplified in phrases such as "on American soil" or demonstrated by the significance to the British of "the white cliffs of Dover" or to Gibraltarians of "the Rock".

In the United States, most States have a State Fossil (Table 3.1). Maine's, for example, is the primitive and rare Devonian plant *Pertica quadrifania,* which was first discovered in the State. Until recently, Dudley in England had the trilobite *Calymene blumenbachii* as its emblem. Raudsep (1994) notes that Estonia is famous for its large erratics, which have a special place in the hearts of Estonians, particularly in the Lahemaa National Park. According to Raudsep (1994, p. 239), "These boulders reflect the history of Estonian land and country; indeed some of them are the bearers of legends and folktales, and many have served as sacrificial places... Almost every family has its own 'babystone', usually behind the house or shed. In children's eyes it had a magic power; they had only to get the parents' permission and knock at the stone and, in time, the stone gave them a little brother or sister".

Coastal communities often feel a strong relationship with, and respect for, the sea and coastal landscapes, frequently due to their dependence on fishing, and a long history of loss of life of both fishing and lifeboat crews. Coastal topography provides

Table 3.1   *Some state fossils of the United States*

State	State fossil	Year adopted
Alabama	*Basilosaurus cetoides* (whale)	1984
Alaska	*Mammuthus primigenius* (woolly mammoth)	1986
Arizona	*Araucarioxylon arizonicum* (petrified wood)	1986
Arkansas	None	
California	*Smilodon fatalis* or *californicus* (sabretooth cat)	1973
Colorado	*Stegosaurus* (dinosaur)	1991
Connecticut	*Eubrontes giganteus* (dinosaur track)	1991
Delaware	*Belemnitella americana* (belemnite)	
Florida	*Eupatagus antillarum* (sea urchin relative)	1979
Georgia	Shark tooth	1976
Hawaii	None	
Idaho	*Equus simplicidens* (Hagerman horse)	1988
Illinois	*Tullimonstrum gregarium* (Tully monster)	1990
Indiana	Limestone	1971
Iowa	Crinoid	
Kansas	None	
Kentucky	Brachiopod	1986
Louisiana	Petrified palmwood	1976
Maine	*Pertica quadrifaria* (plant)	1985
Maryland	*Ecphora gardnerae* (marine snail)	1984
Massachusetts	Dinosaur tracks	1980
Michigan	Petoskey stone (fossilised coral)	1966
Minnesota	None	
Mississippi	*Zygorhiza kochii* (whale)	1981
Missouri	*Delocrinus missouriensis* (crinoid)	1989
Montana	*Maiasaura peeblesorum* (dinosaur)	1985
Nebraska	Mammoth	1967
Nevada	*Shonisaurus popularis* (dinosaur)	1977
New Hampshire	None	
New Jersey	*Hadrosaurus foulkii* (dinosaur)	1991
New Mexico	*Coelophysis* (dinosaur)	1981

navigational landmarks such as headlands and islands, and thus increases the sense of place. Some societies have complex relationships with the rivers on which they are sited, depending on them for water for agricultural and domestic use but fearing the impacts on life and property of flood events. Similarly, those living on the slopes of Mount Etna in Italy value the fertile volcanic soils but fear the often very real effects of the periodic eruptions.

Communities may have more subtle relationships with their physical environment. Adams (1996, p. 72) argues that landscape is a cultural construction: "The various physical and biological elements exist, of course, but the way in which we understand them, and the way we describe them is the result of values and ideas that we associate with them, not their innate properties". Loss of landscape therefore creates powerful emotions in those who have developed such cultural values. Adams (1996, p. 78) gives the example of the M3 motorway cutting through Twyford Down in Hampshire, England: "the value of the place was obvious: the majesty of the landscape with its sweep of chalk hill, its importance as an open space for the people of Winchester......". This is an example of the fact that in England "Landscapes have been progressively simplified, and areas of habitat have shrunk and deteriorated. It is the response to this erosion of diversity on the part of ordinary people that empowers and gives life to conservation" (Adams, 1996, p. 79). Developing some ideas outlined by Gray (1997a), Adams (1998, p. 168) argues that societies may see landforms as valuable "because they form a basic constituent of landscape, and hence share the values associated with its cultural construction, related, for example, to 'distinctiveness' or 'familiarity'". Landscape gives identity and sense of place to our surroundings. As Swanwick & Land Use Consultants (1999, p. 1) state, "It provides the setting for our day to day lives, gives identity to the places where we live, work and visit, and supplies us with enjoyment and inspiration. We all have special associations with landscape because it triggers responses, memories and emotions that are very personal to us. Every landscape is important to someone, whether it be a small patch of wasteland or a great wilderness".

Furthermore, part of the value of "natural" landforms comes from the slowness of some landforming processes relative to human life spans and from the fact that they reflect common social values. Communities may therefore value their local landscape and be resistant to changes to it. A related point is made by Bak (1997, p. 4–5): "we do not live in a simple, boring world... The surface of the earth is an intricate conglomerate of mountains, oceans, islands, rivers, volcanoes, glaciers and earthquake faults, each of which has its own characteristic dynamics. Unlike very ordered or disordered systems, landscapes differ from place to place and from time to time. It is because of this variation that we can orient ourselves by studying the local landscape around us". The importance of landscape diversity in allowing people to locate where they are, their direction of travel or their altitude on a hillside is another important element of the sense of place.

3.4 Aesthetic Value

The aesthetic value of geodiversity is a rather more tangible concept. It refers quite simply to the visual appeal (and those of other senses) provided by the physical environment. This may be through landforms at all scales from mountain ranges to local ponds, from coastlines to river banks, but all have value because of the diversity of topography they provide for residents or travellers (Bourassa, 1992). English Nature (2000a) also notes the psychological and physiological health benefits of having access to natural areas.

Natural landscapes have not always been regarded as aesthetically pleasing. For example, during the sixteenth and seventeenth centuries, wild landscapes were shunned, the English Lake District, for example, being described as full of "dreadful fells, hideous wastes, horrid waterfalls, terrible rocks and ghastly precipices" (Thomas, 1983). This was a time when the formal English garden was seen as an improvement on the wildness of nature. The Romantic movement of the late eighteenth and nineteenth centuries in Europe and North America promoted the inherent or intrinsic worth of nature. Landscape gardeners such as Capability Brown, Sir Humphrey Repton and others adopted less formal lines in the design of the grounds and lakes of English country estates, and eventually wildness and wilderness came to be valued for their own sake. Rather than seeing the Lake District as "hideous", "horrid" and "ghastly", it became the focus and inspiration for the Romantic Lakeland poets such as William Wordsworth and Samuel Coleridge.

Thus, in the nineteenth century, there was the flowering of the idea that nature was worthy of protection for its own sake. In fact, as we shall see in Chapter 5, this Romantic movement led on to the creation of the first national parks and the nature conservation movement in general. Lyrical writings, like those of John Muir in America, inspired the public and the politicians to value the impact of nature on all the senses:

"Climb the mountains and get the good tidings. Nature's peace will flow into you as sunshine flows into trees. The winds will blow their own freshness into you, and the storms their energy, while cares will drop off like autumn leaves... Nature's sources never fail. Like a generous host, she offers her brimming cups in endless variety, served in a grand hall, the sky its ceiling, the mountains its walls, decorated with glorious paintings and enlivened with bands of music ever playing... Fears vanish as soon as one is fairly free in the wilderness"

(Muir, 1901, p. 56).

3.4.1 Local landscapes

Many landscapes have an aesthetic appeal. Daly (1994) argued that peatlands are major landscape features in countries like Finland, Ireland, Scotland and Poland. "As such, they are not only of geomorphological importance but are part of the beauty and scenery of these countries...". With reference to limestone pavements, Goldie (1994, p. 220) believed that "the urge to conserve these landforms for the common good derives largely from local and personal appreciation of their scientific and aesthetic worth". Jarman (1994, p. 41) argued that "Human perception values variety, intricacy, pattern and regional character.... The contribution of landform type and wealth of surface detail to the popularity of tourist areas such as the Yorkshire Dales or residential areas (Kent v Essex), is under-rated. Erosion of this character and detail will reduce the attractiveness of an area".

Norton (1988) refers to this as "amenity value" where the existence of a natural feature improves our lives in some non-material way. This is certainly true of many physical features like mountains, beaches, cliffs, rivers, lakes, glaciers and waterfalls. The National Trust recently bid £1.5 million to buy 600 ha of land at Divis Mountain and Black Mountain, which dominate the skyline of Belfast, northern Ireland. They have been closed to the public, but the National Trust proposes to open them to public access. Landform is undoubtedly undervalued as an element of landscape and it is the ever-changing nature (diversity) of the natural world that creates much of the visual interest and beauty. And aesthetic value also bestows economic value and social status. As Jarman (1994, p. 42) argues "Advanced societies will pay a premium for landscape attributes of a property, starting with a sea view". Hill-top locations often acquire more than topographic height since they are associated with greater social and economic status than adjacent lowland.

Rocks also provide landscape diversity where they outcrop. Red sandstones are distinctive not just in their natural setting but also by giving a distinctive colour and texture to buildings and villages when used as a local building stone (see below). Since the soils of these areas are also typically red, there is a sense here of communities rising out of the land.

3.4.2 Geotourism and leisure activities

There is also an increasing market for geotourism either separately or linked with eco-tourism (Hose, 1998; Larwood & Prosser, 1998; McKirdy *et al.*, 2001; McKeever & Gallagher, 2001). First, there is the general value of scenery, wilderness and environment, often promoted as part of national tourism campaigns (Fennell, 1999). There is increasing interest in touring and walking holidays and the general attraction of rural landscapes for day trips and short breaks by increasingly urban populations. This is in addition to the worldwide attraction of beach visits and holidays.

Secondly, there are specific geological/geomorphological wonders that are highly attractive to tourists. Examples include the Grand Canyon, Niagara Falls, Old Faithful geyser in Yellowstone National Park, the Norwegian fjords, Uluru/Ayers Rock in Australia, geothermal lakes in Iceland, and so on. Sometimes the use of scenery in films can lead to major boosts for tourism. For example, *Lord of the Rings,* filmed in New Zealand has stimulated tourist interest there (see Fig. 3.7). The countless "Westerns" filmed in Monument Valley on the Utah/Colorado border, USA, continue to draw tourists. A recent survey by the BBC of 20,000 viewers, concluded that the top place "to see before you die" is the Grand Canyon, while others places in the top 50 include Uluru (Australia), the Matterhorn (Switzerland), Lake Louise (Canada) and no less than four waterfalls (Angel Falls, Niagara Falls, Iguaca Falls, Victoria Falls). The list suggests that abiotic features are able to attract tourists at least as much as biotic ones and that the value of geotourism as opposed to ecotourism ought to be better understood.

Thirdly, there may be local geological activities that can be used to attract tourists (e.g. fossil hunting, lapidary, geological trails, museums and visitor centres). And voluntary environmental work sometimes brings the volunteers into contact with the geological world, for example, in the construction of footpaths and steps, the repair of drystone walls, or creation of ponds and ditches. In turn, this leads to greater awareness and appreciation of the character and properties of geomaterials and processes.

Finally, there are recreational activities such as skiing, caving, canyoning, glacier hiking, white-water rafting and climbing that require specific landscapes or geological environments. Climbers, in particular, value the diversity of rock types and structures for the variety of challenges they bring, from granite slabs to sandstone pinnacles (Mellor, 2001).

Figure 3.7 *Tourist advertisement linking the film Lord of the Rings with the New Zealand landscape (Reproduced by permission of the New Zealand Tourist Board)*

Larwood & Prosser (1998, p. 99) conclude that "Tourists, whether they are aware or not, will in some way all be geotourists". It is also an activity with huge economic potential.

3.4.3 Artistic inspiration

As Bennett & Doyle (1997) point out, landscape is an extremely important source of inspiration for artists, musicians, poets, writers and others. The Lakeland poets were inspired by the scenery of the English Lake District. After seeing Fingal's Cave on the Isle of Staffa in Scotland, Turner painted it, Jules Verne used it as a location for a short story, Strinberg set a scene in *Dreamplay* on the island and Mendelssohn composed the *Hebrides Overture*. Novelists like Thomas Hardy and Jane Austen founded their writings within distinctive local landscapes and were clearly inspired by them (Hardyment, 2000). In Hardy's case, the landscape was the rolling downlands and vales of Wessex (Hampshire to Devon) in England, and in Box 3.4, he gives an evocative description of the aesthetic qualities of the landscape structure, colour, atmosphere and local distinctiveness of the "Vale of Blackmoor". The English poet Lord Byron, on the other hand, preferred the landscape of the Scottish Highlands:

England! Thy beauties are tame and domestic
To one who has roved o'er the mountains afar;
Oh for the crags that are wild and majestic
The steep frowning glories of the dark Loch na Garr.

Lord Byron (1807) *Lachin Y Gair.*

Box 3.4   Example of Thomas Hardy's Landscape Description

"The village of Marlott lay amid the north-eastern undulations of the beautiful Vale of Blakemore or Blackmoor...an engirdled and secluded region, for the most part untrodden as yet by tourist or landscape-painter.... It is a vale whose acquaintance is best made by viewing it from the summits of the hills that surround it...This fertile and sheltered tract of country, in which the fields are never brown and the springs never dry, is bounded on the south by the bold chalk ridge that embraces the prominences of Hambledon Hill, Bulbarrow, Nettlecombe-Tout, Dogbury, High Stoy and Budd Down. The traveller from the coast, who, after plodding northwards for a score of miles over calcareous downs and corn-lands, suddenly reaches the verge of one of these escarpments, is surprised and delighted to behold, extended like a map beneath him, a country differing absolutely from that which he has passed through. Behind him the hills are open, the sun blazes down upon fields so large as to give an unenclosed character to the landscape, the lanes are white, the hedges low and plashed, the atmosphere colourless. Here, in the valley, the world seems to be constructed upon a smaller and more delicate scale; the fields are mere paddocks, so reduced that from this height their hedgerows appear a network of dark green threads over-spreading the paler green of the grass. The atmosphere beneath is languorous, and is so tinged with azure that what artists call the middle distance partakes also of that hue, while the horizon beyond is of the deepest ultramarine. Arable lands are few and limited; with but slight exceptions the prospect is a broad, rich mass of grass and trees, mantling minor hills and dales within the major. Such is the Vale of Blackmoor".

Thomas Hardy (1891) *Tess of the D'Urbervilles*

3.5   Economic Value

Economists have attempted to put a financial value on all environmental assets (Foster, 1997), but many geological materials have more than a theoretical economic value. Rock, minerals, sediment, soil and even fossils, all have economic value, though this varies depending on the nature of the material involved. The "Millennium Star", a 203-carat diamond, and 11 rare blue diamonds, were together valued at over £200 million when attempts were made to steal them from the Millennium Dome in London in November 2000 (*The Guardian,* 9 November 2001). On the other hand, a tonne of gravel may be worth only a few pounds.

The usual classification of economic mineral resources is into mineral fuels (e.g. petroleum and coal), industrial, metallic and precious minerals (e.g. metal ores and gemstones) and construction minerals (e.g. sand and building stone), but the economic value of the abiotic environment should also include fossils, other forms of energy and indeed soil and landscape resources. The distinction is also blurred in that most single mineral rocks like limestone, chalk, gypsum and silica sand have both constructional and industrial uses (Woodcock, 1994). The diversity of these resources has been exploited with ingenuity over the centuries and has given societies the huge range of materials they have needed to progress to their modern sophistication.

3.5.1 Mineral fuels

The three main types of mineral fuels are coal and peat, petroleum and uranium.

Coal and peat

These are the products of accumulation of mainly terrestrial plants in a humid environment. Peat that has accumulated over the last few thousand years has undergone only limited compaction and alteration so that it is still moist and relatively spongy. But in its compressed form, it is the traditional fuel source in parts of Finland, Sweden, Ireland, Scotland, Russia, Belarus, Ukraine, China, Indonesia and the Falkland Islands (Asplund, 1996), and in some places it is still dug, dried and burnt today. The International Peat Society (IPS) estimates that the world's peatlands cover an area of almost 4 million km^2, representing about 3% of the Earth's land surface, and contain 5000 to 6000 Gt of peat. In addition, there are over 2.4 million km^2 of wetlands where the amount of accumulated peat or other organic matter is unknown (Lappalainen, 1996). Some countries have large areas of peatland. Finland, for example, has 10 million ha covering 30% of the land area and both Finland and Ireland use peat for electricity generation (Kelk, 1992; Asplund, 1996). Other economic uses of peat include agriculture (Okruszko, 1996), forestry (Päivänen, 1996), horticulture (Schmilewski, 1996), health therapy (Korhonen & Lüttig, 1996) and a diverse range of minor, but locally important, uses (Mutka, 1996).

There is great diversity in peat deposits depending on the types of source water and organic materials. For example, Daly (1994) and Maltby & Proctor (1996) describe raised bogs, blanket bogs, tropical forest peats and fen peats. The first two of these form ombrogenous bogs in cool temperate climates where the only water source is rainfall. Consequently, they form deep acid peats of bulk density of 0.1 to 0.2 and low mineral content. Fen peats, on the other hand, are fed by mineral rich springs or other sources and have near-neutral pH, high cation concentrations, and higher bulk densities and mineral content. The Geological Survey of Finland maintains a peat data bank with information on all bogs covering more than 20 ha and includes over 200 variables of bog and peat characteristics (Kelk, 1992).

Coal is the compacted and chemically matured form of peat in which increasing pressure and temperature produce physiochemical changes, which in turn define a continuous spectrum of coal maturity or rank (see Fig. 3.8). This spectrum extends from lignite (soft brown coal) through sub-bituminous coal (hard brown coal), bituminous coal, anthracite to graphite. This succession is produced in anaerobic conditions by progressive loss of water, carbon dioxide, methane and volatiles. During the process of conversion, oxygen and hydrogen content decrease and the percentage of carbon increases until it reaches 100% in the case of graphite. The calorific value also increases, bringing greater economic value to the higher-ranking coals such as anthracite. Coal may contain a variety of impurities, including clay minerals (which

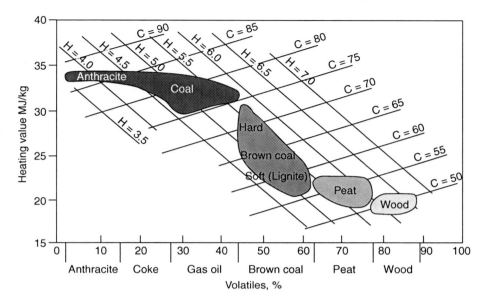

Figure 3.8   *Diversity of fuel characteristics from wood to anthracite (Modified after Asplund, D. (1996) Energy use of peat. In Lappalainen, E. (ed) Global Peat Resources. International Peat Society, Finland, 319–325, by permission of International Peat Society)*

increase ash residues), pyrite (which gives emissions of sulphur dioxide) and sodium chloride (which accelerates corrosion of boilers) (Woodcock, 1994), so diversity is not always beneficial. Rarer types of coal include sporinite (coalified spores and pollen) and alginite (coalified algae) also present in oil shales (Evans, 1997; Thomas, 2002).

Petroleum

This term covers a diverse range of natural, solid (e.g. bitumen), liquid (e.g. crude oil) and gaseous (e.g. natural gas) hydrocarbons. For an economically exploitable resource to occur, there must be a coincidence of appropriate source rock lithologies, processes and rock geometries. The source rock must have a significant organic content (0.5–10%), as would occur through the steady accumulation of dead organisms (algae or plankton) in lake or sea-bed muds that by diagenesis are transformed into shales. Other source rocks include limestones such as the Jurassic Hanifa Formation of Saudi Arabia and the Cretaceous La Luna Formation of Venezuela. In the early stages of burial in anaerobic conditions, the organic matter is converted to kerogen but the type of kerogen produced will have different proportions of hydrogen, oxygen and carbon depending on the type of source organic matter (see Fig. 3.9). In turn, kerogen type controls the type of hydrocarbon produced, but in general terrestrial organic matter leads to kerogen Type III which produces mainly natural gas, whereas marine algae leads to kerogen Type I which produces light oil (Woodcock, 1994). Impurities such as sulphur may be present but there is a high demand for low sulphur fuels (sweet crudes) (Evans, 1997).

The processes are also important in producing diverse end products. At low pressures and temperatures, the decay of organic matter results in methane, much of which

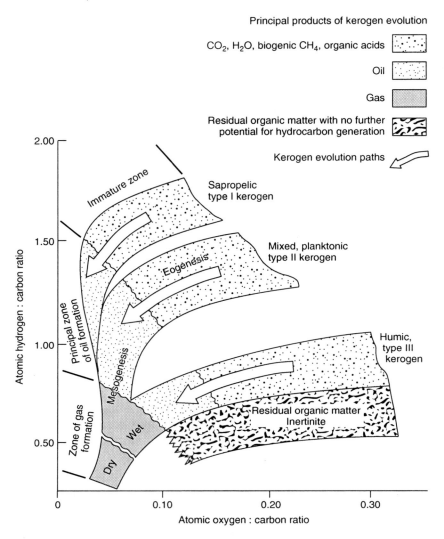

Figure 3.9 *Diversity of kerogen types and fuel paths during diagenesis (After Evans, A.M. (1997) An Introduction to Economic Geology and its Environmental Impact. Blackwell Science, Oxford, by permission of Blackwell Science Ltd)*

escapes to the atmosphere, though some is trapped by frozen ground in permafrost areas, by water pressure on the deep sea bed or within coal beds. It is sometimes exploited for local heating as in Anchorage, Alaska, USA. The more economic natural gases as well as oils are produced at temperatures of 50 to 200 °C, with crude oil, thermogenic wet gas and thermogenic dry gas being released successively as the temperature rises. Oil will itself be transformed to gas if heated further, and the transformations can occur at slightly lower temperatures if timescales of millions of years are involved.

Once released, the fluids will migrate along pressure gradients, usually towards the surface, and will escape if nothing prevents them from doing so. Thus, in Venezuela and some other oil-rich areas, it is not unusual to find "tar ponds" where liquid hydrocarbons are seeping out at the surface. However, in some cases, the rock geometry has been such that low-permeability barriers (caprocks or seals) trap the upward migration of oil and gas. Shales and evaporites form most of the oil and gas seals and Fig. 3.10 illustrates the diversity of some stratigraphic and structural traps. Below these traps, the oil will accumulate in suitable reservoir rocks (predominantly sufficiently thick sandstones and limestones) that are those that have high porosity or permeability or preferably both. High porosity is important since it provides the pore space in which the oil and gas can accumulate, while high permeability is important in allowing fluids to migrate freely into the reservoir and out again via extraction boreholes.

Along with the crude oil and natural gas mentioned above, a range of other hydrocarbons occur, reflecting the diversity of source materials and maturation processes. These include heavy oils, such as those that occur in the Orinocco Oil Belt of Venezuela, which are the result of low maturity or alteration processes within the reservoir. Bitumen has an even higher viscosity than heavy oils and acts as tar cement within sandy reservoir rocks. The bitumen is extracted either by *in situ* steam or gas injection or, in the case of the famous Athabasca Tar Sands in Canada, by mining and heating in retorts. Shale oil occurs where shallow burial has resulted in low-maturity kerogen that has not migrated from the source shale. Other forms of natural gas occur either as "geopressure natural gas", in deeply buried source rocks, or as "tight-reservoir gas", in low-permeability sedimentary rocks.

This diverse range of hydrocarbons is valuable in providing a huge range of end uses though about 90% is used as fuel. Crude oil is distilled to produce fractions such as bitumen for road surfacing, wax and naptha for chemical manufacture (detergents,

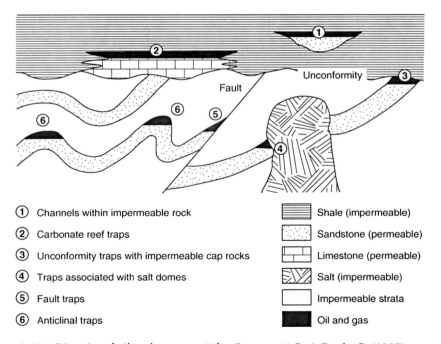

① Channels within impermeable rock	▤ Shale (impermeable)
② Carbonate reef traps	▦ Sandstone (permeable)
③ Unconformity traps with impermeable cap rocks	▦ Limestone (permeable)
④ Traps associated with salt domes	▩ Salt (impermeable)
⑤ Fault traps	▢ Impermeable strata
⑥ Anticlinal traps	■ Oil and gas

Figure 3.10 *Diversity of oil and gas traps (After Bennett, M.R. & Doyle, P. (1997) Environmental Geology. Wiley, Chichester, by permission of Wiley)*

textiles, rubber, agrichemicals, plastics, cosmetics), fuel oil and kerosene/paraffin for heating and aircraft fuel, gas oil for diesel engines, petrol/gasoline for cars and liquid petroleum gas for a variety of fuel uses. Natural gas is used for electricity generation, domestic and industrial cooking and heating and chemical manufacture (e.g. methanol, ammonia, fertilisers). Waxes and heavy oils can be converted to petrols by cracking, thus giving flexibility in meeting geographical or temporal demands.

Uranium

This is a different type of mineral fuel that releases energy not by combustion but by radioactive decay. The average concentration of uranium in the upper continental crust is 3 ppm, but for commercial exploitation, the concentration must be at least 100 times this. A number of geological processes are capable of achieving this, mainly as the mineral uraninite (U^{308}). First, primary uranium deposits form in magmatic veins from residual magmatic liquids, or in hydrothermal veins from hot magmatic waters flushing through joints and fault zones. Subsequent uplift and erosion may expose the uraninite to surface erosion and transport, and because of its high density it becomes concentrated into sedimentary placers, particularly in the low-oxygen environments of the Precambrian. Uranium also goes into solution and may be precipitated within sediments and today is found in black shales, coals and phosphates (Woodcock, 1994).

Renewable energy

These are not mineral fuels, but many are related to the abiotic environment and it is convenient to consider these briefly here. Geothermal energy relies on the internal heat of the Earth and has been utilised for some time in active volcanic provinces like Iceland and New Zealand. More recently, schemes that circulate water into the crust to be heated at depth and pumped to the surface have been explored. Hydroelectric power relies upon a vertical fall of water and therefore requires favourable topographic conditions and water supplies, including the use of glacial meltwater in Norway. Similarly, wave and tidal power can exploit particularly coastal locations where conditions are favourable, and wind power is greatest in upland, coastal and offshore situations.

3.5.2 Industrial, metallic and precious minerals

These minerals form by a wide variety of mineralisation processes (Woodcock, 1994; Evans, 1997), some of which mirror those discussed above for uranium (Fig. 3.11). For example, endogenic mineralisation may occur by early crystallisation and differential settling of heavy minerals within a magma chamber resulting in economically exploitable concentrations of, for example, magnetite, ilmenite and chromite. In other cases, immiscible liquids may develop in these magmas with the separation of, for example, nickel sulphide liquid. On crystallisation, pods of pyrite (iron sulphide), chalcopyrite (copper sulphide) and, sometimes, gold and platinum may form economic concentrations. Metamorphism associated with heat and water transfer from acidic magma intrusions may cause mineral deposits to be concentrated in the surrounding country rock. Magnetite, haematite, cassiterite, pyrite, chalcopyrite, galena, sphalerite and molybdenite all form in this way. Hydrothermal precipitation occurs when super-heated water travels through joints and fissures carrying dissolved compounds, which are then precipitated and concentrated as the water cools to form mineral veins. On the

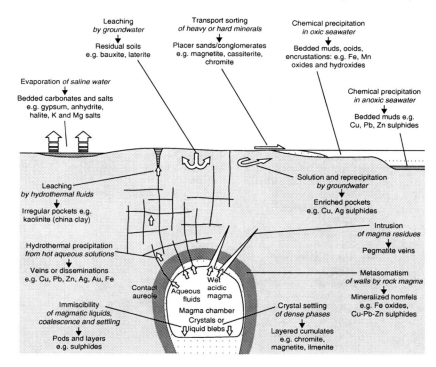

Figure 3.11 *Diversity of mineralisation processes (After Woodcock, N. (1994) Geology and the Environment in Britain and Ireland. UCL Press, London, by permission of UCL Press)*

other hand, the movement of cooler hydrothermal waters may chemically leach rocks leaving residual deposits such as kaolinite (china clay). Mineralisation also occurs by igneous and hydrothermal processes (e.g. black smokers) on the ocean floor or by the flow of saline water through basin floor sediments (Bennett & Doyle, 1997).

Exogenic processes also result in economic concentrations of minerals, including fluvial or coastal transport, density sorting and deposition in placers of heavy metals and some gemstones such as diamonds (Sutherland, 1984). The recently discovered mineral sands of the lower Murray-Darling Basin in Australia occur on an ancient shoreline and comprise one of the richest mineral deposits on Earth at 45 million tonnes including ilmenite (an important source of titanium), rutile, zircon and monzanite. Surface weathering and leaching may result in an economic residue (e.g. bauxite) or precipitation of a leached mineral at or below the water table (e.g. copper). Evaporation of surface fresh or saline water has often resulted in thick mineral deposits including salt and gypsum. Chemical precipitation may also occur within water bodies as exemplified by sea-floor manganese nodules. These and other processes (Woodcock, 1994, Fig. 9.3) result in a very wide range of industrial, metallic and precious minerals. Table 3.2 lists some of the major minerals, their uses, major world producers and reserves (Cutter & Renwick, 1999).

Industrial minerals

These are valued for their particular chemical or physical properties. There is a very diverse range of such minerals and an even more diverse range of uses in the manu-

Table 3.2 *Major minerals, their uses, major producing nations and reserves (after Cutter &* *Renwick, 1999)*

Mineral	Major uses	Major producing nations	Reserves (yrs)
Antimony	Flame retardants, transportation, batteries	China, Russia, S. Africa, Bolivia	39
Arsenic	Wood preservatives, glass, agricultural chemicals	China, Chile, France	20
Asbestos	Insulation, friction products, gaskets	Russia, Canada, Kazakhstan, China	Large
Barite	Well drilling fluids, chemicals	China, USA, India, Morocco	40
Bauxite	Packaging, building, transportation, electrical	Australia, Guinea, Jamaica, Brazil	211
Beryllium	Alloys for aerospace & electrical equipment, computers	USA, China, Brazil, Russia	n/a
Bismuth	Pharmaceuticals, chemicals, machinery	Peru, Mexico, China	34
Boron	Glass, agriculture, fire retardants	Turkey, USA	57
Bromine	Fire retardants, agriculture, petroleum additives	USA, Israel, UK	n/a
Cadmium	Coating & plating, batteries, pigments	Japan, Canada, Belgium, USA	29
Caesium	Electronic and medical applications	Canada, Namibia, Zimbabwe	n/a
Chromium	Metallurgical & chemical industries, refractory industry	S. Africa, Kazakhstan, India	349
Cobalt	Superalloys, catalysts, paint driers, magnetic alloys	Canada, Zambia, Russia, Australia	205
Columbium	High-strength low-alloy steels, carbon steels, superalloys	Brazil, Canada	259
Copper	Building construction, electrical & electronic products	Chile, USA, Canada, Russia	32
Diamond	Machinery, abrasives, stone & ceramic products, minerals	Australia, Russia	20
Diatomite	Filter aid, fillers	USA, France, former USSR	533
Feldspar	Glass, pottery	Italy, USA, Thailand, S. Korea	Large
Fluorspar	Metal processing	China, Mexico, S. Africa	52
Gallium	Optoelectronic equipment, integrated circuits	Germany, Russia, Japan	n/a
Garnet	Petroleum industry, filters, transportation	USA, Australia, China, India	Moderate
Germanium	Fibre-optics, infrared optics, detectors	USA	n/a
Gold	Jewellery, electronics, medicine	S. Africa, USA, Australia	20
Graphite	Refractories, brake linings, packings	China, S. Korea, India	29
Gypsum	Cement retarder, agriculture, plaster	USA, China, Canada, Iran	Large
Helium	Cryogenics, welding, pressurising, controlled atmospheres	USA, former USSR, Algeria	n/a

Table 3.2   (*continued*)

Mineral	Major uses	Major producing nations	Reserves (yrs)
Ilmenite	Titanium pigments	Australia, S. Africa, Canada	82
Indium	Coatings, solders, alloys, electrical, semiconductors	Canada, Japan, France	17
Iodine	Animal feeds, catalysts, chemicals	Japan, Chile, USA	n/a
Iron ore	Steel	China, Brazil, Russia, Australia	150
Kyanite	Refractories	S. Africa, France, India	Large
Lead	Batteries, fuel additives	Australia, USA, China, Peru	24
Lime	Steel furnaces, water treatment, construction, agriculture	China, USA	Moderate
Lithium	Ceramics, glass, aluminium, lubricants	Chile, Australia, Russia	349
Magnesium	Metal refractories, aerospace, auto components	USA, Canada, Norway, Russia	Moderate
Manganese	Construction, machinery, transportation	S. Africa, China, Ukraine	93
Mercury	Electronic and electrical applications, paints, chemicals	Spain, China, Algeria	42
Molybdenum	Machinery, electrical, transportation	USA, China, Chile, Canada	47
Nickel	Metal alloys, stainless steel	Russia, Canada, New Caledonia	51
Perlite	Building construction, filters, horticulture	USA, Turkey, Greece	412
Phosphates	Fertiliser	USA, China, Morocco, W. Sahara	80
Platinum	Automotive, electronic, chemical, jewellery	S. Africa, Russia, Canada	431
Potash	Fertiliser	Canada, Germany, Belarus, Russia	321
Pumice	Building blocks	Italy, Turkey, Greece	n/a
Quartz	Electronics	USA, Brazil	Large
Rare earths	Petroleum catalysts, metallurgical, ceramics	China, USA, former USSR	Large
Rhenium	Petroleum catalysts, super alloys	USA, Chile, Peru, Canada	86
Rutile	Titanium pigments, titanium metal, welding	Australia, S. Africa, Sierra Leone	83
Scandium	Metallurgical research, halide lamps, lasers	China, Kazakhstan, Madagascar	n/a
Selenium	Electronics, glass, chemicals, pigments	Japan, USA, Canada, Belgium	37
Silicon	Metal alloys, stainless steel	China, USA, Russia, Ukraine	Moderate
Silver	Photography, electronics	Mexico, Peru, USA, Australia	20
Strontium	Colour TV tubes, magnets	Mexico, China, Turkey, Iran	36
Sulphur	Fertiliser, chemicals	USA, Canada, China, Mexico	27
Talc	Ceramics, paint, paper, plastics	China, USA, Japan	Large
Tantalum	Electronics, machinery	Australia, Brazil, Canada	65
Tellurium	Iron & steel, catalysts, chemicals	Canada, Japan, Peru	n/a

(*continued overleaf*)

Table 3.2   (*continued*)

Mineral	Major uses	Major producing nations	Reserves (yrs)
Thallium	Electronics, pharmaceuticals, alloys	USA, Canada	25
Thorium	Nuclear fuel, electrical	Australia, Brazil, Canada, India	n/a
Tin	Cans & containers, electrical, construction	China, Indonesia, Brazil, Bolivia	39
Titanium	Aerospace, chemicals	Japan, Russia, Kazakhstan	4
Tungsten	Lamps, electrical, metalworking	China, Kazakhstan, Russia	105
Vanadium	Transportation, machinery, tools, building	S. Africa, Russia, China	286
Yttrium	Television monitors, lasers, alloys, catalysts	China, former USSR, Australia	699
Zinc	Metal plating, alloys	Canada, Australia, China	20
Zirconium	Foundry sands, refractories, ceramics	Australia, S. Africa, Ukraine	36

n/a = not available

facture of products and materials (Fig. 3.12) and even as healing tools employed by "crystal therapists". Figure 3.13 shows Finland's industrial mineral extraction sites.

Metallic minerals

These occur in relatively pure forms such as gold and silver (native metals) or as compounds with other elements (ores). About 75% of all elements are metals and there is therefore a huge diversity of metallic compounds.

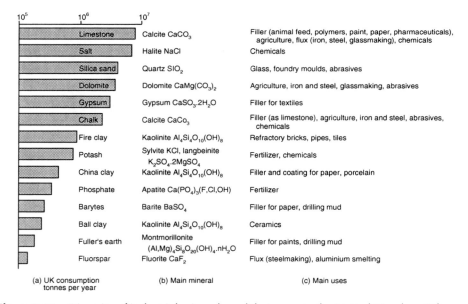

Figure 3.12   *Diversity of industrial minerals and their uses in the United Kingdom (After Woodcock, N. (1994) Geology and the Environment in Britain and Ireland. UCL Press, London, by permission of UCL Press)*

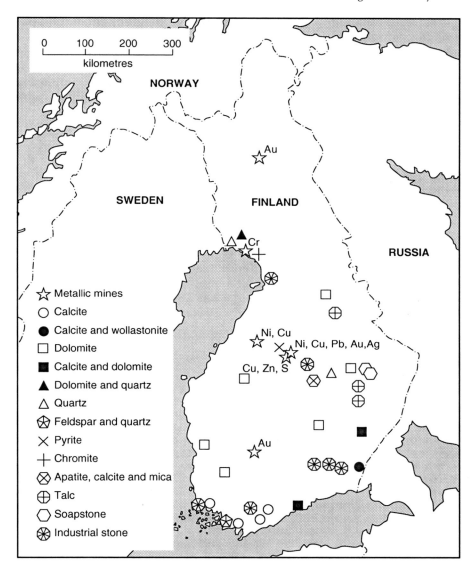

Figure 3.13 *Diversity of industrial and metallic mineral extraction in Finland (Reproduced with permission from Geological Survey of Finland)*

Metals have been valued and utilised for millennia. Gold was probably discovered over 8,000 years ago and its pliable properties enabled it to be moulded into jewellery or hammered into thin decorative sheets. Native copper was found to be stronger than gold and could be fashioned into weapons, tools and ornaments. It was discovered that molten metals could be combined into alloys with new properties, such as bronze alloyed from copper and tin. And although the extraction of iron was more difficult, it allowed superior tools and weapons to be produced and gave rise to a revolution in metal technology during the Iron Age. The Romans used lead for water pipes and appreciated its malleable properties.

This search for new alloys, metal properties and uses has continued to the present day, and it is no exaggeration to say that modern society could not exist without its diversity of metallic artefacts, from aluminium with its strength and lightness to titanium with its high melting point, toughness and corrosion resistance for use in rockets and aeroplanes. Iron has by far the greatest production volumes and its use in the production of steel also utilises manganese, chromium, nickel and molybdenum for particular applications (Woodcock, 1994).

Gemstones

These are valued for their rarity, durability and aesthetic beauty, though some have more mundane uses (e.g. as abrasives or drill-bits). There is, in fact, a very wide range of gemstones, displaying a huge diversity of chemical composition, crystal structure, colour and light properties. O'Donoghue (1988) and Read (1999), for example, list about 200 gemstone types but there is diversity even within each gemstone type (see Box 3.5). An important characteristic of gemstones is their inclusions, whose composition allows geologists to predict further occurrences and gemmologists to say where a particular stone has come from. "Not only can characteristic inclusions identify a natural stone, distinguish it from its synthetic counterpart and provide valuable information on the formation conditions of an unknown specimen, they can also in

---

Box 3.5   Gemstone Diversity

Beryl has the formula $Be_3Al_2Si_6O_{18}$ with some Fe, Mn, Cr, V, and Cs. It displays a wide range of colours from green in emerald, blue to greenish-blue in aquamarine, golden yellow, pink or peach in morganite and an orange- to near ruby-red colour in bixbite from Utah, USA. Beryl can also be colourless (goshenite). The colour in emerald arises from replacement of aluminium by chromium and/or vanadium. The colour of green aquamarine is due to iron, while morganite is believed to acquire its pink colour from manganese. A dark blue beryl (Maxixe or Maxixe-type) was discovered in Brazil in 1917, and it derives its colour from nitrate or carbonate impurities. Beryl is found in a variety of geological environments and rock types around the world. It is found in mica schists (particularly emerald), metamorphic limestones (emerald) and hydrothermal veins. It is common in granitic rocks, particularly pegmatites. The finest emeralds come from Colombia and were known to Colombian Indians as far back as A.D. 1000.

Corundum (Al_2O_3 with some iron, titanium and chromium) comes in a number of forms, but the best known are ruby and sapphire. Ruby colours range from a strong purple-red to orange-red. The world's finest rubies come from Burma and show a strong red fluorescence, whereas those from Thailand, which make up 70% of the world's fine ruby production, are less intensely red due to the presence of iron. Sapphire comes in a range of blues, greens, oranges and purples. Blue sapphires of industrial and gem quality were mined from dykes in Montana, USA, with 25 million dollars being recovered from the Yogo Gulch mines between 1865 and 1929. Placer sapphires have also been quarried from Missouri River terrace gravels where they occur along with mined gold deposits. Some corundum shows 12-pointed or 6-pointed stars due to rutile inclusions.

Quartz (SiO_2) comes in diverse forms and colours including amethyst, rock crystal, milky, smoky and rose quartz, chalcedony, cornelian, onyx, jasper and agate, but each type itself exhibits a diversity of forms and colours due to impurities and processes of formation (O'Donoghue, 1988).

many cases establish exactly where the stone came from – sometimes down to the actual mine" (O'Donoghue, 1988, p. 3). Burmese rubies and Kashmir blue sapphires are regarded as high quality and therefore more valuable than other sources, but certification can only be given by a gemmologist on the basis of the characteristic inclusions.

Gemstones are used in an infinite and dazzling variety of metallic and other settings. For example, Fabergé made several ornamental Easter eggs for Tsar Nicholas II, the most expensive of which is embellished with more than 3000 gemstones and was recently sold for c. £4 million. The engraved rock crystal shell sits on a rock crystal base in the form of a melting ice block with platinum-mounted rose diamond rivulets (*The Independent*, 21 March 2002). René Lalique's Dragonfly Pectoral made in 1897–1898 is a masterpiece of design and colour in gold, enamel, chrysoprase, chalcedony, moonstones and diamonds, and can be viewed in the important Lalique collection in Lisbon's Gulbenkian Museum. Gemstones are, in fact, one of the most remarkable demonstrations of the principle and economic value of geodiversity.

3.5.3   Construction minerals

The greatest volume of geological materials is used in construction work and there is a very diverse range of geomaterials used in a huge number of applications. Although bio-materials are used in building construction (e.g. timber frames, thatched roofs, wooden cladding), urban environments are dominated by bulk minerals in ways that should make the value of geodiversity very evident to their inhabitants. However, most urban dwellers probably never give their mineral surroundings a second glance or a second's thought.

Construction minerals are usually classified into

- building stone – quarried blocks, rough or cut, used in wall construction or as facing slabs or roofing tiles;
- aggregates – either coarse clastic sediments or crushed stone used in concrete manufacture and many other uses;
- limestone – used in the manufacture of cement and other materials;
- structural clay – used in the manufacture of bricks and tiles;
- gypsum – used in plaster;
- sand – for glass manufacture;
- volcanic products;
- bitumen – used in asphalt (see petroleum above).

In some cases, rock is used *in situ* to provide shelter (as in the case of cave dwellings) or form the walls of a building (as in the Temppeliaukio Church in Helsinki, Finland), but in most cases, the geological material is excavated and processed in some way.

Building stone

This has a long history of use dating back to at least 8000 BP when stone was built into the walls of Tell-es-Sultan near Jericho (Prentice, 1990). Not all stone is suitable for use as load bearing or facing material. The five key factors that determine whether a rock type is viable as a building stone are (Bennett & Doyle, 1997)

- structural strength (load-bearing capacity – determined by mineralogy and presence/absence of discontinuities);
- durability (ability to withstand exposure, weathering, salt crystallisation, etc.);
- appearance (e.g. colour and texture);
- ease of working (e.g. whether the rock can be worked into precisely shaped blocks);
- availability (e.g. cost of transport).

Sedimentary rocks are often thinly bedded and in these circumstances may be usable when thin sheets or slabs of rock are required, for example, roofing, floor tiles or cladding. Where sedimentation has been continuous, there may be thicker beds suitable for load-bearing wall construction. Intrusive igneous rocks, such as granite batholiths,

(a)

Figure 3.14 *Diversity of wall materials: (a) mixed, rounded glaciofluvial cobbles, brick and metamorphic stone cladding in juxtaposition, Banff, Canada and (b) mixed stone rubble and brick patterns, Porvoo church tower, Finland*

(b)

Figure 3.14   *(continued)*

often give massively jointed rocks suitable for cutting into variously shaped and sized blocks. Metamorphic calcareous rocks are relatively easily re-crystallised to produce massive marble deposits, whereas granitic rocks require high-grade metamorphism to be converted to usable gneisses (Prentice, 1990). Together, these different rock types produce a dazzling diversity of building materials (e.g. Fig. 3.14).

Several types of building stone can be recognised. Dry stone walls are constructed from roughly broken rock and rely on the skill of the builder to produce an interwoven structure. The diversity of the United Kingdom's dry stone walls is brilliantly illustrated by the 19 sections of the Millennium Wall, at the National Stone Centre in Derbyshire. Rough-cut blocks of various sizes and shapes are referred to as "rubble" though the final effect in buildings is usually very attractive. "Dimension stone" (ashlar to the architect) is shaped and dressed into regular sized and shaped blocks, slabs or tiles, which are then used to build load-bearing walls, cladding for walls, roofing materials (Fig. 3.15), floor tiles, pavement flagstones or road sets and cobblestones. This type of use has a long history, and it reached its peak in the late nineteenth century when a host of public buildings and private houses were constructed in stone. In developed countries its use declined in the twentieth century as cheaper, composite materials became available. However, in most parts of the world, the vernacular architecture can be seen to reflect the local geology, and there has been a welcome recent revival in the use of building stone. (see Section 6.5.4). The local variations in building stone often reflect the geological diversity of particular countries and nowhere is this relationship between geology and traditional building stone clearer than in Britain (see Box 3.6). In general, local building stones have been used in traditional buildings because of the high cost of transport. The major exception is Caen stone, which was easily shipped to south-east England from northern France.

Figure 3.15 *Diversity of roof tiles used to create patterns in the roof of Prague Cathedral, Czech Republic*

"But usually no materials look so well as the local ones, which belong organically to their landscape, harmonize with neighbouring buildings and nearly always give the best colours. There are still many old houses scattered over our countryside which convey the impression of having grown out of the soil untouched by the hand or mind of Man" (Clifton-Taylor, 1987, p. 23–24).

Box 3.6   Building Stone in the United Kingdom

In Scotland, grey granites dominate in Aberdeen where local sources have been actively quarried, for example, at Rubislaw Quarry. In Edinburgh and Glasgow, on the other hand, Carboniferous and Devonian sandstones dominate the traditional buildings. "To see the best of all Britain's Carboniferous sandstones, it is generally considered necessary to go to Edinburgh" (Clifton-Taylor, 1987, p. 132). McMillan *et al.* (1999) list 97 sandstone quarries in Scotland and northern England that supplied stone for the city's buildings and each has its own distinctive rock texture, colour and structure that provide a rich and diverse heritage of stone buildings.

Throughout Britain there are examples of villages where distinctive Old Red Sandstone dominates the buildings and field walls (e.g. East Lothian and Devon). Clifton-Taylor (1987, p. 138) describes the diversity of the stone as follows:

"They may be a dense and even rather depressing red; but they can also be pink, purple, brown, greenish-grey, pure grey, or grey with a blush of pink, a delicious colour. The tint may change slightly almost from block to block, and sometimes within the single block of stone, yielding effects of much charm".

Continued on page 101

Continued from page 100

However, the most intensively worked sandstone in Britain has been the Carboniferous buff-coloured "Yorkstone" of the Pennine region of west Yorkshire. The term "Yorkstone" covers a variety of types. Some are massively bedded with blocks weighing several tonnes offered by some quarries. Others are finely bedded at spacings of 2 to 3 cm, making these types ideal for splitting into flagstones, copings or sills. The mica grains in many Yorkstone types give them an attractive sparkle.

From Humberside to Dorset, Jurassic limestones are used as a major building stone and have been quarried in the Cotswolds, Bath area and Isle of Portland. Many public buildings in London were constructed of Portland stone, including the British Museum. Limestone and sandstone are also used as a roofing material, for example, in the Cotswolds, where the fashion is to place the largest tiles at the base and decrease the size with each course towards the ridge. In the Lake District, Wales and south-west England, Lower Palaeozoic slates and slatey mudstones are extensively used as a building material for walls, floors and, particularly, roofs. About 20 standard sizes of roofing slates are recognised, many named after the female aristocracy. Therefore, we have princesses, duchesses, marchionesses, countesses, viscountesses and ladies, not to mention the politically incorrect wide countesses and narrow ladies (Clifton-Taylor, 1987)!

Eastern England is generally poor in building stone, though locally, Ragstone, Greensand, Chalk and Carstone have been used (Clifton-Taylor, 1987). There is also extensive use of flint derived from the Chalk as a walling material, but "not used for building in any other country on so extensive a scale" (Clifton-Taylor, 1987, p. 193). In the Neolithic age, there were 200 flint mines up to 12 m deep at Grimes Graves in Norfolk, England, exploiting flint seams in the Chalk. It can be worked in various ways, for example, as rounded facing pebbles derived from local beaches or glaciofluvial deposits or as knapped or squared flints producing "flushwork" blocks in churches or important buildings. It is sometimes mixed with chalk or brick rubble in walls.

The local distinctiveness of vernacular architecture in Britain has much to do with the diversity of geological materials in use in different parts of the country (Fig. 3.16). A superbly illustrated map of building stones and quarries of the UK, compiled in collaboration with various other groups, has been published recently by the British Geological Survey (British Geological Survey, 2001).

In developed countries, dimension stone is now rarely used in a load-bearing role, but instead there has been a major expansion in the use of stone as facing material on steel-framed or concrete buildings. It is not just major companies that have become multinational; their geological edifices provide dazzling displays of colourful facades and interiors assembled from around the world. This includes polished cladding slabs for use as external or internal facing materials, service counters or wall and floor tiles. As Robinson (1996, p. 39) puts it: "all hell broke loose in the post-war years as buildings became clad in Namibian gneiss or Brazilian yellow granite. The important fact remains, that geology is all around us if we care to look". Robinson (1984, 1985) has described geological walks in London to view the geological materials but there have been many further additions in the last 20 years. Stone has become "the symbol of prestige, and it is not surprising that it finds favour with the banks and financial houses of today, as it did with the church builders of medieval times, in conveying a sense of permanence and solidity" (Prentice, 1990, p. 43).

For example, in Peterborough, England, the Queensgate shopping centre is dominated by polished Jurassic limestones from Italy and Germany and orbicular granite from Finland (Larwood & Prosser, 1996). Larvikite, a distinctive, iridescent, blue, coarse-grained syenite from southern Norway, has been used extensively in European cities.

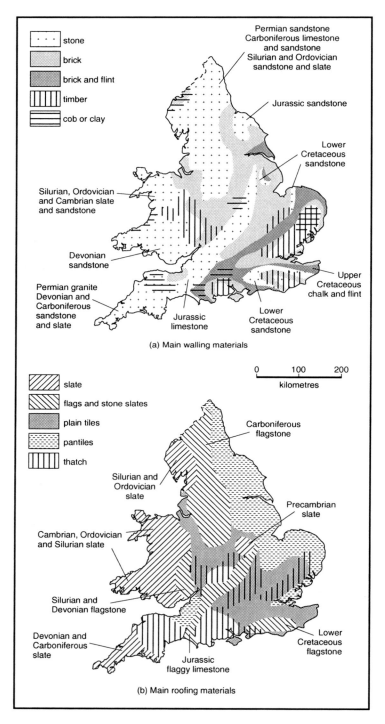

Figure 3.16   *Diversity of main building stones used in the United Kingdom, (a) main walling materials and (b) main roofing materials (After Woodcock, N. (1994) Geology and the Environment in Britain and Ireland. UCL Press, London, by permission of UCL Press)*

The character of the rock stems from the presence of large crystals of the feldspar labradorite, which contains abundant reflective cleavage plains. The marble quarries at Borba and Estremoz in eastern Portugal produce a range of white marble tiles for use in Iberia and beyond. The Finnish company Tulikivi produces a brochure illustrating almost 100 cladding stone types, mostly igneous but with some metamorphic rocks.

The use of the attractively coloured and textured geological materials led to their early use in artistic contexts. The Romans used the diversity of stone colour to create beautiful mosaics. In Mediterranean countries, small stone blocks are used extensively as paving materials and there is a tradition of laying these in patterns often with contrasting colours. For example, in Lisbon and Funchal, Portugal, a white durable limestone is used in conjunction with a black basalt to produce an impressive set of geometric pavement patterns or pavement pictures (Fig. 3.17).

Figure 3.17   *Diversity of pavement patterns produced from black basalts and white limestones*

Figure 3.17   *(continued)*

In some cases, a great variety of ornamental stones have been used in building construction. For example, Robinson (1987) gives a detailed account of the 10 main building stones used in the construction of the Albert Memorial in London in the 1860s. This includes one Irish and two Scottish granites as well as Campanella Marble from Italy. In addition, in the upper parts, there are bronze statues, several other stones and what the architect George Gilbert Scott described as "an elaborate fretwork of exquisite workmanship, and inlaid with polished, gem-like stones. These gems and inlays are formed of vitreous enamel, spar, agates and onyxes, upwards of 12,000 in number; of these, 200 are real onyxes, many of which are 0.75 inches in diameter" (quoted in Robinson, 1987, p. 30).

Stone in garden walls and rockeries also has a long history in many parts of the Europe and the Far East, though different approaches are used. In Europe, a common aim has been to use natural stone to mimic the natural bare rocky habitats of alpine plants. Weathered surface rocks such as limestone pavements have been very popular, but this has resulted in significant damage to these outcrops (see Box 4.1). An increased use of water worn cobbles and pebbles is also of concern if derived from active beach or river environments. Gravel spreads, through which plants grow, have become popular as a low-maintenance garden solution. In the Far East rock colour, structure and texture are used together with diverse aggregate mixes to give artistic and structural qualities to gardens.

Finally, armourstone is used to protect sea walls, breakwaters and vulnerable structures from wave impact and erosion. In high-energy environments, very large blocks (up to 20 tonnes) are necessary to withstand wave attack and therefore require durable rock types with widely spaced joints. In the East of England, an accessible source of such blocks is by boat from Scadinavian coastal quarries.

Aggregates

These are collections of rock particles, produced either by natural processes of erosion, transportation and deposition, or by the mechanical crushing of larger rock masses. Aggregates vary in particle size from sand to cobbles and are by far the greatest volume of exploited material in most countries of the world.

Natural sand and gravel aggregates come from either active or inactive sources and, of course, sediment can be recycled through various geological deposits. Glacial, fluvial, coastal, and aeolian processes all produce aggregates, but the major sources are from present day, active river or beach environments or from unconsolidated Quaternary glaciofluvial sources, either quarried onshore or dredged offshore. Subsidiary quantities may come from older, weakly cemented rocks such as the Cretaceous Greensand of south-east England and Triassic Bunter sandstones and pebble beds of the English Midlands (Woodcock, 1994). Glacial till is rarely used as a gravel aggregate because of its high fine sediment (clay and silt) content, but it is widely used as a fill material, for example, in dam or road construction (see Box 3.7). The use of clay in brick and tile manufacture is discussed below. Finally, we must mention the use of mud and clay mixtures for the construction of humbler buildings in many parts of the world.

Crushed aggregates are usually coarse, angular limestones, granites, dolerites or other igneous rocks. In the United States, 70% of crushed rock is limestone, 20% is granite or basalt and 10% is sandstone and quartzite. There is an increasing use of coastal superquarries to provide large quantities of easily transported (by boat) crushed aggregate and armourstone blocks. Glensanda, a granite quarry in the western Highlands of Scotland is currently the only UK superquarry (but see Box 6.1), and several exist in Scandinavia. In Ireland, a coastal quarry at Arklow Head, south of Dublin, produces dolerite for export to Germany as well as armourstone for coastal defences in Wales.

Box 3.7   Road Construction

Roads have been constructed since Roman times, but it was John McAdam who developed the modern bitumen or "black-top" road in the nineteenth century. Modern roads are still constructed in layers, though the details vary from country to country and depend on the type of traffic loading expected. The sub-base comprises a drainage layer of unbound coarse aggregate. This is overlain by a road base that is generally sand and gravel, compressed to give a firm bed, followed by a base course that is usually a bitumen-bound coarse, crushed aggregate. Finally, the wearing course is an asphalt with 30% aggregate, sometimes with a top-dressing of fine gravel chippings (see Fig. 3.18, Woodcock, 1994). This wearing course is the most important since it must be able to withstand frictional wear, provide grip for tyred vehicles and result in low surface noise. Much testing of particle size, shape, strength and durability is carried out to ensure that only suitable materials are used (Bennett & Doyle, 1997). In Finland, studded tyres are permitted in winter, but this imposes special demands on the quality of the aggregates in the asphalt (Kelk, 1992).

Concrete roads are still constructed but are criticised for their surface noise. In 2001, the UK government stated its intention to remove all concrete sections of trunk roads within 10 years. Gravel roads are also common in some countries, for example, USA, Canada, Norway and Sweden, and a variety of aggregate materials is used in their construction, including glacial till (Knutz, 1984). The Denali Highway in central Alaska, USA, is about 200 km of mainly glacial till-surfaced road open only in summer. The till matrix makes a good compacted road surface, but the clasts are often loose on the

Continued on page 106

Continued from page 105

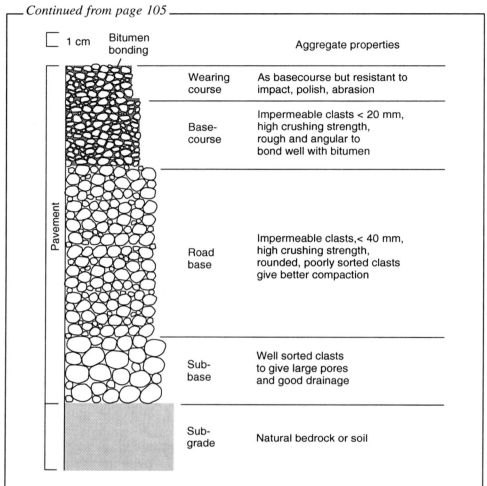

1 cm	Bitumen bonding		Aggregate properties
		Wearing course	As basecourse but resistant to impact, polish, abrasion
		Base-course	Impermeable clasts < 20 mm, high crushing strength, rough and angular to bond well with bitumen
		Road base	Impermeable clasts,< 40 mm, high crushing strength, rounded, poorly sorted clasts give better compaction
		Sub-base	Well sorted clasts to give large pores and good drainage
		Sub-grade	Natural bedrock or soil

Figure 3.18 *Diversity of aggregates and other geomaterials used in road construction (After Woodcock, N. (1994) Geology and the environment in Britain and Ireland. UCL Press, London, by permission of UCL Press)*

surface with the result that most drivers allow at least five hours to drive it! Scoria, a natural, fused-volcanic cinder is extensively used as a road base for minor roads in recently active volcanic areas, for example, Cascade Mountains and Hawaii, USA and Iceland.

Once quarried, these materials are sorted into different size fractions, which are then used in a diverse series of applications. For example, sand is used in the production of cement, concrete and mortar, as a base in road construction or pipe laying and in the manufacture of bricks, tiles and drainage pipes. Because the packing behaviour of grains is important, grain size and shape analysis are important in assessing the suitability of sand for particular uses (Prentice, 1990; Evans, 1997). So-called "soft sands" often from marine or fluvial sources have rounded grains and are more easily worked than "sharp sands", for example, from fluvioglacial sources or crushed rock that have rougher-textured particles are more acceptable in concrete or road aggregate.

Gravels of various sizes are also tested for their suitability for various functions. If they are to be combined with cement or bitumen, they must react favourably with these materials to ensure the structural stability of the end product in its construction position. Resistance to heavy loads, high impacts and severe abrasion may be important as, for example, in its use as road stone. Strong, durable, crushed aggregates are commonly used as the bed for rail track sleepers. On the other hand, crushed aggregates result in less easily worked and denser concrete mixes, and natural aggregates are usually preferred. However, not all coarse aggregates are suitable for concrete manufacture since some lithologies with high clay or mica content display high shrinkage properties during drying, leading to structural weaknesses in the set concrete. Porous particles make the concrete prone to frost weathering. The most suitable aggregates are hard, quartz-rich lithologies such as quartzite and durable sandstones. Aggregate is also used in roughcast or pebble-dash wall renders, which can be produced in a diverse series of colours and textures.

Limestone

This is used in cement and concrete manufacture and is a major quarrying industry in most countries. Evans (1997, p. 230) refers to it as "probably the most important industrial rock or mineral used by man" and lists the uses of the material (Table 3.3). Prentice (1990, p. 171) believes that "It is difficult to visualise what our present-day urban landscape would be like without concrete; or to imagine how today's major engineering structures – bridges, tunnels, roads, high-rise buildings – could have been constructed without it". Over 1 billion tonnes of cement and concrete are produced annually and they are clearly of major importance to both the global construction industry and the global economy.

Cement is a mixture of 75 to 80% crushed limestone (calcium carbonate) and 18 to 25% clay or ash (silica and alumina), which is fired in a kiln, cooled to produce clinker and finally mixed with about 5% gypsum. Cement, with sand and water added, is worked as a floor screed or wall render, which sets hard when dry. In the United Kingdom, Lower Carboniferous or Jurassic limestones and Cretaceous chalks are frequently used to provide the carbonate content, while Jurassic, Cretaceous and Tertiary clays and shales or pulverised fly ash supply the silica and alumina. In some countries, for example, Italy and the Rhine rift valley, volcanic ash has a long history of use. Some rock lithologies, for example, cementstones, come "ready mixed" in the correct proportions or can be readily made up to them. The Rosendale Formation on the Hudson River upstream from New York contains this type of mixture and has been mined and quarried since 1820 to supply New York, particularly its concrete sidewalks (Marshak, 2001). At Dunbar in Scotland, Carboniferous limestone and shale are interbedded so that quarry design has allowed the correct mix to be produced on site. Most countries have limestone and clay that can be utilised for cement production even if the carbonate has to come from Precambrian marble, Holocene coral sands, ultra-basic igneous rocks (Greece) or carbonatite intrusions (Malawi) (Prentice, 1990). Mixed with sand, and sometimes lime, cement becomes mortar for use in bricklaying.

When mixed with a 60 to 80% gravel aggregate and water, cement becomes concrete. The cement minerals react with water to form interlocking crystals of hydrated silicates and aluminates, binding the aggregate into a solid material (Woodcock, 1994). As

Table 3.3 *Diversity of limestone uses based on product size (after Evans, 1997)*

Size	Use	Main requirements
>1 m	Cut and polished stone ("marble")	Mineable as large blocks free of planes of weakness. Consistent white or attractive colouration patterns. Low porosity and frost resistance.
>30 cm	Building stone	Wide spacing of bedding planes and joints. Low porosity and frost resistance. Consistent appearance. High compressive strength.
>30 cm	Rip-rap or armourstone	Wide spacing of bedding planes and joints. High compressive and impact strength. High density. Low porosity and frost resistance.
1–30 cm	Kiln feed stone for lime kilns	Chemical purity. Degree of decrepitation during calcination and subsequent handling. Burning characteristics of stone.
1–20 cm	Aggregate, including concrete, roadstone, railway ballast, granules, terrazzo, stucco	Impact strength. Abrasion resistance. Resistance to polishing. Alkali reactivity with cement. Soluble salts. Tendency to form particles of particular shape.
0.2–0.5 cm	Chemicals and glass	Chemical purity. Organic matter. Abrasion resistance.
3–8 cm	Filter bed stone	Chemical purity. Compressive strength. Moisture absorption. Abrasion resistance. Crust formation.
3–8 mm	Poultry grit	Chemical purity. Shape of grains
<4 mm	Agriculture	Chemical purity. Organic matter
<3 mm	Iron ore sinter, foundry and non-ferrous metal flux stone, self-fluxing iron ore pellets.	Chemical purity.
<0.2 mm	Filler and extender in plastics, rubber, paint, paper, putty	Chemical purity. Whiteness and reflectance. Oil, ink and pigment absorption. pH. Nature of crushing and grinding circuits.
<0.2 mm	Asphalt filler	
<0.2 mm	Mild abrasive	Near white colour. Low quartz content.
<0.2 mm	Glazes and enamels. Funcicide and insecticide carrier	Chemical purity. Near white colour. Organic matter.
<0.1 mm	Flue gas desulphurisation	Chemical purity. Surface area. Microporosity.
Various	Bulk fill	Customer requirements. Size gradation.

stated above, the characteristics of the aggregate are very important to the strength of the concrete. There have been some notable failures due to aggregate shrinkage during drying (see above) or adverse reactions between alkali solutions from the cement and water, and silica or silicates in the aggregate, particularly when crushed types have been used (Prentice, 1990; Bennett & Doyle, 1997). Failure may also occur where reinforcing rods rust or salt crystallisation occurs within the concrete. Many examples of the latter have been experienced in the Middle East owing to rising saline capillary waters in coastal locations (Cooke & Doornkamp, 1990).

Structural clay

This is a clayey sediment used in brick or tile making and occurs as diverse types (Table 3.4). The processes have a long history, stretching from the time of the Egyptians, through the impressive innovations by the Romans, to the sophisticated variety of modern bricks and tiles (Prentice, 1990). Brick clays are usually sedimentary deposits though some are of residual origin. But whatever their source, they are a diverse mixture of four clay minerals: kaolinite, illite, smectite and chlorite, each of which reacts differently to the various stages of brick making and many interact in complex ways. Quartz sand or silt grains may also be present or are added as "grog" (inert additive) in order to provide a more open texture, assist in the processes and provide strength and durability. In fact, quartz can make up to 90% of the total. Micas, vermiculite and sepiolite may also be present in brick clays to affect their properties.

Clifton-Taylor (1987) describes brick colours in England and believes (p. 236) that "There is scarcely a limit to the colours that have been obtained". The diversity of colours is partly related to the mineralogy of the clay and the nature of the firing, but nowadays, mineral colourants such as ochres are often used. The iron minerals, such as haematite, limonite, pyrite, siderite and magnetite, will provide the natural red colour of most bricks with depth of colour increasing at firing temperatures of over 1000 °C. Blue colours, such as Staffordshire Blue bricks made from the iron-rich Etruria marl of the West Midlands of England, are produced by producing reducing conditions, either by setting the bricks close together in the kiln, so reducing the passage of air around the bricks, or by controlling the fuel supply so that all the oxygen is burnt. "In a reducing atmosphere, the iron combines with the silicates in the clay to form ferrous silicates, which, unlike haematite, become liquid at kiln temperatures, thus forming a dark blue skin on the surface on cooling (Prentice, 1990, p. 149). Black colourations can be produced by creating reducing conditions in the interior of individual bricks. This is done by sealing the brick with a high-temperature pulse, thus producing a vitrified external skin. This prevents escape of gases, which burn internally, producing a black core, which may extend in patches to the brick surface. However, this colour effect weakens the brick. Pale colours (buff or yellow) can be produced by a complex series of processes, which suppress free haematite, or by adding calcite to produce the lighter colours of iron carbonates such as ankerite, or dicalcium ferrite. However, control of clay mineralogies and firing conditions are not always easy; hence precise control of colour is, in fact, extremely difficult (Prentice, 1990).

Brick texture also varies. In handmade bricks, sometimes creases or wrinkles may be seen on the surface where the clay has been pushed together. Texture is often added by dusting with sand and modern bricks can be patterned or glazed in a variety of colours. Engineering bricks are produced for their durability and resistance to frost and damp.

Table 3.4 *Diversity of clay types, mineralogy and uses (after Evans, 1997)*

Type	Mineralogy	Uses
Kaolin (also called *china clay*)	Kaolinite of well-ordered crystal structure plus very minor amounts of quartz, mica and sometimes anatase. Highest quality material will be almost pure kaolinite and will contain >90% of particles <0.002 mm. Low to moderate plasticity.	Paper filler and coating; porcelain and other ceramics; refractories; pigment/extender in paint; filler in rubber and plastics; cosmetics; inks; insecticides; filter aids.
Ball clay (plastic kaolin)	Kaolinite of poorly ordered crystal structure, with varying amounts of quartz, feldspar, mica (or illite) and sometimes organic matter. Kaolinite particles usually finer than in kaolin. Moderate to high plasticity.	Pottery and other ceramics; filler in plastics and rubber; refractories; insecticide and fungicide carrier.
Halloysitic clay	Halloysite with varying amounts of quartz, feldspar, mica, carbonate minerals and others. Similar particle size distribution to ball clay. Moderately plastic.	Pottery and other ceramics; filler in plastics and rubber.
Refractory clay or fire clay	Structurally disordered kaolinite with some mica or illite and some quartz. Very plastic.	Refractories (firebricks); pottery; vitrified clay pipes.
Flint clay	Kaolinite and diaspore. Plastic only after extended grinding	Refractories (firebricks); pottery.
Common clay and shale	Kaolinite, and/or illite, and/or chlorite, sometimes with minor montmorillonite, quartz, feldspar, calcite, dolomite and anatase. Moderately to very plastic.	Bricks; tiles; sewer pipes.
Bentonite and Fullers' earth	Montomorillonite (smectite) with either Ca, Na or Mg as dominant exchangeable cation. Minor amounts of quartz, feldpar and other clay minerals. Forms thixotropic suspension in water. Forms a very plastic, sticky mass.	Iron ore pelletising; foundry sand binder; clarification of oils; oil well drilling fluids; suspending agent for paints; adhesives; absorbent; landfill site lining.
Sepiolite and attapulgite	Sepiolite or attapulgite with very minor amounts of montmorillonite, quartz, mica, feldspar. Thixotropic and plastic as montmorillonite clays.	Absorbent; clarification of oils; some oil well drilling fluids; special papers.

Clay minerals are produced by the weathering of alumino-silicate minerals, and since these are present in almost every rock, it follows that clays are very abundant either as residual weathering products or as transported, sorted and re-deposited clay sedimentary units. Clays are also produced by volcanic activity or released from pre-existing rocks, for example, by weathering of chloritic shales in Scandinavia. It is therefore possible to find clays and make bricks in almost every country in the world, though the type of clay and brick varies greatly. Dry climates favour weathering to smectite, whereas illite forms under wetter conditions and kaolinite and vermiculite under very wet conditions. Gibbsite and kaolinite are principal constituents of soils in the humid tropics.

Most commercial brick making clays are post-Devonian because of the degree of alteration of older clays. This explains why there is relatively little brick making in the western United States but major industries in the central and eastern states (Prentice, 1990). In the United Kingdom, nearly half of all bricks are produced from the 70 m thick, Upper Jurassic Oxford Clay, which has a relatively high hydrocarbon content. This enhances the plasticity of the clay and reduces the cost of firing by as much as 75% (Prentice, 1990). In Belgium, there are thick and extensive deposits of clays, which form the basis of a large-scale industry. Included are the late Oligocene "Boom clay", a 50 m thick deposit of smectite–illite rich rhythmites. Quaternary rythmites (varves) and wind-blown loess are also used in brick making. Indeed, the wind-blown deposits of South-East England are often referred to as *brickearths* and produce the yellowish-grey bricks (due to high calcite content) common in London. However, the most common source of brick clays are recent alluvial sediments where for centuries the deposits of the Yangste Kiang, Huang Ho, Indus, Ganges, Mekong, Nile and others have been extracted from shallow excavations to produce bricks, albeit of relatively low strength (Prentice, 1990).

For underground pipes that must be able to withstand attack by internal fluids, external groundwater and overlying loads, particular clay qualities are required. In western Europe, the kaolinite–smectite Westerwald clays from Germany are often used and exported overseas, while in North America, carefully blended coal-measure shales are commonly utilised. The high strength is achieved by firing to 1200 °C.

Floor, roof and wall tiles are also made from clay, though often of a high quality, more carefully prepared and baked harder (Clifton-Taylor, 1987). Like bricks, the colour of traditional tiles is linked to clay composition and, in fact, a great diversity of clays have been used in ceramic tile production. The German Westerwald clays are certainly used as are the ball-clays of South-West England. A greater variety of clays is used in Spain and Portugal, the latter having a famous hand painted, azuleju tile tradition. In the United States, extensive kaolinite deposits occur in South Carolina and Georgia where they are derived from Cretaceous erosion of earlier weathered rocks. The clays were deposited in a series of lakes and deltas and subsequently had iron removed by leaching processes during the Tertiary so that the clays are very pure. Tile form also varies with flat floor and roof tiles, half-cylinder tiles, pantiles and wall-hung tiles of various shapes being popular in different parts of the world.

Porcelain and other ceramics are made from kaolinite or china clay, while high-grade ceramics use ball clay (plastic kaolin) or halloysitic clay (Table 3.4). To produce porcelain, the clay is partially melted at temperatures above 1250 °C. Just as it begins to melt, the potter cools it quickly to create a translucent, vitreous appearance. South-West England is the major source of china clay, which along with ceramics is also used in paper, plastics and paint industries throughout Europe (Bristow, 1994).

Gypsum

This is the main raw material used in the production of plaster for use as a finishing render on internal walls and ceilings. It originates as marine or terrestrial evaporite deposits but nowadays is also produced as a by-product of the flu-gas desulphurisation process in coal-burning power stations, where a single large plant can produce a million tonnes of gypsum in a year.

Plaster of Paris gets its name from the Tertiary gypsum deposits around the city and there are also extensive deposits in Italy, Spain and many other Mediterranean countries derived from the time during the Miocene when the Mediterranean was landlocked. Other evaporites include the Cambrian of Pakistan, Silurian of New York and Ohio, USA, the Carboniferous of Utah, the Permian deposits of New Mexico and Texas, USA, and the famous Permian deposits of the Zechstein Sea in Holland, Germany and Poland. Plaster is produced by dehydrating gypsum at about 107 °C and mixing the end product with other materials to provide a range of plaster types with different uses. One is render for walls and ceilings, with both base-coat bonding and finishing plaster being produced. Fine plaster can be worked into intricate raised ceiling or wall patterns, the latter being referred to as *pargeting* in England. Other uses include plasterboard and for the setting of broken bones.

Glass sand

This is a particularly pure type of sand deposited in iron-free environments or subsequently iron-leached. The sands should also ideally be clean (clay free) and well size-sorted to ensure even melting during production. Such sands are found extensively in the United States, Holland and Argentina and in smaller quantities in many other countries. Very pure sand is needed because impurities would produce discolourations (iron) or imperfections (e.g. zircon, chromite) in the glass. Optical glass must be low in alumina (Evans, 1997). But even the purest quartz sands usually have some impurities and these are removed by washing, sieving or flotation methods. Additives may include boron (from evaporites) or feldspar (from igneous rocks) as flux materials, lithium (from the mineral spodumene) for use in ceramics, fluorite or arsenic for heat-resistant glass and strontium (from the mineral celestite) for use in television tubes where it reduces radiation. Another product is glass fibre, used as insulation or incorporated with resin to produce fibreglass.

The ideal conditions for deposition of pure, iron-free sand occur where granitic rocks have been weathered (to reduce the silicates to clays), eroded (to produce the quartz) and very well sorted (to remove the iron clays). Such conditions occurred during the Upper Cretaceous at Loch Aline in Scotland and Provodin in the Czech Republic and during the Tertiary in Guyana, where very large reserves exist.

Volcanic products

We have already mentioned the use of volcanic ash to provide the clay fraction in cement production in, for example, Italy. Other volcanic rocks used in construction include pumice, a highly porous, solidified volcanic froth. Moulded with cement and cured slowly, it produces a lightweight block with good insulating properties. Pumice is worked on Lipari in Italy, the Greek islands and Turkish mainland, and in Oregon, California, Arizona and New Mexico, USA. Large reserves occur on the Caribbean islands. Other low-density volcanic materials with similar properties include perlite and vermiculite.

3.5.4 Fossils

Fossils can have significant commercial value if they are rare, well preserved and/or well known. Dinosaur's fossils, for example, can command large sums given their popularity with the general public. A complete skeleton of *Tyrannosaurus rex* ("Sue") collected by a commercial company on federal lands in South Dakota and subsequently confiscated by the FBI was sold at an auction by Bonhams in New York for $8.36 million (Fiffer, 2000). This escalating commercial market for fossils has driven some to undertake weekend or summer excursions in the United States to find their million-dollar dinosaur!

Details of recent US Natural History auctions are given by Forster (2001). The Staatliches Museum in Stuttgart, Germany offered £180,000 for the world's earliest fossil reptile ("Lizzie"), discovered at East Kirkton Quarry in Scotland. The *Caloceras* beds at Doniford Bay in Somerset, England are valued at £825 per m^3, but 65% has been removed by irresponsible collecting (Webber, 2001). Even already collected material can be sold off for commercial gain. The famous collections of the Moscow Palaeontological Museum have been broken up and are turning up in the commercial markets of Europe and the Americas (Karis, 2002).

The market for fossils is not just scientific. So-called "décor fossils" are sold to a general market on the basis of their aesthetic appeal, an example being fossils encased in amber, iridescent ammonites ("Ammolite" jewellery) and the small oval trilobite *Elrathia kingii*, which is used in necklaces, tie-pins and so on. In fact, the value of most fossils is assessed according to aesthetic appeal or rarity rather than on scientific grounds (Norman, 1994). Forster (1999) lists fish, crustaceans, crinoids, ammonites, trilobites and large leaves as the most popular specimens, with prices ranging from £50 to £5,000 dependent on size, colour and quality of preservation. Fossil and mineral shops are increasing to cater for public demand (Fig. 3.19).

Fossiliferous ornamental stone is also common, coralline and crinoidal rocks being particularly popular. In Scandinavia Ordovician, red limestones with straight nautiloids provide an attractive stone, while is southern Europe, Cretaceous limestones containing rudists are commonly polished as a facing stone (Fortey, 2002).

Figure 3.19 *Fossil and mineral shop, Banff, Canada. Also note the geological diversity in the stonework*

Fossils *in situ* have great economic value in stratigraphic correlation and thus in oil and mineral exploration worldwide. Particular use is made of a diverse range of microfossils (e.g. conodonts, ostracods, foraminifera, coccoliths, pollen and spores).

3.6 Functional Value

Functional value has rarely been discussed in nature conservation, but it is clear that soils, sediments, landforms and rocks all have a functional role in environmental systems, both physical and biological. In turn, we can recognise two sub-divisions of functional values. First, there are utilitarian values to human society of geodiversity *in situ*, as opposed to the extracted value described above. Secondly, geodiversity has a functional value in providing the essential substrates, habitats and abiotic processes, which maintain physical and ecological systems at the Earth's surface and thus underpin biodiversity.

3.6.1 Utilitarian functions

Platforms

The land surface provides a platform or foundation on which, and into which, development and all human activities take place. This gives land a functional and economic value. Particular combinations of landforms, bedrock types, or soils make some areas best suited for agriculture, others for urban development, still others for hydroelectric power. Airport runways require flat sites, whereas skiing requires slopes of varying steepness and complexity depending on the ability of the skier. Geodiversity therefore results in a diversity of utilitarian functional values of different parts of the landscape. While we do not always use land in ways to which it is best suited, it would be better if we did. Soil provides the growing medium for plants and is therefore the basis of all life. Diversity of platforms is therefore important.

Storage and recycling

The physical earth also acts as a store of various types. Particularly important is the role of soil and peat as carbon stores. The total carbon stored in world soils amounts to 2100 to 2400 Gt, about 25% of this being held in peatland soils, which cover only 3% of the land area. The peatland carbon store "is probably more than three times the carbon stored in the world's tropical rainforests, and in the region of a half of the estimated total carbon in all terrestrial biota" (Maltby & Proctor, 1996, p. 17). Research is also being carried out in Europe, North America and Australia on the possibility of storing carbon dioxide from power plants by forcing pressurised gas into capped permeable rocks such as offshore oil and gas fields or deep saline aquifers.

Other examples of the abiotic environment acting as important stores include water stored in sub-surface aquifers or surface lakes, oil and gas stored in geological traps and mineral resources stored in ore bodies, sediments, and so on (see above).

The physical environment also plays a role in recycling earth materials. For example, leaf fall leads to humus production and recycling of nutrients through the soil and tree roots. At the global scale, the hydrological and other cycles operate.

Burial

Human burial has long used the physical resources of the land by placing bodies either into the land (as in graves) or constructing monuments above ground (as in the pyramids or, at a smaller scale, in cairns or dolmens). In Johannesburg, South Africa, consideration is being given to turning disused gold mines into catacomb-style cemeteries to accommodate the increasing number of people dying from AIDS. About 25% of the 3 million population are infected with HIV and the city needs to find places to bury over 20,000 people per year.

A diverse variety of rock types are also used by modern stone masons for making gravestones (Fig. 3.20), though an important property here is durability, particularly in retaining inscriptions. Older gravestones in sandstone or softer limestones are sometimes unreadable, and the preferred rock types are durable limestone, marble, slate, granite, gabbro and other igneous rocks. In Copenhagen, Denmark, the most suitable headstone materials have traditionally been the rounded erratic metamorphic and igneous boulders found in the glacial tills of the area. This local style has been preserved even today when rectangular blocks imported from around the world are fashioned into rounded boulders (Parkes, 2002). In Iceland, sections of columnar basalt are frequently used as headstones. In the United Kingdom, while some families choose black gabbros or dark grey slates as the traditional funereal colour of death, others prefer white limestones or marbles. Portland limestone was almost universally adopted for the graves of the British Commonwealth war dead (Bennett & Doyle, 1997). Some writers mourn the diversity of stone used in modern graveyards. Clifton-Taylor (1987, p. 146), for example, believes that "there can be little room for doubt that....no stone has contributed so much as polished granite to converting many English churchyards, once so attractive, into ghoulish eyesores". Some church authorities, such as the Norfolk

Figure 3.20   *Diversity of gravestone rock types in an Irish cemetery*

Diocese, England, agree and have banned the use of stones other than sandstones and limestones.

Waste materials are also buried in the ground (landfill sites) or above ground (landraising). Sites have been selected because of the *in situ* properties of local rocks for waste disposal. This includes municipal waste disposal based on clay soils like the London or Oxford Clays in England, though usually with a reworked floor or lining system of plastic and/or bentonite. The low permeability of glacial till means that it can be used as an aquaclude for use in lining landfill sites or reservoirs, but its fissured nature and sand lens inclusion means that some reworking is generally necessary (Fredericia, 1990; Gray, 1993). The burial of sheep and cattle in the United Kingdom following the 2001 foot-and-mouth disease outbreak occurred very rapidly and may not have fully assessed the geological properties of the burial sites.

Radioactive waste disposal is an increasing problem, with many countries stockpiling waste in temporary surface stores. One solution is to use underground chambers as retrievable locations for this waste, and the United States has recently decided to use an underground site in dry rhyolite in the Yucca Mountains of Nevada to store its nuclear waste. Several other countries are exploring similar options (National Research Council, 2001) and while some are in quite stable parts of the world (e.g. Finland and the Canadian Shield), others are in tectonically active zones (e.g. Japan and Switzerland). Other disposal suggestions include use of subduction zones where the waste will be carried down into the Earth, the interiors of low-permeability salt domes or the use of underground clay formations, since clay can more readily absorb and trap radioactive products.

Pollution control

Soil, sediment and rock perform a role in attenuating polluting substances and therefore in helping to maintain the quality of surface and groundwaters. This attenuation can occur by adsorption, ion exchange, microbial decomposition or dilution, but the effectiveness of these processes is variable. For example, the thickness, composition and structure of the soils, sediments and rocks will strongly influence the susceptibility of an area to groundwater pollution. A thin capping of soil and sediment above a water table will provide little protection from surface pollution by agrochemicals or spills, whereas a thick cover of massive clay will generally be very effective in attenuating pollution. The value of the latter is not always recognised.

Water chemistry

Bedrock geology affects the chemistry of water circulating at depth. This affects the chemistry of springs, some of which are utilised for bottled mineral water. Malvern "English natural mineral water", for example, advertises itself as "naturally filtered through Precambrian granite and sourced from deep beneath the Malvern Hills".

Other drinks are made from surface or groundwaters and have variations in taste or other attributes as a result of geological diversity (Maltman, 2003). For example, Lloyd (1986) examined the relationship between hydrogeology, brewery location and beer taste in England. He found that English breweries were mainly concentrated on the Triassic Sandstone and Chalk aquifers though there were important breweries on other major aquifers. He also found that calcium, bicarbonate and sulphate are important in the brewing process with the latter ion having the most direct impact on taste. He quotes sulphate values of 820 mg/L for the Triassic Sandstone but only 65 mg/L for the

Box 3.8   The Whiskies of Islay

Islay is known for its heavy, pungent, phenolic whiskies, but in fact the whiskies of Islay vary depending on the geology and local source of water. The geology of Islay is shown in Fig. 3.21 but, in general terms, is an anticline dipping to the north-east. Quartzites and greenstones form the limbs of the anticline in the north and south-east with older sedimentary rocks from the core of the anticline exposed by erosion in the centre of the island.

Three distilleries, Laphroaig, Lagavulin and Ardbeg, are situated in bays on the south coast of the island. The water used in distilling originates as rainfall on the quartzite hills to the north. The water descends to the coast crossing phyllites and greenstones and their peaty cover on the way, but may be intercepted by small natural or dammed lochs. The quartzite rocks of the hills and the peaty land make the waters highly acidic, but because they are surface waters, they have a low mineral content. These are the heavy whiskies of

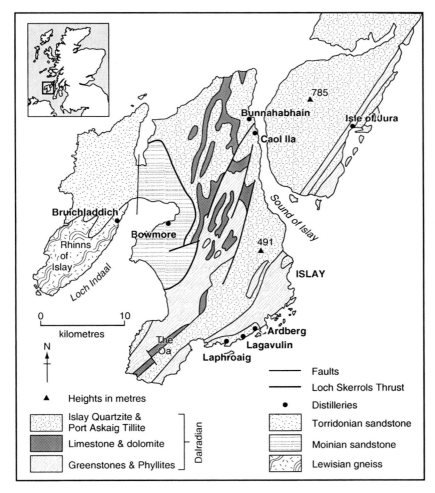

Figure 3.21   *Diversity of rock types and water chemistry on Islay, Scotland, helps to explain the diversity of malt whiskies from the island (Modified from Whittow, 1992)*

Continued on page 118

—— *Continued from page 117* ——

Islay, though their phenolic taste derives more from the peat smoke allowed to permeate the barley prior to distilling.

On the north-east coast of Islay, the distilleries of Caol Ila and Bunnahabhain are situated on older rocks in the core of the Islay anticline. The waters here are derived from springs from the Port Askaig tillite (Caol Ila) and dolomite (Bunnahabhain) and the whiskies are light and clean.

At Bowmore, in the centre of the island, the water is extracted from the River Laggan having flowed westwards towards Loch Indaal over quartzite hills, limestone outcrops, glacial deposits, alluvium, peaty lowlands and grey Moinian sandstones. The water is therefore a blend of Islay's geology and the whisky likewise is intermediate between those of the south and north-east coasts (Cribb & Cribb, 1998).

Chalk, with Jurassic Limestone, Magnesian Limestone and Coal Measures at intermediate values.

Whisky is also made from water, and the diversity of Scottish malt whiskies (of which there are about 100) is partly due to the different sources of water (see Box 3.8)

Health

About 17 elements are thought to be essential for plant and animal life. Along with the obvious elements of carbon, hydrogen and oxygen, they include nitrogen, phosphorous, potassium, calcium, magnesium and sulphur, as well as trace amounts of iron, manganese, boron, molybdenum, zinc, copper, chlorine and cobalt. A balanced intake of these and other minerals and nutrients are essential for human health but some are toxic at concentrations even slightly above environmental norms. Magnesium plays a role in the production and transport of energy in the body, zinc is important in the proper functioning of the immune system, and calcium helps to build dense bones and prevent osteoporosis. Some trace minerals are thought to offer protection from heart disease or diabetes; others may be antioxidants that reduce the risk of cancer.

An example of the latter is selenium where there is a narrow range between deficiency and toxicity (40–400 μg/day). Recent studies in America, for example, have indicated that people with relatively high selenium levels have a 50% lower chance of dying of cancer than those with low levels, and there are concerns about the low selenium concentrations in European wheat (*The Observer*, 20 October, 2002). Selenium deficiency leads to Keshan Disease, named after a Chinese district where it was discovered in 1935 after causing damage to the heart muscles and eventually death of many local people. Selenium supplements, fertilizers or crop sprays can be used to counter deficiency (Fordyce & Johnson, 2002).

The recommended intake of iodine is c. 100 μg/day and iodine deficiency causes a swelling of the thyroid known as goitre, which affects 190 million people worldwide. In pregnant mothers, it can also lead to impaired brain function in children, and the World Health Organisation estimates that 100,000 sufferers are born every year, particularly in developing countries.

Chromium is a trace mineral that is essential for good health and is also believed by many to be usefully supplemented for protection against or treatment for Type 2 diabetes. Adequate chromium is necessary for insulin to be effective and some studies have shown that chromium supplements may help those with sluggish insulin response to process blood sugar more quickly.

Clays have been used for a variety of purposes. For example, in clay and mud packs as skin cleaners and toners. But they are also taken internally, for example, in kaolin and morphine medicine for diarrhoea and indigestion. Soil eating by humans is quite common in some African countries (Uganda and Kenya), perhaps because it contains magnesium, iron and zinc, which are essential during pregnancy, early childhood and adolescence (Smith, 2002). Some South American Indians neutralise poisons in wild potatoes by cooking them with clay. Many animals exhibit similar behaviour. For example, parrots have been observed to gorge themselves on river clay in the Amazon rainforest and this is thought to be a way of neutralising their diet of poisonous berries and fruits (*The Guardian,* 4 April, 2002). Many gorillas and chimpanzees act similarly. Other animals visit "salt licks" but at the same time consume insoluble clays and minerals (Smith, 2002). Elephants seek out sodium-rich rocks in caves at Mount Elgon in Kenya, which they grind and swallow to stimulate bodily defences against toxins in the plants they eat (Engel, 2003).

Soil functions

The usual perception of soil is as a medium for plant growth, and there is no doubting the importance of this function. As Rachel Carson (1962) wrote: "The thin layer of soil that forms a patchy covering over the continents controls our own existence and that of every other animal of the land. Without soil, land plants as we know them could not grow, and without plants no animals could survive". Plants need sunlight and carbon dioxide from the atmosphere but they rely on soil for the nutrients and water that are essential to life. Soil is therefore an absolutely vital part of terrestrial environmental systems, though it's role has usually been undervalued (Thompson & Bullock, 1994). In fact, a more diverse series of important functions of soil is increasingly being recognised (Royal Commission on Environmental Pollution, 1996; Taylor, 1997; DEFRA, 2001; Brady & Weil, 2002). For example, they

- interact with many other parts of the environment (atmosphere, hydrosphere, biosphere, lithosphere);
- act as a habitat for soil biota and as a gene reserve;
- filter and bind substances from water and receive particulates from the atmosphere;
- act as a store of water and carbon and recycler of organic matter;
- support ecological habitat and biodiversity;
- regulate the flow of water from rainfall to watercourses, aquifers, vegetation and the atmosphere;
- act as a growing medium and nutrient supply, for food, timber and energy crops and the basis for livestock production (see above);
- act as environmental archives.

However, soils are not uniform and the value of soils functional diversity must be recognised. As Bridges (1994) states, soils are the vital link between the inanimate world (geosphere) and the living world (biosphere).

- *Agriculture* About 90% of our food comes from the land and only 10% from water and air. Although population has been increasing at a very high rate, food production has been increasing even faster owing to expansion of cropland areas, drainage and irrigation, and increased use of machinery, fertilisers, pesticides and high-yield varieties (Mather & Chapman, 1995). At the same time, there has been concern over whether these increases represent sustainable agricultural solutions or are, in fact, a threat to soil resources (Mather & Chapman, 1995; Trudgill, 2001; see Chapter 6). There has been an increasing production of cash crops for urban markets or export and there is now a very large trade in food around the world amounting to about 15% of total production, vital to the economy of many countries. One only has to visit the fruit and vegetable displays (or wine counters) in a modern superstore in Europe or North America to appreciate the variety of produce from around the world. There has also been an increase in crops grown for animal feed, reflecting a rise in livestock products.

  Soil diversity is important since different crops favour different soils. For example, in the East Anglia region of England, the sandy loam soils are favoured by root crops and sugar beet, potatoes, parsnips and carrots are grown in rotation with cereals and vining peas. Rotation is essential for soil health and irrigation is required because of the light soil and low summer rainfall. On the medium soils, winter wheat is grown in rotation with oilseed rape and a legume crop. On the heavy clay soils derived from the glacial till, wheat is grown and land is also left fallow under a government "set-aside" scheme to encourage wildlife. Thus, farmers have learnt to utilise soil diversity for sustainable crop production.

- *Viticulture* Along with grape variety, climate and manufacturing processes, geodiversity accounts for some of the diversity of wine. The most important variables are as follows:

  - Rock and soil mineralogy, which control the nutrient and mineral supply to the vines. Potassium is particularly important and is found in the potash feldspars of many igneous and metamorphic rocks, in arkosic and glauconitic sands and in illitic and chloritic clays (Selley, 2002). Geology also strongly influences soil acidity. Vines grow in the basaltic soils of the Portuguese island of Madeira, but the soil is too acidic to produce high-quality table wines. Instead, the grape crop is largely used to produce the fortified wine, Madeira. Trace minerals give wine diversity of taste but it has not yet proved possible to identify, for example, a Maastrichtian wine, from its taste.

  - Rock and soil structure/texture, which control water content. Vines do not thrive in either drought or waterlogged soils, but they do require a steady supply of water. This is best supplied where the soil is well drained, but the roots are able to penetrate deep into the rock via fractures. An example of this occurs in the Douro Valley east of Oporto, Portugal, where the vine roots penetrate deep into the schists for coolness and moisture. The English and French Chalk also provide good conditions derived from their porosity and fracture characteristics.

  - Topographic variables are important in providing appropriate growing conditions. Altitude and aspect affect the degree of solar warmth. In marginal climates for vine growth like Britain, south-facing slopes at low altitude will maximise solar warmth, particularly where there is a lake or river at the base of the slope to reflect the sunlight. In hot climates like Spain or Italy, vines often grow better at higher and therefore cooler elevations.

- *Forestry* About 4,000 million hectares of the Earth are covered in forest, though this is declining at an alarming rate (10–12 million hectares per year). About 60% of the resource is in America or the former Soviet Union and only 29% is in Africa and Asia. These last two continents have high populations and a long history of use of wood for fuel. In fact, fuel wood is now in short supply in several countries. Other uses of forests include wood pulp for papermaking, the manufacture of wooden furniture, cladding, and so on, and woodland recreation ecotourism, though the exact mix of products varies greatly from place to place. There is a growing area of plantation forest not only in the temperate developed world but also in countries such as Brazil and Chile. These usually have much higher levels of productivity than natural forests due to careful selection of species and intensive management. Both natural and plantation forest products are increasingly traded, though mainly within the developed world. However, this is likely to change given the potential for plantation growth in the tropics. Forestry also protects soil from erosion, particularly in areas of high rainfall and sloping terrain.

3.6.2 Geosystem functions

River channels perform the function of transporting water and sediment from land towards the sea and their capacity is adjusted to the stream discharge. The form of this channel affects the flow of water in it and, through erosion and deposition, the flow modifies the form. Interference with the channel can cause effects elsewhere in the drainage basin. Similarly, floodplains perform a water storage function in times of flood. They process large fluxes of energy and materials from upstream areas. Beaches act to protect the coastline from erosion under normal conditions and they perform the dynamic function of allowing sediment movement along the coast. Salt marshes are efficient in trapping fine sediment, which has helped them accrete vertically and laterally in response to the Holocene sea-level rise (French & Reed, 2001). Many of these physical systems are in dynamic equilibrium and their continued functioning is important to environmental systems.

3.6.3 Ecosystem functions

The physical environment generally plays a huge role in providing diverse environments, habitats and substrates that create biodiversity, yet this appears to have been rarely recognised by ecologists. Warren's (2001) irritation at this state of affairs is very apparent: "Even general discussions of patterns of diversity.... and authoritative statements about landscape ecology....where the main cause of variation is blatantly geomorphological, almost wilfully ignore the pattern and dynamics of the substrate" (Warren, 2001, p. 39). "At worst, the physical environment is seen as an unchanging backdrop to the development of the ecosystem..." (Warren & French, 2001, p. 1). Fortunately, Warren & French note that the situation is changing and engineers and ecologists have become aware of the need to understand the processes and patterns in the physical environment if they are to manage processes and habitats successfully (see Chapter 7). For example, "ecologists working in tropical rainforests have suddenly become aware that slope dynamics have a vital role to play in the patterning and dynamics of the ecosystem, yet have only just begun to collaborate with geomorphologists" (Warren & French, 2001, p. 3). Similarly, in the case of the management of coastal dunes, a highly protectionist philosophy among ecologists and engineers is giving way to one that recognises the importance of allowing the operation of shifting sands and dynamic physical processes; an understanding of the latter is critical to an understanding of the population dynamics of plant and animal species, and both need attention if dunes are to be managed in a way that conserves their diversity (Warren & French, 2001). Bartley & Rutherfurd (2001, p. 15) believe that "Biologists increasingly acknowledge that geomorphological surfaces form the template for development of both flora and fauna communities" (Fig. 3.22).

MacArthur and Wilson's (1967) famous prediction that species numbers were strongly related to land area is clearly undermined by the important variable of topography. McCoy & Bell (1991) and O'Connor (1991) found that heterogeneous habitats have an important effect on measurements of species diversity, and it is often geomorphology, geology and soil diversity that provide habitat heterogeneity. For example, in Rhode Island, USA, geomorphology is a more important control on plant species diversity within woodlands than woodland size (Nichols *et al.*, 1998). Warren (2001, p. 41) believes that this type of topographical control is even clearer in very dry areas where "the slightest hollow can support many times more plant production than a nearby slope". Warren also argues that geomorphology is more of an ecological factor in high- than low-energy landscapes.

There are similar relationships with animal populations. Warren (2001) quotes the case of the Bay Checkerspot butterfly in California, USA, which needs warm slopes where larvae can develop quickly and contiguous cool slopes where the females can

Figure 3.22   *Diversity of slope angle and hydrology creates habitat and species diversity*

emerge and reach diapause before their host plants become senescent as summer proceeds (Murphy *et al.*, 1988). Moreover, in wet years, more of these butterflies survive
on warmer slopes, while in dry years, there are more on the cooler slopes. Similarly, Belsky's (1995) study of butterflies in East Africa demonstrated that they needed
different facets of the landscape at different times of the year. He concluded that there
was greater animal species diversity and biomass where there was more landscape
diversity. Thus, the influence of geodiversity on biodiversity is increasingly being
recognised by biologists (Burnett *et al.*, 1998) as well as geoscientists. The degradation of landforms, soils and surface waters will inevitably adversely impact on the
biological species and communities living in or on them.

Among the physical factors that influence biodiversity, altitude and aspect must rate
as two of the most important (Huggett, 1995). Most mountains display an altitudinal
zonation of natural or semi-natural vegetation, where this exists, linked to soil and
climatic gradients. Aspect, by influencing topographic shelter and shade, can have an
important effect on plant distributions, by affecting, for example, insolation, temperature, evapotranspiration, soil moisture and duration of snowlie. But many other physical
factors influence biodiversity. Warren (2001) gives examples from Brazil and Niger. In
the central Brazilian plateau, a strong relationship between facet type and vegetation
was established by Furley (1996). The upper slopes are covered in arboreal savanna
(*cerrado*), the side slopes support seasonally wet grassland (*campo limpo*) and the colluvium and alluvium on the valley floors support gallery forest (*mata ciliar*). But at the
same time, these slopes function geomorphologically with weathering and breakdown
of rock on the upper slopes, movement of this material downslope through the soil
and sub-soil and eventually reaching the stream. The whole natural environment therefore functions as an integrated geomorphological and ecological system (Furley, 1996).
In semi-arid areas of south-west Niger, the "tiger stripe" patterns of vegetation visible
from the air occur where trees and shrubs are arranged in bands parallel to the contours
with almost bare ground between. The low-angle surfaces are impermeable and plants
can only grow where sufficient runoff from upslope has been collected. However, the
growth of vegetation then cuts off the supply to the zone immediately below where the
process of downslope water collection begins anew (Thiéry *et al.*, 1995). The clear relationship between ecology and geomorphology is very evident from these two examples.

Other environments can also be cited. Salt marshes provide the saline conditions
enjoyed by several specially adapted plants. In the intertidal zone, several species of
mollusc and seaweed cling to rock and boulders, sometimes deriving nutrients, and
always deriving physical support, from their rocky home. Mudflats and shallow ponds
provide excellent environments for wading birds that often have long beaks adapted
for probing into the mud for food.

Acid peatlands provide the homes for many flowering plants (e.g. *Calluna vulgaris*),
mosses (e.g. *Sphagnum*) and even insectivorous plants (e.g. *Sarranaceniaceae*). Fenlands, on the other hand, are dominated by grasses or sedges often with rich herbaceous
floras and colonised by fen-carr shrubs (Maltby & Proctor, 1996). According to Daly
(1994, p. 18), peatlands are unique ecosystems "representing extreme habitats where
waterlogging and, in many cases, restricted nutrient supply are important features, and
they support unique combinations of plants and animals adapted to these environmental
conditions". They are internationally important for many bird species, and peat swamp
forests are the home of the orang-utan. The United Kingdom's National Vegetation
Classification illustrates the relationship between woodland and scrub vegetation on

Table 3.5 *Extract from an analysis of the relationships between National Vegetation Classification communities and the soils and rocks that support them (after Rodwell, 1991; Usher, 2001)*

NVC number	NVC name	Notes on habitats and soils
W1	*Salix cinerea – Galium palustre* w	Wet mineral soils on the margins of standing or slow-moving open waters and in moist hollows
W2	*Salix cinerea-Betula pubescens-Phragmites australis* w	Topogenous fen peats, especially flood plain mires
W3	*Salix pentandra-Carex rostrata* w	Peat soils kept moist by moderately base-rich and calcareous groundwater
W4	*Betula pubescens – Molinia caerulea* w	Moist, moderately acid, though not highly oligotrophic, peaty soils
W5	*Alnus glutinosa – Carex paniculata* w	Wet or waterlogged organic soils, base-rich and moderately eutrophic
W6	*Alnus glutinosa – Urtica dioica* w	Eutrophic moist soils
W7	*Alnus glutinosa – Fraxinus excelsior-Lysimachia nemorum* w	Moist to very wet mineral soils, only moderately base-rich and usually only mesotrophic
W8	*Fraxinus excelsior – Acer campestre-Mercurialis perennis* w	Calcareous soils with mull humus
W9	*Fraxinus excelsior – Sorbus aucuparia-Mercurialis perennis* w	Permanently moist brown soils derived from calcareous bedrock and superficial deposits
W10	*Quercus robur – Pteridium aquilinum-Rubus fruticosus* w	Base-poor, brown soils
W11	*Quercus petraea – Betula pubescens-Oxalis acetosella* w	Moist, but free-draining, and quite base-poor soils
W12	*Fagus sylvatica – Mercurialis perennis* w	Free-draining, base-rich and calcareous soils
W13	*Taxus buccata* w	Moderate to steep limestone slopes with shallow, dry rendzina soils
W14	*Fagus sylvatica – Rubus fruticosus* w	Brown earths, with low base status and slightly impeded drainage
W15	*Fagus sylvatica – Deschampsia flexuosa* w	Base-poor, infertile soils

the one hand with soils and geology on the other (Rodwell,1991; Usher, 2001) and an extract is shown in Table 3.5.

In the permafrost of Siberia and Canada, northern species such as the Dahurian larch (*Larix dahurica*) and Siberian dwarf pine (*Pinus pumila*) are adapted to the shallow depth of soil available with spreading root systems and no taproot. On the other hand, taproots sent down to the water table from low points in the desert surface are one method by which plants can survive in tropical deserts. Other adaptations in these environments include direct nourishment of plants by rain, which consequently spend most of their lives in a dormant state, and water storage in succulent leaves and stems.

Zovodovski Island in the South Sandwich Group, Antarctica, has the world's largest colony of chinstrap penguins. Two million of them nest on the island, and this is attributed to the high geothermal heat flux from the island's volcano. The penguins can lay their eggs on the snow-free, warm soil, thus giving them an advantage over most of the other land and penguin colonies of the Antarctic. On the other hand, it is quite common for snakes, reptiles, and so on to escape the heat of the day by sheltering under rocks or in rock crevices. Female turtles come ashore to lay their eggs in the beach sands in places like Ascension Island in the Atlantic, Costa Rica in the Caribbean and Crab Island of the Queensland coast in Australia. In summer, Beluga whales rid themselves of their old skin by vigorously rubbing themselves against the gravel beds of Arctic estuaries, and walruses do the same on coastal rocks. The spawning grounds of the Atlantic salmon are well-oxygenated, pool-and-riffle sequences in gravel bed rivers. Sea birds like the gannet, guillemot and puffin nest on ledges on mainland or offshore island sea cliffs. The gannet (*Sula bassana*) gets its name from the Bass Rock, a small island off the coast of East Lothian, Scotland, where there is a colony of 20,000 birds.

Burrowing animals such as badgers, moles and rabbits rely on the soil and sub-soil for their living quarters. Huggett (1995, p. 149–151) gives examples of animal preferences for different substrates. For example, four species of pocket mice live in Nevada, USA. One lives on fairly firm soils of slightly sloping valley margins; the second is restricted to slopes where stones and cobbles are partly embedded in the ground; the third is associated with the fine, silty soil of the valley floor and the fourth can live on a variety of substrates. The soil itself provides a living environment for thousands of smaller faunal species. As Bridges (1994, p. 12) puts it "… not only does soil support life, it is itself an ecosystem teeming with living creatures belonging to many different phyla. All these different forms of life, from bacteria to mammals, participate to some degree in the recirculation of chemical elements in which the soil plays a central role...It provides all terrestrial ecosystems with physical support, nutrients and moisture...". Bats, on the other hand, find amenable living conditions in caves.

Adams (1996, p. 165) argues that "The link between process and ecology is particularly clear in the case of rivers, where characteristic species and communities are maintained within different parts of the channel and the floodplain by processes of erosion and deposition, and by patterns of over-bank flooding and groundwater recharge. If those processes are changed, ecological changes are likely to follow". Hughes & Rood (2001, p. 105–107) believe that "Floodplains are unique linear landscapes. They are among the most important ecosystems on the planet because they are highly productive and have a high species diversity.... Floods promote regeneration of plants... by creating open, moist sites necessary for many riparian species". Both Hughes & Rood (2001) and Richards *et al.* (2002) believe that flood attenuation has led to a loss of biodiversity on floodplains. In parts of the Amazon basin, it has been suggested that the high species diversity is the result of long-term disturbance through flooding and channel migration (Salo *et al.*, 1986). Floodplains also act as biological corridors through landscapes that would otherwise be inhospitable to both plant and animal species (Hughes & Rood, 2001).

Lithology and geochemistry also play a vital role. Calcareous rocks provide the special environments for a range of plants and animals, including limestone pavements (Webb & Glading, 1998) (see Box 3.9). The mortar-filled joints of walls also provide a calcareous habitat for plants, while "on walls built of mixed materials...every change in the substrate is reflected by a corresponding adjustment in the covering of lower plants (lichens and mosses)" (Gilbert, 1996, p. 7). In the United Kingdom, heavy metal mineralisation produces toxic conditions for most plants, yet a small number of

---

Box 3.9   Influence of Geodiversity on Biodiversity, Ingleborough, England

The Yorkshire Dales area of England is an example of an area where geology can be seen to be an inseparable part of the natural world. The development of vegetation on the limestone rock faces, crevices and screes is influenced by the stability of the rock, slope angle, aspect and shelter. The less stable and more exposed the habitat, the more specialised the flora.

Limestone pavements sustain a rich range of habitats for botanical diversity (Ward & Evans, 1976). The crevices (grykes) in the Ingleborough limestone pavements provide the right sheltered and shady conditions for relatively rare plants like purple saxifrage, yellow saxifrage, alpine-meadow grass, hoary whitlowgrass, lesser meadow-rue, wall lettuce and baneberry. The microclimate in grykes varies depending on their orientation (Burek & Legg, 1999) but, in general, is more like that of a woodland than an exposed rock surface. Snails and butterflies including some rare fritillaries live in these habitats and birds such as wheatear and wren sometimes nest in limestone pavements.

Elsewhere in the area, metallic ores, including mining waste, provide the conditions for a distinctive range of species able to cope with high concentrations of heavy metals.

---

rare and local plants are more or less restricted to such situations, including Leadwort (*Minuartia verna*), Alpine Penny-cress (*Thlaspi alpestre*) and several rare mosses and lichens (Hopkins, 1994). A type of mountain pink (*Lychnis alpina* var. *Seprenticola*) is adapted to growing on nickel-rich serpentine and can be used as a nickel ore indicator. Similarly, distinct plant communities occur on serpentine rocks in many parts of the world including the Klamath-Siskiyou Ecoregion on the Oregon/California border, USA. Of the 200 endemic plant types in the ecoregion, 141 are either rare or uncommon and have evolved in the toxic and dry conditions. Huggett (1995) refers to these as "lithobiomes" and gives several examples.

Coastlines and streams, bogs and moors, deserts and mountains, glaciers and volcanoes: the infinite variety of life on earth is adapted to its physical environment and these diverse physical systems therefore have a functional value for biological systems and biodiversity. As Warren (2001, p. 60) puts it "Plant and animal species have adapted to and therefore now need the spatial variety and scale that geomorphological processes provide". It is now fairly clear as a generality that areas of high geodiversity lead to high biodiversity (Hopkins, 1994), though the reverse is not always the case, and certainly more research is needed (see Chapter 7).

3.7   Research and Educational Value

Although this is the final value to be reviewed, it is in many ways the most important. The physical environment is a laboratory for future research, and "it is often only field sites which provide a reliable test of many geological theories" (Bennett & Doyle, 1997, p. 161). In this section, the need to conserve geology and geomorphology for research and education purposes will be assessed by citing some examples of the contribution of past research. Damage to physical systems and sites inevitably damages our ability to undertake research and teaching on the physical environment. Consequently, "we should maintain the means to seek knowledge in the future" (Nature Conservancy Council, 1990, p. 17).

3.7.1 Processes and scientific discovery

Dynamic sites are important in allowing research aimed at understanding the processes at work in active systems and has implications for pure and applied research. As Ellis *et al.* (1996, p. 8) have noted, "by studying the dynamics of natural systems, such as rivers and coasts, it may be possible to predict how land and coastal processes will operate in the future. This will aid flood prediction and management, the mapping of physically hazardous areas and coastal management" (see Box 3.10). For example, to understand the impacts of sea-level rise, an understanding of coastal processes is crucial.

Box 3.10   Fluvial Research

J. Hooke (1994) recognises the value of sites in the middle sections of rivers, for increasing scientific understanding of

- links between upland and lowland systems and therefore between the zones of maximum runoff and sediment generation and the zones of transportation and deposition;
- dynamics of flooding;
- erosion processes;
- sediment transport mechanics and sediment loads;
- depositional processes and floodplain development;
- mechanism and rates of short-term changes and adjustments of channels;
- long-term changes and variations in behaviour of channel systems.

She believes that such sections "are essential to understanding natural processes and trends so that effects of modification elsewhere can be assessed and management practices can be instituted which work with nature rather than against it. .. ... An essential point about these sites, though is that they are active and should be kept so" (J. Hooke, 1994, p. 115).

Along with research on earth surface processes, other scientific research includes testing geomaterial properties, understanding landform sensitivity and geoforensics. The latter involves linking suspects to crime scenes through the distinctiveness of soil on their clothing, footwear, car tyres, and so on. This generally involves searching for unusual minerals in the samples and linking these with the distinctiveness of the soil mineralogy at the crime scene (Donnelly, 2002).

3.7.2 Earth history

The study of the geological record has enabled geologists to reconstruct in considerable detail the history of the Earth over the last 4,600 Ma. It is a record of amazing complexity and a tribute to the meticulous work of thousands of geologists over a long period of time. It has been deciphered from rock outcrops and boreholes in all countries of the world and it continues to be refined by further research. Major discoveries are still being made, particularly in the less well studied parts of the planet, but even where intensive studies have been made, the geological record needs to be conserved for future study using new techniques and approaches and to allow findings to be checked and reinterpreted (Page, 1998). This geological rock record therefore has enormous research value.

In the case of Britain, Ellis *et al.* (1996, p. 8) argue that "Natural rock outcrops and landforms, and artificial exposures of rock created in the course of mining, quarrying and engineering works, are crucial to our understanding of Britain's Earth heritage. Future research may help resolve current geological problems, support new theories and develop innovative techniques or ideas only if sites are available for future study".

The fossil record contained in the rocks has demonstrated the evolution of species from the simplest unicellular organisms to the early history of humans. "Fossils are not only records of evolution...they also allow us to have a look at the construction of living matter of past biospheres" (Wiedenbein, 1994, p. 118). For example, the Rhynie Chert in Scotland contains some of the oldest known fossils of higher plants, insects, arachnids and crustaceans in such a well-preserved state that microscopic detail and cell structures can be studied. Such sites are a rare and irreplaceable part of the world's geological heritage. The history of the dinosaurs and their extinction has caught the public imagination. Research on the triggers for mass extinctions, including the idea of extraterrestrial impacts, has also been stimulated by the nature of the fossil record, which therefore has great research and educational potential and therefore value.

Geological research has also enabled the reconstruction of the changing geography of the planet as the supercontinent of Pangaea initially fractured and then drifted apart on huge tectonic plates driven by mantle convection currents. Research has demonstrated the close link between tectonic plate margins and volcanic and earthquake activity, and study of the pattern of past natural hazards helps us to predict the location and timing of future disasters.

Other researchers have deciphered the history of the Quaternary Ice Age through studies of geology, geomorphology and biostratigraphy. The record is one of glacials and interglacials and shorter climate changes during which the ice sheets and glaciers advanced and retreated on many occasions. Figure 3.23 shows how geological diversity has enabled reconstruction of directions of ice movement due to the distinctiveness and limited outcrop of the source rocks.

3.7.3   History of research

Ellis *et al.* (1996, p. 7) have drawn attention to the fact that many British sites have value because they "have played a part in the development of now universally applied principles of geology". Examples include Hutton's unconformity at Siccar Point in Scotland (Fig. 5.22) and his section on Salisbury Crags in Edinburgh, Scotland, where he deduced that the dolerite forming the crag must have been molten. Many of the names of periods of geological time are derived from Britain while many others serve as international reference sections. These sites have become international "standards" that "must be conserved so that they can continue to be used as the standard references" (Ellis *et al.*, 1996, p. 7).

Of course, it is not just Britain that has such sites or where such arguments for conservation apply. For example, the history of fossil hunting is important in several parts of the world, including the dinosaur-rich areas of the western fringes of the American and Canadian Rockies (e.g. Wyoming, Montana and Alberta) (Horner & Dobb, 1997; Gross, 1998). Such sites, where important discoveries have been made in

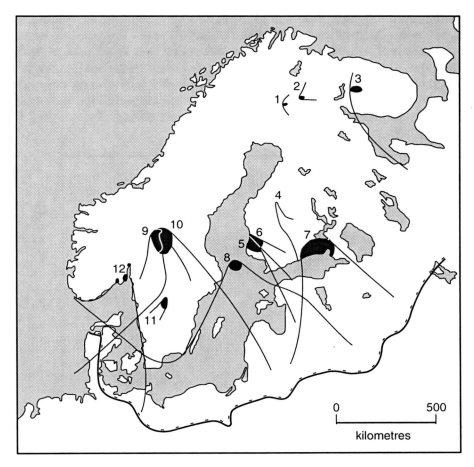

Figure 3.23 *Geological diversity enables directions of ice movements to be established from studying erratic dispersal in Scandinavia. Erratic sources are 1. Jatulian sandstone and conglomerate; 2. Nattanen granite; 3. Umptek and Lujarv-Urt nepheline syenite; 4. Lappajarvi impactite; 5. Vehemaa and Laitila rapakivi granite; 6. Jotnian sandstone (Satakunta); 7. Viipuri rapakivi granite; 8. Aland rapakivi granite; 9 Jotnian sandstone (Dala); 10. Dala porphyries; 11. Cambro-Silurian limestone; 12. Smaland granite; 13. Oslo rhomb porphyries and larvikite (Modified after Donner, J.J. (1995) The Quaternary history of Scandinavia. Cambridge University Press, Cambridge, by permission of Cambridge University Press)*

the past, are worthy of conservation to illustrate and illuminate the history of scientific research.

3.7.4 Environmental monitoring

The record of sediments in lakes, bogs and ice cores also provides records of the effects of human activities on the environment through pollution, vegetation clearance, soil erosion, and so on. These records are valuable not only in reconstructing the past human impacts on the environment and the history of human use of the land but also in assessing the effects of current and potential future impacts. For example, research in the 1980s based on the analysis of diatoms, trace metals and fly-ash particles in lake sediment records provided definitive evidence for the causes of lake acidification in the United Kingdom (Battarbee *et al.*, 1985; Battarbee, 1992). This research contributed significantly to UK government decisions to introduce sulphur emission reduction policies in the late 1980s leading to the establishment of the UK Acid Waters Monitoring Network in 1988.

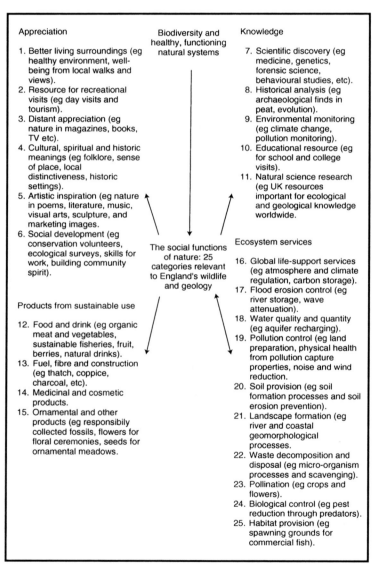

Figure 3.24   *English Nature's summary of the "social functions of nature: 25 categories relevant to England's wildlife and geology" (After English Nature, (2002) Revealing the Value of Nature. English Nature, Peterborough. by permission of English Nature)*

The climate history of the Earth of the last few million years gives us many clues as to the causes of climate change, the role of humans in altering climate and how physical systems will respond to future changes. The evidence for climate change comes from a very wide variety of sources, but they include the records preserved in peat bogs, cave sediments and ice sheets. Records like this are important in reconstructing recent climate changes and correlating these with other terrestrial and marine locations. Study of such sites can also provide empirical data on which global models of climate change can be developed and tested, and may help predict future changes to the global climate system.

3.7.5   Education and training

The geological record has huge research value, but it also has a role in education and training. Students and teachers need sites and areas that they can use to demonstrate geological principles and processes in the field. Trained geologists, geomorphologists and pedologists are needed to locate and utilise mineral resources, predict natural hazards and ensure the sustainable use of land. Rock exposures, fossil sites, landforms, soil sections and active processes play a valuable role in the education of children, the training of the next generation of geologists, and amateurs with an interest in their environment and the geological history of the planet. We should not lightly allow them to be destroyed.

3.8   Conclusions

The combined values of geodiversity are considerable and it is useful here to provide a synthesis and summary. De Groot (1992) proposed 25 categories of the social functions of nature and this was modified by English Nature (2002) to reflect the English situation (Fig. 3.24). However, it mainly applies to biological values and, in particular, omits most of the geological/geomorphological and other geovalues and products discussed in this chapter. Table 3.6 is an attempt at a similar synthesis reflecting the full range of values of geodiversity discussed in this chapter and amounting to over 30 distinctive values. Guthrie (2003) has outlined a research project in the United Kingdom aimed at quantifying all the social, economic and cultural values both nationally and locally that arise from geological sites.

Table 3.6   *Summary of geodiversity values*

Intrinsic value	1. Intrinsic value	Abiotic nature free of human valuations
Cultural value	2. Folklore	Giant's Causeway, UK; Devil's Tower, USA
	3. Archaeological/ Historical	Petra, Jordan; Stonehenge, UK; local tools and artefacts
	4. Spiritual	Uluru, Australia; N.American Indian sites
	5. Sense of place	White Cliffs of Dover, UK; Rock of Gibraltar; local places
Aesthetic value	6. Local landscapes	Sea-views; countryside walks; vernacular buildings
	7. Geotourism	Grand Canyon, USA; Norwegian fjords; Canadian Rockies

(*continued overleaf*)

Table 3.6   *(continued)*

	8. Leisure activities	Rock climbing; caving; whitewater rafting; fossil hunting
	9. Remote appreciation	Nature in magazines and TV; "Walking with dinosaurs"
	10. Voluntary activities	Wall repairs; footpath construction; mine restoration
	11. Artistic inspiration	Literature (Hardy); music (Sibelius); painting (Turner)
Economic value	12. Energy	Coal & peat; oil & gas; uranium; geothermal; hydro-electric; tidal
	13. Industrial minerals	Potash; fluorspar; kaolinite; rock salt
	14. Metallic minerals	Iron; copper; chromium; zinc; tin; gold; platinum
	15. Construction minerals	Stone; aggregate; limestone; brick clay; gypsum; bitumen
	16. Gemstones	Diamond; sapphire; emerald; onyx; agate
	17. Fossils	Tyrannosaurus "Sue"; fossil and mineral shops
	18. Soil	Food production; wine; timber; fibre
Functional value	19. Platforms	Building and infrastructure construction
	20. Storage & recycling	Carbon in soil & peat; oil & gas in traps; hydrological cycle
	21. Health	Nutrients and minerals; therapeutic landscapes
	22. Burial	Human burial; landfill sites; underground nuclear chambers
	23. Pollution control	Soil & rock as a water 'filter'; landform screens
	24. Water chemistry	Mineral water; whisky
	25. Soil functions	Agriculture; viticulture; forestry
	26. Geosystem functions	Continued operation of fluvial, coastal, aeolian processes, etc.
	27. Ecosystem functions	Biodiversity
Research &	28. Scientific discovery	Geoprocesses; geotechnology; geoforensics
Education value	29. Earth history	Evolution; geological history of Earth; geoarchaeology
	30. History of research	Early identification of unconformities, igneous activity, etc.
	31. Environmental monitoring	Ice cores; sea-level change; pollution monitoring
	32. Education & training	Field studies; professional training

These values sometimes overlap. Fossils have both an aesthetic and economic value; similarly with landscapes, which are in essence free and attractive in many senses yet can also generate economic prosperity through geotourism. It will also be very clear that values may conflict. The economic values of geological materials and the need for modern society to exploit them often conflict with the aesthetic values of the landforms and landscapes that would be impacted by quarrying.

The following chapter deals with the threats to geodiversity from mineral exploitation and a range of other human activities, and later chapters address the issue of how decisions are made as to what values and elements of geodiversity are sufficiently important to warrant protection.

4

Threats to Geodiversity

4.1 Introduction

There is perhaps a general tendency to think of the biological world as fragile and vulnerable and therefore in need of conservation, whereas the abiotic world of mountains and rock is seen as stable, static and much too prolific ever to be endangered. This is a gross oversimplification, and many threats to the geodiversity of the planet or local areas are comparable to those facing biodiversity. Furthermore, geoconservation is not just about protecting the static elements of the landscape. It is also about allowing dynamic processes to continue operating within the historical range of natural rates. It should also be noted that disturbances to geological, geomorphological and soil processes can be produced in ways that are not always local or obvious. For example, individually or cumulatively, vegetation clearance, agriculture, water diversion, forestry and urbanisation can all have profound impacts on river landforms and sediments by changing runoff rates and magnitudes, sediment loads, and so on. There are many significant, real and potentially damaging activities that ought to be better understood if we are to properly conserve and manage geodiversity. As Sir David Attenborough (1990, p. 5) puts it:

> "Increasingly quarries, gravel pits, old mines and caves are pressed into service as waste disposal sites. Eroding cliffs with their extensive exposures of rock sections, the source of sand and shingle beaches elsewhere, are rendered invisible and geomorphologically impotent behind concrete. Rocky mountain crags are shrouded by conifers. Even the shape of the land itself is changing as features are levelled and whittled away to feed the insatiable demand for development land and construction materials".

The number of these threats is great (Glasser, 2001; Gordon & MacFadyen, 2001), though only a few are likely to apply in most locations. Many threats are the result of natural processes, for example, coastal or river-bank erosion or weathering of fossils (Koch *et al.*, 2002), but this book concentrates on human impacts. The types of artificial activities that will degrade geodiversity depend on the types of values concerned (see Chapter 3). Furthermore, the impact of an operation will vary depending on the sensitivity, stability or robustness of the site in question (Schumm, 1979; Brunsden & Thornes, 1979). An operation that would have a devastating effect in one area may be more acceptable in another, more robust, location. This is because some systems are capable of repairing themselves in a relatively short time due to the continued operation of natural processes (e.g. reformation of gravel bar following a flood), whereas other changes are irreversible because the processes no longer operate or the change to the landscape is fundamental (e.g. removal of an esker by quarrying or loss of soil

Geodiversity: valuing and conserving abiotic nature Murray Gray
© 2004 John Wiley & Sons, Ltd ISBNs: 0-470-84895-2 (HB); 0-470-84896-0 (PB)

cover). This concept of landscape sensitivity is a fundamental one in understanding the threats to geodiversity (Werritty & Brazier, 1994; Gordon *et al.*, 2001; Haynes *et al.*, 2001; Werritty & Leys, 2001; Sharples, 2002a), and is considered further in Section 4.16.

In general terms, threats to geodiversity are the result of development pressures and land-use change (Gordon & MacFadyen, 2001), but others may result from natural processes or from human-induced change (e.g. climate change and sea-level rise), though it is often difficult to separate these effects (Harrison & Kirkpatrick, 2001). The human impacts on geodiversity can be summarised as

- complete loss of an element of geodiversity;
- partial loss or physical damage;
- fragmentation of interest;
- loss of visibility or intervisibility;
- loss of access;
- interruption of natural processes and off-site impacts;
- pollution;
- visual impact.

Some of these impacts affect specific sites of geoconservation value while others impact widely across large land areas, but all may lead to loss of or damage to elements of geodiversity (e.g. see Fig. 4.1).

The United Kingdom's Wildlife and Countryside Act (1981) allows for the specification of a standard list of "Potentially Damaging Operations" (PDOs) and this is reproduced in Table 4.1. Although some are exclusive to biodiversity interests, many can and do affect geodiversity. Table 4.2 is a more explicitly earth science–related list of threats and will be expanded upon in this chapter. English Nature (1998) has also recognised many threats to the country's geoheritage (termed Operations Likely to Damage – OLDs) and these have also been included in the following discussion.

Figure 4.1   *Graffiti on a roadside hoodoo, Montana, USA*

Table 4.1 *List of potentially damaging operations (PDOs)*

Reference no.	Type of operation
1*	Cultivation, including ploughing, rotavating, harrowing and reseeding
2*	Grazing and changes in the grazing regime (including type of stock or intensity or seasonal pattern of grazing and cessation of grazing)
3	The introduction of stock feeding and changes in stock feeding practice
4	Mowing or other methods of cutting the vegetation
5	Application of manure, fertilisers and lime
6*	Application of pesticides, including herbicides (weedkillers)
7*	Dumping, spreading or discharge of any materials
8*	Burning or changes in the pattern or frequency of burning
9	The release into the site of any wild, feral or domestic animal, plant or seed
10	The killing or removal of any wild animal, including pest control
11*	The destruction, displacement, removal or cutting of any plant or plant remains, including tree, shrub, herb, moss, lichen, fungus or turf
12*	The introduction of tree and/or woodland management, including afforestation and planting
13a*	Drainage (including moor gripping and the use of mole, tile, tunnel and other artificial drains
13b*	Modification of the structure of watercourses (e.g. rivers, streams, springs, ditches, drains) including their banks and beds, as by realignment, regrading and dredging.
13c	Management of aquatic and bank vegetation for drainage purposes
14*	The changing of water levels and tables and water utilisation (including storage and abstraction of existing water bodies)
15*	Infilling of ditches, drains, pools or marshes
16	The introduction of, or changes in, freshwater fishery production and management including sporting fishing and angling
17*	Reclamation of land from the sea
19*	Erection of sea defences or coastal protection works, including cliff or landslip drainage or stabilisation measures
20*	Extraction of minerals including peat, shingle, sand, gravel, topsoil and subsoil
21*	Construction, removal or destruction of roads, tracks, walls, fences, hardstands, banks, ditches or other earthworks, or the laying, maintenance or removal of pipelines and cables, above or below ground
22*	Storage of materials on any part of a beach complex
23*	Erection of permanent or temporary structures, or the undertaking of engineering works, including drilling
24*	Battering or grading of sand dunes
26*	Use of vehicles or craft likely to damage or disturb features of interest
27*	Recreational or other activities likely to damage landforms or features of interest
28	Changes in game or waterfowl management and hunting practice

*PDOs that are applicable to earth science interests.

Major reviews of the human impact on the environment have been undertaken by several authors and readers are referred to Detwyler (1971), Turner *et al.* (1990), Goudie & Viles (1997) and Goudie (2000) for fuller information.

4.2 Mineral Extraction

Extraction of minerals from quarries and other open excavations, including building stone, rock aggregate, metallic ores, sand and gravel, beach shingle, peat and soil is

Table 4.2   *Threats to geoheritage (modified after Gordon & MacFadyen, 2001)*

Threat	Examples of on-site impacts	Examples of off-site impacts
Mineral extraction (includes pits, quarries, dunes & beaches)	Destruction of landforms and sediment records Destruction of soils, soil structure, soil biota May have positive benefits in creating new sections	Contamination of watercourses Changes in sediment supply to active systems Extraction from rivers and beaches, leading to erosion and scour
Landfill & quarry restoration	Loss of exposures Loss of natural landform & soil disturbance Detrimental effects of leachate & landfill gases Habitat creation	Contamination of surface watercourses Contamination of groundwater
Land development and urban expansion	Large-scale damage and disruption of landforms and soils Changes to drainage systems Creation of slope instability	Changes to processes downstream Contamination of watercourses
Coast erosion & protection	Loss of coastal exposures Loss of active and relict landforms Disruption of natural processes	Changes to sediment supply & processes downdrift
River management, hydrology & engineering	Loss of exposures Loss of active and relict landforms Disruption of natural processes	Changes to sediment movement and processes downstream Change in process regime; drying of wetlands
Forestry, vegetation growth & removal	Loss of landform and outcrop visibility Physical damage to small-scale landforms Stabilisation of dynamic landforms Soil erosion Changes to soil chemistry and soil water regime	Increase in sediment yield and runoff during planting & deforestation Changes to groundwater and surface water chemistry
Agriculture	Damage or loss of small-scale landforms through Ploughing, ground levelling and drainage Soil compaction, loss of organic matter & soil biota Changes to soil chemistry from fertilisers Effects of pesticides on soil biota Soil erosion	Changes in runoff arising from drainage Episodic soil erosion by wind and water Pollution of surface and groundwater from excess agrochemical application
Other land management changes	Loss or degradation of exposures and landforms Loss or contamination of soils Changes to soil-water regime	Changes in runoff and sediment supply
Recreational/ tourism pressures	Physical damage to small-scale landforms & soils Localised soil erosion Damage to cave systems	

Table 4.2 (*continued*)

Threat	Examples of on-site impacts	Examples of off-site impacts
Removal of geological specimens	Loss of fossil record Loss of mineral specimens	
Climate & sea-level changes	Changes in active process systems Coastal erosion and inundation	Changes in flood frequency Changes in rates of geomorphological processes
Fire	Loss of organic soils Loss of vegetation leading to soil erosion	
Military activity	Loss and damage to soils and small-scale landforms by vehicles Production of craters by bombing	
Lack of information/ education	Loss or damage to active processes or static features through ignorance of values	

clearly necessary to modern society. And while these activities have frequently resulted in important geological exposures, many of which are subsequently protected, there is also always some loss of geodiversity as the landscape and topography are disturbed and sediments or rocks below are removed. This may not be significant where the resource being quarried is large, but it will become problematic where rare soils, important landforms, limited rock outcrops or important fossil-bearing strata are removed during mineral extraction and there are many examples of, for example, eskers being quarried away for aggregate.

Geological resources are, by-and-large, non-renewable. Woodcock (1994) argues that oil is being used at least a million times faster than it is being created and most other use of geological materials is unsustainable, that is, all these materials will become exhausted eventually but the timescales involved will be very different for different materials. Kelk (1992, p. 34) noted that in the United Kingdom, "the annual usage of sand and gravel is equivalent to 50 to 60 barrowloads for every man, woman and child. ... Most of the geological resources, such as coal and hydrocarbons, are in limited supply, whereas water – perhaps the most important natural resource of all – is both limited in quantity and increasingly threatened in quality". The issue of exhaustion of mineral supplies is considered in Sections 6.5.2 and 7.3.3 as part of the issues of sustainable use of geomaterials and geological extinction.

"Mining can be a nasty business; it can be one of the most environmentally damaging activities undertaken by humans. ... Today, more land is devastated because of the direct effects of mining activities than by any other human activity" (McKinney & Schoch, 1998, p. 266). This may not be entirely accurate since estimates are that this disturbance amounts to only about 1% of the Earth's land surface, though this is likely to be an underestimate, given off-site impacts like acid drainage and the fact that large areas have been disturbed by past mining activities. However, other impacts such as agriculture, deforestation and development may be much more spatially significant.

Underground mining has a less serious impact on the surface, but may still involve subsidence of several metres, for example, as a result of coal mining (English coalfields), iron mining (north-east France), brine pumping (e.g. Lüneburg, northern Germany), water abstraction (Mexico City) or oil extraction (Wilmington, California, USA). It is also more expensive and not feasible for many materials like aggregate. Visual intrusion of the opencast quarries is therefore a major issue. Blunden (1991) and Jones & Hollier (1997) point out that the extent of visual intrusion from mining depends to a large degree on the local topography, but some like the Bingham Canyon copper mine in Utah are so large (4 km wide and 1 km deep) that they can be seen from outer space. An important example of unsustainable rock exploitation is the removal of limestone pavements in parts of the United Kingdom and Ireland (Box 4.1).

Box 4.1   Removal of Limestone Pavements in the United Kingdom and Ireland

Limestone pavements constitute "a unique and finite part of Britain's physical landscape" (Bennett & Doyle, 1997, p. 106). They "are limestone outcrops which have been stripped of any pre-existing soil or other cover by some scouring mechanism, generally but not exclusively, glacial scour" (Goldie, 1994) and subsequently subject to joint weathering and solution, producing a variety of surface morphologies including clints (abraded and weathered surface blocks) and grykes (weathered joints). In Europe, they occur predominantly in the Carboniferous limestones of Upland Britain and Ireland, though there are many Alpine sites and patches in Sweden. In North America, limestone pavements in Devonian limestones occur in New York State, for example, around Syracuse and the Finger Lakes (Burek & Conway, 2000), and on the top of the Niagara Escarpment on the Bruce Peninsula, Ontario, Canada.

Serious loss of these pavements by unscrupulous operators has occurred in Yorkshire and Cumbria in Britain and the Burren in Ireland to satisfy the demand for weathered limestone in garden rockeries and walls where their weathered characteristics are valued by horticulturalists. "Clints have been sold at garden centres to purchasers who are probably ignorant of their beauty and interest *in situ.*... Pavements thus damaged have a much altered geomorphology, a messy, ugly, broken surface with much loose debris and rough remnant clint tops, lacking attractive runnelling, and with a much depleted, or even totally destroyed, flora" (Goldie, 1994, p. 216). Blocks of pavement can sell for over £100 per tonne, but broken up into hand-sized pieces, the limestone can reach £360/tonne (Webb, 2001). The market demand far exceeds the legal consequences if caught (Bennett & Doyle, 1997).

Unfortunately, the strengthening of the UK legislation (see Limestone Pavement Orders, Section 5.10.5) has meant that unscrupulous dealers have turned their attention to Ireland where protection is less stringent. A research investigation into the trafficking of water-worn limestone, including under-cover surveillance, has revealed that pavements are being removed from The Burren and other Irish localities by at least 13 traders, with almost 10,000 tonnes per year being imported into the United Kingdom, as well as quantities going to Germany, Holland and Belgium.

It is estimated that the cumulative world use of sand for mineral extraction in the 25 years between 1976 and 2000 was 37,000 km^2 (Evans, 1997) with the greatest proportion being in the developed world. Thus, there is a threat to local landscapes, for example, where sand and gravel for use in glass production, road construction or concrete manufacture are excavated. Stürm (1994) notes that the impressive moraine landscape around Zurich in Switzerland is under pressure because of the demand for sand and gravel to satisfy the building boom in that area. In Cheshire, England, English Nature (1998) sees an important threat from "the extensive winning of construction sand from the Delamere Forest so significantly modifying the undulating landscape", which is glaciofluvial in origin.

The excavation of marine, beach and river aggregate also needs careful assessment in order to avoid damage to functioning geomorphological systems. Masalu (2002) describes illegal sand mining along the beaches and rivers of the Tanzania coast. This is causing local coastal erosion, threats to coastal properties and instability of bridges due to alteration of stream-bed morphology. Offshore dredging for sand and gravel alters the shape of the sea bed and hence may impact on current patterns, possibly aggravating coastal erosion. At a smaller scale, problems have arisen at Crackington Haven in Cornwall, England, where the beach comprises dark grey rounded shale pebbles with attractive quartz veins. These have become the target of collectors as a result of television gardening programmes promoting the use of rounded cobbles and pebbles in gardens. According to Anon (2000, p. 6), "pebbles are being removed by the bucket load", and there have even been incidents reported of car and trailer loads being taken from the site that is a protected area. The threat is not just the loss of the beach but also to the protection it gives to the cliffs and local settlement.

Thorvardardottir & Thoroddsson, (1994, p. 228) describe the threat to volcanic landscapes in Iceland. "If an area is not protected. ... Landowners are permitted to mine gravel, rock, scoria, or pumice on their estate. This has led to the production of many mining pits, especially where lava and pumice are found, as these have thin or no vegetation cover, are lightweight, porous, without frost activity and give good insulation". They are therefore used for road, footpath and driveway construction. Thorvardardottir & Thoroddsson (1994, p. 228) argue that the many abandoned mining pits in Iceland "look like battlefields in the landscape". They believe that many interesting volcanic formations have been lost, although nowadays mining is planned with more care to avoid craters, in particular. In New Zealand, Buckeridge (1994) notes that all 48 volcanic cones around Auckland have suffered some damage, and over half have been quarried away or covered over. Similarly, Rosengren (1994) expresses concern about the impact of quarrying on Australia's important Late-Cenozoic volcanic province. Although the quarries often reveal important internal structures, contacts and stratigraphies,

> "scoria, and tuff are non-renewable resources and the configuration of the eruption points cannot regenerate or recover until there are more eruptions. Quarry operators may not recognise the significance of uncovered material or structures and may unknowingly destroy specimens or exposures, fill in craters or bury material with overburden or stockpile. Outcrops or topography significant in displaying volcanic history may be modified, buried or removed. . . quarrying produces synthetic landforms. Holes and overburden mounds alter the form and slope angles of cones and mounds and may confuse future interpretation of original eruption topography and products" (Rosengren, 1994, p. 109).

This extract has been quoted at length because it neatly describes a series of threats to an important geological and geomorphological heritage, from a quarrying operation and thus implicitly makes the case for control that takes account of the heritage interest.

The impacts on the landscape are greatest where area strip mining is involved (US Department of the Interior, 1971). This method is commonly used where there are gently dipping beds or other resources within a few tens of metres of the surface, for example, Carboniferous coals, Jurassic ironstones or sedimentary kaolins. Examples of the latter occur in Georgia and South Carolina in the United States, while Ohio, Pennsylvania, West Virginia, Kentucky, Illinois, Wyoming and Missouri are States with significant coal strip mines. The method is to work a continuous swathe across an area, extracting the mineral and backfilling with the stripped overburden and other wastes. Box 4.2 gives another example from the Czech Republic.

Box 4.2   Impact of Brown Coal Mining in the Czech Republic

Domas (1994) describes the impact of brown coal extraction in the Czech Republic in the 1960s and 1970s. According to Domas (p. 93) "The government decided to obtain energy from brown coal, regardless of the impact". Opencast mines were used so that the impact came from both the pit and the spoil. Pits were 1 to 5 km long, 1 to 3 km wide and up to 150 m deep. "Each coal quarry was required to deposit the overburden on its outer reaches at the beginning of mining operations, in the process burying many square kilometres of original landscape... the new profiles which result dominate the landscape".

In the North Bohemian Brown Coal Basin, some 260 km² of land surface has been influenced by opencast mining, 80 km² of which are spoil dumps (see Fig. 4.2). More

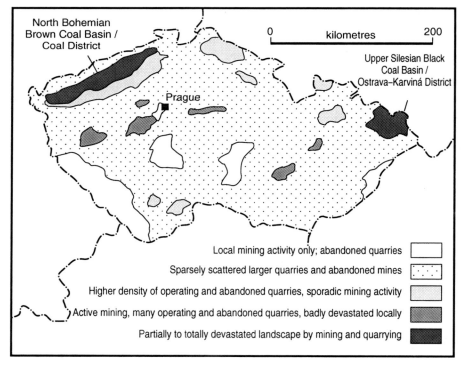

Local mining activity only; abandoned quarries

Sparsely scattered larger quarries and abandoned mines

Higher density of operating and abandoned quarries, sporadic mining activity

Active mining, many operating and abandoned quarries, badly devastated locally

Partially to totally devastated landscape by mining and quarrying

Figure 4.2 *Impact of opencast coal mining in the Czech Republic (After Domas, J. (1994) Damage caused by opencast and underground coal mining. In O'Halloran, D., Green, C., Harley, M., Stanley, M. & Knill, J. (eds) Geological and Landscape Conservation. Geological Society, London, 93–97, by permission of The Geological Society)*

_____ Continued on page 141 _____|

Continued from page 140

than 70 villages and settlements were destroyed by the advancing quarries. The mining policy of the government at the time was to exploit this resource to exhaustion of the basin's reserves, but a subsequent change of policy imposed obligatory spatial limits on the operating mines, which must not be overstepped by either quarrying or spoil deposition.

In the Upper Silesian Black Coal Basin, mining is mainly by underground workings, sometimes on up to 10 floors. This has caused significant land subsidence of up to 40 m in places. The original glacial and periglacial geomorphology of the landscape has been totally changed, with large areas now being flooded. In addition to the constant need to transfer and uplift roads and railways, more than 2,500 houses have been destroyed. Settling pits infilled by waste from coal preparation plants contain several metres of thick black mud with admixtures of organic and inorganic pollutants. With no lining systems, polluted water migrates from the base of these ponds into surface and groundwaters. Methods to restore landscapes are described in Section 6.3.1 but as Domas (1994, p. 97) states "Elimination of the mining impact, so devastatingly and carelessly caused in the past, will certainly not be easy".

Box 4.3   Peat Extraction in Ireland

About 16% of the land surface of Ireland is covered in peat over 1 m thick. This occurs either as blanket bog 1 to 6 m deep along the western coastal strip and mountain areas throughout Ireland where there are high rainfall amounts and relatively impermeable rock types, or as raised bog up to 15 m thick over the low-lying, poorly drained midland counties.

After World War II, the Irish government established a large-scale peat excavation project both for domestic use and for electricity generation. A state company (Bord Na Mona (BNM)) was established to extract the peat from 90,000 ha of bog over 80 years, 1950 to 2030. Average production is about 3 to 6 million tonnes/year. However, in recent decades there has been concern from environmentalists about the impact of this extraction, particularly on the raised bogs, 95% of which have been lost, fragmented or affected by human activity. Thus, there is currently a dilemma about the exploitation of Irish peat bogs. On the one hand, the material is valuable as a fuel and horticultural material as well as providing local employment. On the other hand, the bogs are part of the Irish landscape and heritage, attracting leisure and tourist activities (Kelk, 1992).

Gunn *et al.* (1994) believe that there are examples in Fermanagh in northern Ireland where commercial peat cutting is being carried out illegally and where enforcement is ignored by the operators.

The extraction of peat for commercial or domestic use as a fuel or garden compost can result in loss of the palaeoecological record of vegetational and faunal change, the archaeological record of human use of the wetland, as well as disrupting wetland landscape, drainage and wildlife (see Box 4.3). Peatlands once covered 8% of western Europe and were a dominant landscape feature. However, they are "extremely delicate and prone to damage. . . (and) their utilization has a considerable economic importance. As a consequence, peat cutting, drainage, afforestation and agricultural development have destroyed most of the peatlands of north-western Europe" (Daly, 1994, p. 17). The Netherlands have lost almost 100% of their peatlands, Britain has lost 98% of its raised bogs and 90% of blanket bogs, and in Ireland the corresponding figures are 94% and 86% respectively (Daly, 1994). In February 2002, the UK government paid the company Scotts £17.3 million to buy out its extraction rights at three of England's most important remaining peat bogs (*The Guardian*, 28 February, 2002).

As indicated by the above examples, mineral extraction often produces waste materials, which themselves impact on the landscape. This may either be due to the need to remove overburden to reach the mineral, or is due to the low concentration of a mineral within an ore body. Mineral ores are rarely greater than 30% pure and can be under 1% pure. For example, at Bingham Canyon, Utah, the copper concentration is about 0.3%. Thus, large quantities of waste materials need to be disposed of. In the case of deep quarries, the backfilling option is excluded by the need to work vertically and not sterilise the remaining resource at depth. Waste material must therefore be dumped beyond the quarry edge, thus increasing the landscape impact. The most serious examples of this occur where there are high quantities of waste relative to mineral recovered, for example, kimberlite diamond pipes. In the United Kingdom, the most serious examples are at the North Wales slate quarries and at the china clay pits in south-west England where about 6 to 8 times as much waste quartz and mica are produced than usable kaolinite (Bristow, 1994).

Silvestru (1994, p. 224) describes the effects of limestone and marble quarries on the karst landscapes of Romania. "... the damage does not consist of mere subaerial artificial cuts... Sometimes... underground drainage is diverted or disrupted, which always results in important regional changes and has long-term consequences. Even more destructive, are the auxiliary works like roads, buildings and everyday activities ...". Silvestru also describes the impact of bauxite mining, which takes place in closed hollows up to 60 m deep. "A closed hollow of this type, takes a very long time to recover, and represents a strikingly unnatural feature in the heart of karstic landforms. In addition, the waste dump left on the slopes is either terra rosa or reddish sediments associated with bauxite, both practically sterile for vegetation. Therefore, the recovery time is even longer". He describes large areas of the northern Apuseni Mountains of north-west Romania as being "a land of desolation" (Silvestru, 1994, p. 224).

In some developing countries, there may be little by way of restoration and landscapes can become devastated by quarrying operations (Hilson, 2002), sometimes carried out illegally. There are several examples of illegal diamond quarries in Africa, many of which are used to fund civil wars in Sierra Leone, Angola and the Democratic Republic of Congo (*The Guardian*, 12 December 2001). Levy & Scott-Clark (2001) refer to the "brutalised landscape" at the jadeite quarries at Hpakant in northern Burma, with "the mountains reduced to rubble". In November 2001, at least 40 gold miners were killed by landslides at illegal surface diggings on a hillside in western Colombia (*The Guardian*, 23 November 2001).

The dumping of metal mine tailings can have serious pollution impacts. When operating in the 1980s, 130,000 tonnes of metal-contaminated tailings per day were dumped into the adjacent river from the Panguna copper mine in Papua New Guinea. This polluted 30 km of river, destroyed all aquatic life and led to civil war. Some elements, for example, cadmium, mercury, antimony and arsenic are very common in small amounts in many polymetallic sulphide ores, but are toxic to living organisms. They may find their way into soils and surface water via mine tailings. An example of this occurred in the Stolberg area of western Germany where wastes from the lead–zinc ore mining in the Carboniferous and Devonian limestones, and the smelting of the ores, led to contamination of soils and grassland. Cattle deaths from lead poisoning in the 1970s brought fears that drinking water and foodstuffs might be contaminated. An extensive series of tests was carried out and various remedial measures put in place, including guidance on the cultivation of vegetables in the cadmium-contaminated soils (Fig. 4.3) (Aust & Sustrac, 1992).

Toxic chemicals such as mercury and cyanide are also used to extract metals such as gold, and can lead to major land or water pollution, for example, at Witswatersrand in South Africa (Evans, 1997), Caroni basin in Venezuela (Cutter & Renwick, 1999) and at several sites in Ghana (Hilson, 2002). A related problem is acid mine drainage resulting from oxidation of sulphide minerals (Blunden, 1985, 1991). Sulphuric acid may be generated, which in turn attacks other minerals to produce acidic water, polluted, for example, by cadmium or arsenic. The problems may only occur after mine closure when the water table rebounds following cessation of pumping (Bell *et al.*, 2002). Hydraulic mining, the washing out of rock with high-pressure water jets, can also be very damaging in silting rivers and lakes downstream or downslope. It is used to extract gold in countries such as Indonesia, the Philippines, Venezuela and Brazil.

Having examined the impacts of mineral extraction, it is important to restate the fact that mining frequently also brings geological benefits by creating exposures of rock and sediment that would otherwise be inaccessible. This often results in valuable evidence of the local stratigraphy, sedimentology or geomorphological origin. In these cases, continued quarrying using traditional techniques and carried out in collaboration with geological and conservation organisations makes available otherwise inaccessible rock or sediment sections. Many disused quarries have become important stratigraphical, mineralogical or palaeontological sites. In assessing mineral extraction proposals, it is important to make a balanced assessment on both the benefits and disbenefits of the proposals and to weigh the arguments that should always include the economic need and geological advantages of extraction as well as the landscape, hydrological and other impacts.

4.3   Landfill and Quarry Restoration

Waste management is a major problem in many parts of the world. The waste material of modern societies is of various types (e.g. domestic, industrial, construction, agricultural, quarry) and some of it may be highly toxic (e.g. hospital waste, radioactive waste, contaminated soils). In many countries, quarries are infilled with domestic, industrial or agricultural waste materials, but this may be a major threat to geodiversity since it obscures the geological interest that has been exposed by the quarrying process. An example is Webster's Clay Pit in Coventry, England, where Westphalian sandstones and mudstones representing the only available exposure of alluvial plain deposits within the Enville Formation were completely destroyed by landfilling in the 1990s. It was the best site in Britain for studying Walchia-like conifers of the Upper Palaeozoic but it is now buried by waste (Prosser, 2003). Similarly, landfilling of Kirkill Quarry in Aberdeenshire, Scotland, buried one of the most important Quaternary sites in Scotland and has prevented further research being carried out on the original exposures (Gordon & Leys, 2001b). Casual tipping of waste can also obscure exposures.

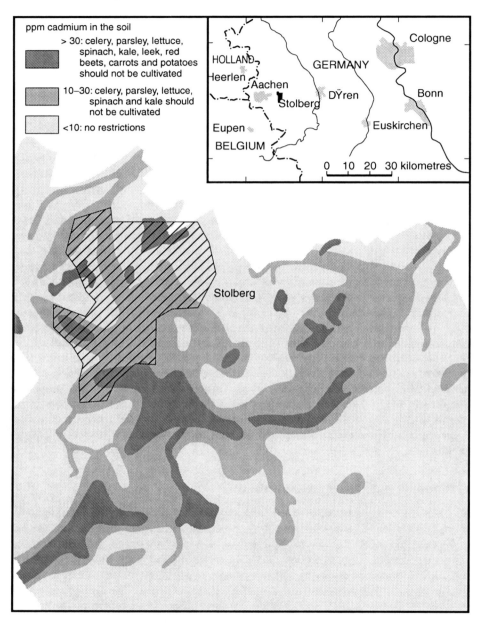

Figure 4.3 *Cadmium distribution in soils derived from mine dumps in the Stolberg area, Germany, and recommendations for the cultivation of vegetables (After Aust, H. & Sustrac, G. (1992) Impact of development on the geological environment. In Lumsden, G.I. (ed) Geology and the Environment in Western Europe. Oxford University Press, Oxford. Reprinted by permission of Oxford University Press)*

"But even when sections of the rock face are left uncovered, accumulation of leachate and landfill gases, slumping of the waste towards the face and difficulties of access often seriously interfere with research or education..." (Nature Conservancy Council, 1990, p. 35). Furthermore, in the absence or failure of a waste containment system, groundwater or surface-water contamination by leachate can occur and landfill gases, including methane and volatile organic compounds, are released to the atmosphere and may contaminate the soil-capping materials (Gray, 1996).

Overfilling of sites or creation of landraised sites will result in the creation of new landforms, many of which are out of character with the local topography (Gray, 1998b, 2002a; see Section 6.4.2). In other cases, natural valleys or depressions may be filled with waste with the resultant loss of landform, rock exposure and soils, and, therefore, loss of geodiversity.

4.4 Land Development and Urban Expansion

New building works can have large impacts on geodiversity by removing topsoil and damaging soil structure and soil biota, removing and re-profiling land surfaces thus leading to loss of landforms, sub-surface sediments and fossils, and obscuring the underlying rocks and sediments. Urban expansion and infilling in cities is leading to the accelerating erosion of undeveloped land, and this is resulting in loss of many important geological sites or semi-natural landscapes. At Gingko Petrified Forest State Park, Washington, USA, construction of a pipeline would have cut through the heart of the important fossil site known for the diversity of its 200 genera and species encased in lava flows. Fortunately, this threat was recognised and the plan was dropped (Gibbons & McDonald, 2001), but thousands of important geological sites have been lost through development.

Infrastructure development also often leads to major landscape and pedological changes, for example, the construction of road or railway cuttings, embankments, dams and reservoirs. Engineering solutions such as the use of gabions are often visually intrusive as is the formation of forest roads across mountain slopes. In the 1960s and 1970s, hundreds of kilometres of vehicular tracks were bulldozed in the Scottish Highlands to improve access to hunting areas and forestry plantations, resulting in unsightly hillside scars (Watson, 1984). Similarly in Iceland, Thorvardardottir & Thoroddsson, (1994, p. 228) describe how the development of hydroelectric and geothermal power plants has affected the landscape. "Even in the preparation stage of a power plant, off-road driving in connection with scientific work to find the best spot for electricity pylons, leaves tracks in sensitive areas....roads and powerlines create scars in the open wilderness landscape ..." In Taiwan, Wang *et al*. (1994, p. 114) describe the damage to the coastal landscape by expansion of the Suhua Highway. "Massive drilling and collapses of slopes have severely scarred the landscape". Kiernan (1996) describes the loss of part of an important section in Tasmania as a result of widening of the Lyell Highway, and Jungerius *et al*. (2002) describe the erosional problems associated with roads in Kenya: "however carefully the measures against erosion are designed, they become rapidly outdated because a new road attracts settlement".

However, as in the case of mineral extraction, road, railway and pipeline cuttings often reveal important permanent or temporary rock exposures (Davies & Pearce, 1993). Where "permanent" sections are left after completion of the project, subsequent maintenance of these rock cuttings can obscure the sections if inappropriate grading or stabilisation techniques such as covering with soil are used. Building development on quarry floors can obscure the faces left by quarrying, while barrages and marinas can threaten coastal rock exposures. Geological inputs to engineering design are therefore extremely important but not always invited.

Related effects may be disruption of local hydrological systems and pollution of watercourses. For example, the construction of low-permeability surfacing in urban areas (tarmac and concrete) leads to lower infiltration rates and faster runoff of storm discharge. In turn, this may lead to increases in flooding and channel erosion down-stream of an urban area, as well as increasing pollution from urban drainage (e.g. oil, salt and pesticides). Pollution from sewage can also occur. Silvestru (1994, p. 222) describes the impact of rural settlement on karst terrain in rural Romania. The karst absorbs all liquid discharges including sewage "so that practically all karstic aquifers in the area are likely to be polluted". Similarly, water extraction can threaten river and lake levels. The Everglades National Park in Florida, USA is threatened with drying up as water is abstracted for urban populations and crop irrigation. Hurlstone & Long (2000) report that of five major geyser fields active in New Zealand in the late 1800s, only Whakarewarewa survives. This is due mainly to hydroelectric developments, though the Rotomahana field was destroyed by a volcanic eruption. Of more than 200 gey-sers active in the 1950s, only 40 remain active today. At Whakarewarewa, there were previously seven large geysers but as a result of more than 800 commercial and domes-tic wells drilled to tap a cheap source of heat, three have stopped erupting (Hayward, 1989; Buckeridge, 1994; Hurlstone & Long, 2000). Similarly, some geologists fear that exploitation of the Island Park Known Geothermal Resource Area (IPKGRA) in Idaho, USA, could affect the famous geothermal activity at nearby Yellowstone National Park in Wyoming (see Box 5.1).

In 1998, the United Nations forecast an increase of more than 50% in world popula-tion between 1998 and 2050. This will almost inevitably significantly increase pressure on the world land resource and geodiversity and the ability to sustain social and eco-nomic development.

4.5   Coastal Erosion and Protection

Coastal erosion is a natural process, but can destroy the geological and geomorpho-logical integrity of sites if important features of limited extent are lost. An example is the loss of Quaternary periglacial structures (ice wedges and patterned ground) and storm surge deposits on the coast of Brittany, France (Regnauld *et al.*, 1998). But on the other hand, the erection of sea defences such as sea walls, cliff stabilisation, rock armouring, flood embankments or coastal slope regrading (Fig. 4.4) can adversely affect or destroy the geological and geomorphological interest of coastlines since "all are designed to counter the natural evolution of the coastline" (Lees, 1997, p. 3). Unfor-tunately, poorly informed, coastal decision-making can increase demands for coastal defence and stabilisation, for example, construction of homes very close to eroding coastlines or agricultural land uses downwind of mobile coastal dunes.

The impacts of coastal defences include

- loss of, or damage to, geological exposures, landforms and habitats;
- stabilisation of active coastal landforms and processes such as dune systems;
- interference with natural interchange of sand between beaches and dunes, reducing sand deposition inland;

Figure 4.4   *Example of a defended coast*

- increased erosion at the flanks of defences;
- reduction in input of sediment to the beach from an eroding coastline, which may exacerbate erosion downdrift.

There is also a loss of habitat since many specialised flora and fauna flourish on eroding coastal cliffs and intertidal zones (Lees, 1997).

   As Hooke (1998) points out, many important geoscience sites for research and teaching occur on the shoreline because of the exposure of geological strata by marine processes and the dynamic landforms created by these processes. "Continued availability of sites is essential to advance knowledge and understanding of both the past and the present" (Hooke, 1998, p. 2).

Even if not permanently destroyed, rock and fossil sites may be damaged by the works and are lost to research and education for the design life of the structures. There are many examples of loss of geological exposures by coastal defence works. At Corton in Suffolk, England, an important Quaternary site is now obscured because of coastal protection measures and related stabilisation of the cliffs. Another example is at Barton-on-Sea in Hampshire, England, where a proposal to extend the coastal defences and drain a slope threatened to severely compromise the integrity of the international Bartonian stratotype (Doyle, 1989). At Burnie in Tasmania, Australia, coastal reclamation has covered natural shore platform exposures of Precambrian dolerite dykes that were amongst the earliest "Geological Monuments" in Australia (C. Sharples, personal communication).

"Coastal defences can isolate landforms such as shingle bars, beaches, saltmarshes and mud flats from the sediment supply that feeds and maintains them, often causing erosion of these features" (Glasser, 2001, p. 893). There is, therefore, often a case for allowing erosion to continue, at least at a slow rate, in order to maintain the exposure. Sea defences such as flood embankments also fix the position of the coastline and may result in "coastal squeeze" on the retreating saltmarsh areas seaward of the embankments. Clearly, any interference with sand dunes will alter the morphology of the dunes as well as affecting the stability and processes affecting the dunes. English Nature has published a report highlighting the degraded condition of the coastline of England, as affected, for example, by coastal defences, port development and recreational pressures (Covey & Laffoley, 2002).

4.6   River Management, Hydrology and Engineering

The human impact on rivers has been extensive, and these have come about from both engineering within the channel and floodplain and from wider land-use changes. For example, deforestation within a drainage basin may increase overland flow and hence the potential erosion and transfer of sediment to the channel. Prosser *et al.* (2001) calculated that at least 127 million tonnes of sediment is delivered to Australia's streams and rivers, and most of this is the result of extensive land clearance, sheep and cattle grazing and mining activities over the last 200 years. The result is higher water turbidity, lower water quality and sand slugs on the river bed that raise bed levels, increase the risk of flooding and smother aquatic habitats (Bartley & Rutherfurd, 2001).

Clifford (2001) classifies the traditional direct engineering impacts on rivers into

- flood prevention and mitigation;
- channel stabilisation;
- flow regulation for navigation;
- water supply and quality;
- effluent disposal.

Many natural river channels have been altered in the past, often as part of flood protection or river management schemes. Rivers have been channelised, straightened, embanked, dammed, diverted, culverted, dredged and isolated from their floodplains

Figure 4.5   *Example of a channelised river between two roads*

(Brookes, 1985, 1986, 1988; Graf, 1992, 1996). In turn, these works have changed the river characteristics including channel roughness and river dynamics, as well as the natural bank, river bed and floodplain habitats (Hooke, 1994; Soulsby & Boon, 2001; Richards *et al.*, 2002). As Clifford (2001, p. 72) puts it, "Channelisation degrades habitat ... by removing bed and bank sediments and vegetation and by altering flow dynamics" (Fig. 4.5, Box 4.4).

But perhaps the most radical impacts are from dam construction where there is wholesale impounding and regulation of flow (Graf, 1996; Clifford, 2001). According to Goudie (2000), engineers had built more than 37,000 major dams (>15 m) around the world by 1990. On a large scale, China's Three Gorges Dam on the Yangtze River is certainly changing the downstream flow regime of the river as well as raising water levels in the gorges themselves and "will forever transform one of the most spectacular river valleys in the world from a natural-flowing stream into a reservoir. . . ." (Cutter & Renwick, 1999, p. 339). The dam is 185 m high, 1.9 km long and impounds a lake 600 km long. Three major dams on the River Indus in Pakistan have led to water deficiencies in the lower reaches of the river and salinisation of soils. It was the flooding of Lake Pedder in Tasmania, Australia (Fig. 4.13) for a hydroelectric power scheme on the Gordon and Serpentine Rivers that is widely regarded as having initiated environmental politics in Australia as well as the awakening of interest in geodiversity issues in Tasmania (C. Shaples, personal communication).

Beavis and Lewis (2001) noted that agricultural expansion and the need for "drought-proofing" led to the building of 300 large dams in Australia between 1940 and 1983, but even more significant effects come from the construction of farm dams of which there are estimated to be nearly half a million in Australia (Beavis and Lewis, 2001). These have a major impact on catchment hydrology and Australian water resources. The "tank" landscape of south-east India is a landscape dominated by small earth dams collecting water from myriads of little streams and areas of overland flow (Spate & Learmonth, 1967; Goudie, 2000).

Interbasin water transfers are also associated with major changes to geomorphological and hydrological systems. For example, the Snowy Mountains Hydro-Electric Scheme in Australia consists of 16 large dams, many smaller diversion structures and hundreds of kilometres of tunnels, aqueducts and pipelines (Pigram, 2000). "Understandably, an undertaking of this magnitude has had far-reaching impacts on the landscape, both in the upland areas immediately affected by construction works and in areas downstream. Perhaps the greatest impact has been on the Snowy River itself." In reaches of the river downstream from the scheme, there are marked deteriorations of streamflow, riparian vegetation and the river channel itself (Pigram, 2000, pp. 343–344). Despite this, major water transfer schemes are still being planned. For example, Spain's National Hydrological Plan includes a scheme to divert water from the Ebro northwards to Barcelona but mainly southwards to the coastal areas of Valencia, Murcia and Andalucia where the tourist complexes, swimming pools, golf courses and vegetable growing consume huge quantities of water. Even bigger is the engineering project recently authorised by the Chinese government to transfer 48 trillion litres of water per year from the Yangtze River to the drier northern provinces including Beijing via three new channels up to 500 km long (*The Guardian*, 27 November, 2002). Previous river diversions in the former Soviet Union have caused the drying of the Aral Sea with profound effects on geomorphological, biological and social systems.

Hughes & Rood (2001, p. 106) see floodplains as among the most abused of ecosystems because of competing interests for resources. Dam construction, channelisation, water removal, gravel and sand extraction and forest clearance all affect the timing and quantity of water and sediment delivery to floodplains". A further issue concerns the impact of river boat wakes on bank erosion, an example being the important Gordon River site in Tasmania, Australia (Bradbury *et al.*, 1995), where natural levees were being eroded at up to 1 metre per year. Some of these changes are reversible by natural or human restoration, but others such as the incorporation of mine tailings into river sediments are more intractable (Balogh, 1997).

Box 4.4   *River Dane, Cheshire, England*

J. Hooke (1994, p. 115) recounts how in 1982 several kilometres of the channel of the River Dane in Cheshire, England, were dredged by dragline to clear out riffles and bars. The material was dumped on the banks. "The purpose of these works was supposedly to increase evacuation of flood waters from the area.... (but) it increased flood peaks at the next urban area downstream, and it was a waste of money because within two years the riffles and bars had reformed". She also believes that "insertion of bank protection is only likely to shift erosion elsewhere and is unlikely to be successful in the long term. The danger is that piecemeal protection is put in, then more added until an artificial, immobile channel is created. Not only are such channels unattractive and represent a loss of habitats but they may cause further problems in the longer term because of the lack of adjustability of the channel".

Water abstraction from rivers or groundwater can also alter river discharge and therefore channel processes and lake levels. For example, Valiūnas (1994) expresses concern about the effect of water abstraction and gravel quarrying at the Trakai Historical National Park in Lithuania, because of the impact on lake levels in the Trakai Lakes, which are a central part of the park's landscape (Fig. 4.6).

The digging of drainage ditches will often alter the geomorphological and hydrological character of the area, for example, by introducing rectilinear drainage lines and accelerating runoff and hence erosion. Daly (1994) gives an example from central Ireland where a raised bog of international importance, traversed by a road, was drained. As a consequence, the peat dried out and its thickness decreased from 12 m to 6 m so that the bog is no longer functioning properly in this area. Drainage of upland peats may result in drying and erosion of peats with resultant loss of palynological records. Bogs can also be affected by water abstraction or arterial drainage at a distance.

4.7   Forestry, Vegetation Growth and Removal

4.7.1   Afforestation

The impacts of afforestation and planting relate partly to potential obscuring of landforms or rocks by blanket woodland so that individual features and landform associations are masked by trees or other planting. The Nature Conservancy Council (1990, p. 40) believed that "Large-scale conifer plantations render location, mapping and interrelation of outcrops virtually impossible" and also encourage lichen growth on outcrops or bury them in leaf litter. Natural vegetation growth may also lead to loss of visibility. For example, the columnar jointing in a young lava flow in the Organ Pipes National Park near Melbourne, Australia, is now largely hidden behind rapidly growing plantings and native vegetation (Joyce, 1999). Larwood (2003) illustrates by photographs how vegetation growth has obscured the visual impact of the geology of Wren's Nest Nature Reserve in the West Midlands, England. In the Yorkshire Wolds, England, English Nature (1998) sees the afforestation of karst valleys as undesirable and in Breckland, England, afforestation or scrub and tree encroachment has lead to the obscuring of fossil pingos. Careful tree removal has been undertaken to re-expose the features whilst avoiding damage to the pingo ramparts.

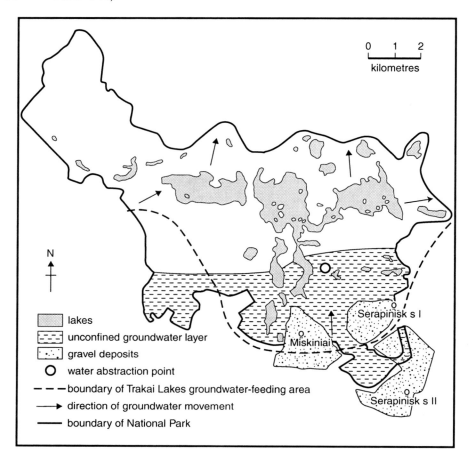

Figure 4.6 *Hydrology of the Trakai Historical National Park, Lithuania (After Valiūnas, J. (1994) Environmental geology maps for national parks and geomorphological reserves in Lithuania. In O'Halloran, D., Green, C., Harley, M., Stanley, M. & Knill, J. (eds) Geological and Landscape Conservation Geological Society, London, 273–277, by permission of The Geological Society)*

   Accessibility of landforms and rock exposures or sediment sections is also likely to be affected by blanket planting, while more limited planting or natural vegetation growth can reduce the visual continuity of landforms or rock exposures. Outcrops on open land or in quarries can quickly become degraded and overgrown by scrub vegetation unless site clearance is regularly undertaken. However, we must also acknowledge the positive role of tree planting in reducing carbon concentrations in the atmosphere or stabilising slopes, sand dunes, and so on.

   Root damage to sensitive geological sections is not unknown. It is therefore important to try to locate new forests and woodlands in less sensitive locations in this respect. For example, in the 1970s, there were plans by the UK Forestry Commission to plant new forest in Glen Roy, site of the famous "Parallel Roads", interpreted as glacial lake shorelines. The shorelines, which are today continuously visible for several kilometres, would have been totally obscured by this plan, which fortunately was abandoned when the impacts were pointed out by the earth science community. Nowadays, much greater account is taken of earth science interests in planning planting schemes in the United Kingdom.

Soil chemistry can also be affected by afforestation, though this will depend on the species planted. Conifers and eucalyptus can induce soil acidification with a decrease in exchangeable cations and increase in aluminium (Aust & Sustrac, 1992).

4.7.2 Deforestation

Deforestation or logging operations may also have major impacts on soils and other earth science interests. Trees and their leaf litter generally have the role of intercepting rainfall and reducing the impact of raindrops on the soil so that erosion rates in forests tend to be low. The removal of trees therefore often results in soil erosion by overland flow, particularly from sloping terrain in high rainfall areas (Douglas *et al.*, 1995; Goudie, 2000). Spectacular examples of gulley erosion following deforestation are found in areas of tropical rain forest areas such as Madagascar. Silvestru (1994, p. 223) describes the effects of local populations felling trees on the karst terrain of rural Romania. "If the present rate of cutting is maintained... the soil will rapidly vanish, leaving a stony desert which would eventually exacerbate the karstification process". Silvestru describes an area in northern Romania where a whole mountain slope has been clear cut for commercial purposes. "... the soil immediately starting to move down the 35° slope, exposing bare limestone to weathering. Moreover, since the limestone dip roughly corresponds to the slope angle, rock slides occurred within four years (after the cutting), because of intense karstification". Valiūnas (1994, p. 275) urges caution about deforestation of fossil sand dunes in the Dzūkija National Park in Lithuania "otherwise there is a danger of renewed deflation, soil degradation, etc."

Commercial felling operations involving mechanical equipment and uprooting of the trees have a major effect in compacting and disrupting the soil. Brady & Weil (2002, p. 141) believe that such practices "can impair soil ecosystem functions for many years. Timber harvest practices that can reduce such damage to forest soils include selective cutting, use of flexible-track vehicles and overhead cable transport of logs, and abstaining from harvest during wet conditions". Kiernan (1991, 1996) describes the devegetated and eroded condition of the soils and landforms in the central part of Tasmania's West Coast Range caused by commercial timber cutting as well as severe air pollution from copper smelting.

Replacement planting operations are also potentially damaging if large-scale ploughing triggers further erosion and silting. It is easy to disrupt the soil structure, alter soil chemistry or change soil biota, and there may be serious effects in areas where important subtle landforms occur.

4.8 Agriculture

In general terms, there is likely to be little damage to geodiversity in areas that have had a long and successful history of cultivation. Farmers who have rotated crops, retained fertility and managed the land to prevent soil erosion have been able to conserve and sustain the soil resource. In some cases, however, particular problems may arise. For example, regular ploughing of steep slopes can affect the physical integrity of the slopes and landforms by increasing erosion and downslope movement of soil. The effects are particularly serious under wet soil conditions. In areas where the landforms are small scale and intricate (e.g. low sand dunes or abandoned channels on river terraces), ploughing can reduce the relief and landform detail. In Breckland, England, much fossil periglacial patterned ground has been ploughed out this century (English Nature, 1998). In Northumberland, England, the wintering of cattle on dune areas has led to damage to the dune vegetation thus increasing the vulnerability to wind erosion. In fact, about a third of the arable area of England and Wales is thought to be at risk from wind erosion, with drained peat soils and Quaternary sands being most at risk. Similar threats have been detected in other areas, for example, on the coastline of north Germany (Aust & Sustrac, 1992). Stocking levels in semi-arid areas can have serious impacts on vegetation cover and thus soil and gulley erosion potential and delivery of sediment to streams (Caitcheon *et al.*, 2001).

But the biggest threats to geodiversity have come from unsustainable land management and the history of land clearance and cultivation is littered with examples of the degradation of soil and other resources (see Box 4.5). Bringing areas into agriculture that have not previously been cultivated will alter many of the soil properties (e.g. soil compaction, organic matter content and distribution, soil biota, soil chemistry from fertilizer or pesticide application) and reduce natural soil diversity. This is particularly true if intensive agriculture is practised year after year, resulting in loss of soil structure and leaching of nutrients. Soil compaction has become an increasing problem as bigger and heavier farm machinery has been introduced. Subsoil compaction may occur below the plough layer, producing a "plough-pan". "Nearer the surface, a combination of compaction and tractor wheel-spin may limit the extent to which plant roots can exploit the soil for moisture and nutrients and the free movement of gases between soil and the atmosphere" (Bridges, 1994, p. 13). Compaction decreases infiltration capacity and porosity, resulting in faster saturation and overland flow. In turn, soil erosion by wind and water, including transport of pesticide residues, siltation of river courses and increased flooding may result. Figure 4.7 shows soil erosion in the United States based on US Department of Agriculture data (Cutter & Renwick, 1999). In the United Kingdom, Boardman (1988) related increases in soil erosion in the 1980s to increased growing of winter cereals and consequent expansion of bare ground in the autumn and winter when rainfall is highest.

Box 4.5   *Agricultural Impacts Around Mexico City and the Great Plains, USA*

Leopold (1959) recorded the sequence of events that led to the degraded land and denuded soils in the foothills north of Mexico City (see also Dasmann, 1984). Early descriptions described a tall, open forest of pine and oak in the area, but as the population grew the demand for farm land increased and extended farther up the slopes of the foothills. The forests were cleared and turned into charcoal, and the land was then planted with corn or wheat. However, rainfall on the sloping ground of the foothills quickly led to soil degradation resulting from topsoil erosion and mineral leaching. In the end, the soil could no longer support a viable corn crop, but maguey, a cactus-like plant grown for its fibres, could be cultivated on the impoverished ground. The open nature of this plant led to further erosion and gulleying until only an impervious hardpan remained, incapable of supporting

Continued on page 155

_____ *Continued from page 154* _____

the maguey. The remaining scant cover of a few desert-tolerant plants could only support the grazing of goats and donkeys until they too were gone and only a wasteland was left. "The story illustrates a process that has been repeated again and again throughout the world, an attempt to find food. . . .resulting in the degradation of the basic and irreplaceable natural resource on which terrestrial life depends" (Dasmann, 1984, p. 179).

Another famous example was "The Dust Bowl" of central USA. In the 1880s, the first settlers began to move into the Great Plains of the United States, an area prone to great fluctuations in rainfall. The natural vegetation was short grassland and droughts did little lasting damage to the underlying soils. However, when the brown soils were cultivated for wheat, they became susceptible to wind erosion under drought conditions. The most serious erosion occurred in 1933 when severe drought brought dust storms that eroded a topsoil that had lost much of its structure due to ploughing and overgrazing. A large area on the borders of Kansas, Oklahoma, Texas and Colorado was affected by a loss of topsoil of between 50 and 300 mm. The area became known as "The Dust Bowl" and was affected by drifting dunes, which buried roads, fences, dwellings and crops in thick deposits of redeposited dust (Dasmann, 1984). Whitney (1994, Chapter 10) describes this as "the most rapid rate of wasteful land use in the history of the world" and uses phrases like "earth butchery" and "predatory agriculture".

The most recent example is occurring in northern China where dust clouds testify to wind erosion of soil brought about by overcultivation, overgrazing, overcutting and overpumping of marginal land. This "Chinese Dust Bowl" is seriously threatening Chinese food supplies and higher world grain prices.

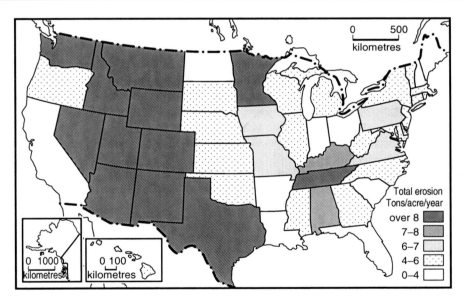

Figure 4.7 *Soil erosion rates (1992) in the United States (After Cutter, S.L. & Renwick, W.H. (1999) Exploitation, Conservation, Preservation: A Geographic Perspective on Natural Resource Use. 3rd ed. Wiley, Chichester. Reproduced with permission from Wiley)*

There has been much debate as to whether soil should be regarded as a renewable or non-renewable resource. Although the organic content of soil, derived from dead and living plants and animals, is at least partially replaceable relatively quickly by pioneer vegetation, leaf falls, colonisation by soil fauna and manuring, the mineral components may take thousands or millions of years to weather from the parent rock. Thus, it is important that soil is regarded as a non-renewable resource in which soil erosion losses are unsustainable even with the addition of fertilisers (Trudgill, 2001).

Ploughing may also mean some loss of surface rock exposures and reduce the accessibility of sites. Off-site effects may include chemical pollution of watercourses or groundwater, more rapid runoff and episodic soil erosion.

The application of fertilizers and pesticides to land has already been referred to and can be very beneficial to the quantity and quality of crop yields. However, residues may remain in the soil and excess amounts may be flushed and leached into surface or groundwaters, thus leading to nitrate and other pollution and water eutrophication. Irrigation can also impact on soil by leaching nutrients, changing groundwater levels and increasing salinity if the water used has a high salt content. Salinity levels can also rise as a result of overabstraction of groundwater in coastal areas. An example is the Llobregat delta near Barcelona in Spain (Goudie, 2000). Irrigation near the Hagerman Fossil Beds in Idaho, USA, has added water to the unconsolidated sediments of the Glenns Ferry Formation, resulting in an increase in frequency and magnitude of landslides and threatening the exposure of the fossil beds.

O'Halloran (1990), Hardwick & Gunn (1994), Gunn (1995) and Eberhard & Houshold (2001) summarise research on the impact of agricultural operations on limestone cave and karst systems. The potential impacts are illustrated in Fig. 4.8 and include the following:

- Farmyard runoff including oil, organic slurry and animal dips can drain into cave systems, causing cave and groundwater pollution.
- Tipping of farm waste, including animal carcasses, may block cave entrances, infill dolines and lead to underground pollution.
- Digging drainage ditches increases peak discharges and may flood caves and erode cave sediments and speleothems. New drainage routes may be opened with consequent impacts on both new and old drainage ways.
- Spoil and sediment from surface excavations, ploughing and erosion may be washed into cave systems.
- The use of agrochemicals (fertilisers, pesticides, etc.) on thin soils overlying porous and permeable rocks often leads to pollution of speleothems, watercourses and cave sediments.

Buffer zones around cave sites have been suggested and are useful, but changes in the whole catchment area can affect caves and cave water in both obvious and more subtle ways. Hardwick & Gunn (1994) describe impacts of ploughing on cave systems in the Castleton area, England. Clastic sediment input may infill cave entrances and internal passages, bury speleothems, generate back flooding and loss or damage to cave features, alter cave microclimates and transport agrochemicals into underground environments.

Changes in agricultural practices can also have an effect on cave hydrology. For example, Gunn *et al.* (1994) found that changes in agricultural grants and subsidies allowed farmers in Fermanagh, northern Ireland, to build access tracks into the hills, undertake drainage works and operate at higher stocking densities. This has increased runoff, leading to accelerated erosion and downstream deposition.

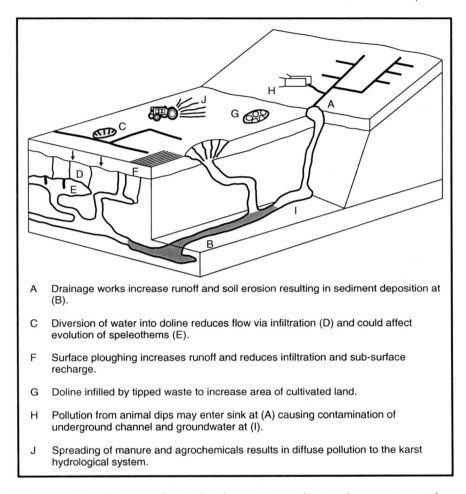

A Drainage works increase runoff and soil erosion resulting in sediment deposition at (B).

C Diversion of water into doline reduces flow via infiltration (D) and could affect evolution of speleothems (E).

F Surface ploughing increases runoff and reduces infiltration and sub-surface recharge.

G Doline infilled by tipped waste to increase area of cultivated land.

H Pollution from animal dips may enter sink at (A) causing contamination of underground channel and groundwater at (I).

J Spreading of manure and agrochemicals results in diffuse pollution to the karst hydrological system.

Figure 4.8 *Potential impacts of agricultural operations on karst and cave systems (After O'Halloran, D. (1990) Caves and agriculture: an impact study. Earth Science Conservation,* **28***, 21–23, by permission of Earth Heritage)*

One of the major impacts of agriculture on the landscape has been the construction of terraces. The function of these is to produce flatter areas of land even on steep slopes, to reduce soil erosion and, in the case of rice cultivation, to retain water. Although some spectacular landscapes have been produced (Fig. 4.9), the result is a loss of the natural landform and change in the soils. Bayfield (2001, pp. 20–21) notes that "Terraces represent a major disruption of the original soils and have a massive landscape impact".

4.9 Other Land Management Changes (e.g. Cutting, Filling, Dumping, Spreading or Discharging Materials)

These activities can have very serious effects on many parts of the physical (and biological) environment. For example, the excavation of materials to create agricultural

(a)

(b)

Figure 4.9 *Impacts of agricultural terracing: (a) vine terraces, Douro Valley, Portugal and (b) rice terraces, Bali, Indonesia*

reservoirs or fishing lakes often results in the dumping or spreading of the subsoil on adjacent land, particularly given transport and other costs or regulations involved in removing the material. However, these processes, in badly managed situations, can lead to soil burial and creation of new, artificial landforms. Similar effects can result from unconsidered disposal of navigation channel dredgings.

The dumping of solid materials can lead to infilling of natural hollows such as kettle holes and may obscure important rock, mineral and fossil exposures. Infilling of ponds and marshes leads to loss of geomorphological and hydrological character and, therefore, habitat. Some geomorphological and biological landscape richness is therefore lost. Significant land remodelling, as often occurs in golf course construction, will also change the local geomorphological character and soil conditions. In sand dune areas such as the Lancashire coast, England, or East Lothian coast, Scotland, golf course construction usually leads to loss of natural dune landscapes (see Section 6.4).

The dumping of quarry and mine wastes has had an important effect on some landscapes, such as the coastline of Durham, England, where waste was tipped over the cliffs, for example, at Easington and Horden. This altered the coastal topography and beach sediments to the detriment of earth science and aesthetic interests (see Section 6.3.3). Deliberate land reclamation will also result in changes to the coastal morphology and can have detrimental effects on shoreline processes by altering the coastal dynamics. The largest-scale dumping has been in the Dutch polders where land reclamation from the sea has provided new agricultural land and flood control measures but has resulted in changes in groundwater characteristics, ground subsidence and foundation damage in adjacent towns (Aust & Sustrac, 1992).

Spreading of manures, sewage sludge or other waste liquids will often provide valuable nutrients and organic matter and reduce the need for chemical fertilizers and disposal sites. However, they may also have adverse effects, such as

- environmental pollution by nutrients within the organic materials;
- environmental pollution by contaminants, including heavy metals (e.g. cadmium, mercury, lead, zinc, chromium, nickel, copper);
- disease risk from contaminants or microbes within the materials;
- physical damage to the soil, such as compaction and waterlogging;
- reduction in soil biodiversity;
- short-term odours and visual impact (DEFRA, 2001).

According to Bridges (1994, p. 13), "Use of phosphatic fertilizers containing cadmium has also been responsible for raising the concentration of this metal in arable lands". Up to a certain threshold, the soil can absorb some of chemical pollutants without harm but "the soil's capacity to store these toxic metals is not infinite. . . . if the pH of the soil is decreased by acidification. . . .. it is possible for some metallic elements to be catastrophically released in what has become known as a 'Chemical Time Bomb" ' (Bridges, 1994, p. 13).

4.10   Recreation/Tourism Pressures

Recreational and tourism is an increasingly important economic activity but one that can lead to damage to biodiversity and geodiversity.

Miller (1998, p. 404) quotes a 12-fold increase in US National Parks between 1950 and 1990 and there has been a further rapid increase since then.

"Under the onslaught of people during the peak summer season, the most popular national and state parks are often overcrowded with cars and trailers and are plagued by noise, traffic jams, litter, vandalism, deteriorating trails, polluted water, drugs and crime. . ... Park Service rangers now spend an increasing amount of their time on law enforcement instead of on resource conservation and management".

It is not surprising that Hunt (1988) expressed the view that in the United States, there is a danger of people "loving their national parks and historic sites to death". While not all of these impacts will directly threaten geodiversity, some of them will. For example, footpath erosion initially tramples the vegetation but ultimately exposes the soil and subsoil to erosion and can lead to significant gulleying (Bayfield, 2001). Dovedale in the English Peak District National Park receives up to a million visitors every year, including many school and university groups. However, the impact of these is leading to the destabilisation of scree slopes and significant footpath erosion. This potential threat and actual damage to the landscape in rural areas has increased with the use of mountain bikes, motorbikes and all-terrain vehicles (ATVs), which often give rise to soil compaction, erosion or gulleying on slopes.

Some areas are particularly vulnerable to recreational pressures including sand dune areas (Fig. 4.10) where worn footpaths and ATVs can instigate serious wind erosion and sand movement (ASH Consulting Group, 1994; Scottish Natural Heritage, 2000c; Catto, 2002). Recreational pressures on beaches can affect the build-up of sand on the dune front and change the natural processes of dune formation and renewal. Beach cleansing can also destroy embryo sand dunes before they can fully develop.

Thorvardardottir & Thoroddsson (1994, p. 229) believe that in Iceland, one of the major tourist impacts on volcanic landscapes is the creation of multiple paths on lava cones. "This is especially striking where mosses have covered the crater's slope, but

Figure 4.10 *Vehicle access to sand dune areas can initiate significant erosion, Donegal Bay, Ireland*

trampling breaks the moss cover, leaving scars open to erosion. Tephra formations and some lava formations are so fragile that they disintegrate underfoot". Off-road driving creates even greater problems: ". . .it seems that tourists, Icelanders and foreigners alike, cannot resist the temptation to try out their four-wheel drive vehicles and their ability to drive, or rather to spin their wheels, on slopes of volcanic craters", although steps have recently been taken to encourage more responsible driving attitudes (Thorvardardottir & Thoroddsson, 1994, p. 228).

Another volcanic landscape affected by pedestrian access is brittle lava, which may be broken underfoot. An example occurs at Craters of the Moon National Monument in Idaho, the only historical lava flows in the lower 48 states of the United States. Here, the glazed black lava crust cracks under visitor weight to reveal an orange lava beneath (see Fig. 4.11). Not far away at Yellowstone in Wyoming, visitors have been found throwing coins, branches, tyres, clothes, and so on into geysers, which can block the geothermal pipes and lead to explosive and damaging events when the pipes are cleared.

Cave systems are another geological environment sensitive to visitor pressure with caving, leading to damage to the speleothems or floor deposits (Wright & Price, 1994; Hardwick & Gunn, 1994). "As soon as a new cave is discovered, there is a gradual, often insidious, degradation of the cave. The rate and degree of this degradation depends on the fragility of the features contained and the care of the cavers involved" (Wright & Price, 1994, p. 196). Touch, breath and light introduced by cavern visitors can encourage algal growth. There are also external threats to caves resulting from human-induced hydrological changes, which may cause erosion or degradation of underground systems (see above).

Rock climbing, canyoning and so on can also damage important geological sections for "as geotourism grows so does the pressure placed on the resource" (Larwood & Prosser, 1998, p. 98). For example, climbers have been ascending Devil's Tower, Wyoming, USA, for over 100 years. Nowadays, over 5000 climbers per year use the

Figure 4.11 *Perceived threats from pedestrian impacts on brittle lava, Craters of the Moon, Idaho, USA*

more than 200 routes identified for scaling the massive columns with the result that the geological integrity of the tower has been affected (Fig. 4.12 and see Section 5.8.1). Particularly damaging has been the permanent placement of bolts and pitons. On the other hand, tourism and outdoor leisure pursuits bring people into contact with natural or semi-natural environment and may lead to a more appreciative public willing to support efforts to reduce the threats and conserve the resource.

Finally, "High-mountain environments are particularly sensitive to the disturbances caused by recreation use. Steep topography, thin soils, sparse vegetation, short growing

Figure 4.12 *Climbers on the Devil's Tower, Wyoming, USA, have damaged the rock through use of bolts and pitons*

seasons and climatic extremes (e.g. heavy precipitation, cold temperatures, high winds) all contribute to the sensitivity of high-mountain environments. Under such conditions, it takes relatively little use to create long-lasting impacts" (Parsons, 2002, p. 363). Impacts of hiking and camping include construction of campgrounds, soil compaction and erosion, inadequate disposal of human waste, blackening of rocks from campfires and movement of rocks to build fireplaces and wind breaks. Parsons (2002) reports that management proposals to restrict recreational use in urban-proximate wilderness in Oregon and Washington, USA, have led to conflicts between hiking groups and conservation groups.

A particular issue in mountain environment is the development of skiing areas. Removal of boulders and smoothing of slopes may impact on geological, geomorphological and pedological features (Bayfield, 2001). Pfeffer (2003, p. 24) states that "in areas above the timber line, geomorphologic features, for example, morainic arcs, rock glaciers or outcropping bedrock, are the main targets to be destroyed in the landscape". She gives the examples of the removal of 50% of an important frontal moraine of the "Gepatsch" Glacier in the Austrian Tyrol by road construction to a skiing centre, and the blasting of a well-developed rock glacier at Sölden to create a steep ski run. "In both cases, evidence of geomorphologic evolution has been destroyed in favour of skiing facilities" (Pfeffer, 2003, p. 25).

4.11   Removal of Geological Specimens

Collecting appears to meet an innate human need. Casual removal of coloured rocks or pebbles from rivers and foreshores may seem inconsequential, but over long timescales depletes the resource. At Yellowstone, USA, at least one petrified tree has been completely lost through gradual removal of pieces by visitors (see Fig. 5.30), and at Fossil Cycad National Monument in South Dakota, USA, 35 years of neglect, unauthorised fossil collecting and unchallenged research collecting led to the near loss of the resource and its removal as a National Monument in 1957 (Santucci & Hughes, 1998).

Fossils, minerals and rocks are often aesthetic specimens and have long attracted collectors in the same way, though not perhaps to the same extent, as some of the Victorian wildlife collectors of, for example, butterflies or birds' eggs. Gemstones and minerals have long had a commercial value, but fossils have acquired significant economic worth only relatively recently (see Section 3.5.4). As Norman (1994, p. 63) points out, "Attaching a monetary value to a natural heritage item such as a fossil raises a spectrum of conflicting assessments. . .". Nowadays, there is even the risk of fossils, including type specimens, being stolen from museums or being forged (V. Santucci, personal communication).

The removal of rock, fossil and mineral specimens is not regarded as significant where the geological resource is extensive (Alcalá & Morales, 1994; Forster, 1999; King & Larwood, 2001). It is also recognised that many fossils are more valuable out of the rock than *in situ*, particularly where they may be vulnerable to erosion, quarrying, burial or other loss. In such cases, proper scientific field recording, curation, housing and accessibility in a suitable museum are essential.

However, there can be serious effects on geodiversity where the resource is extremely limited (e.g. within a cave or river channel deposit) or where there are very rare or scientifically valuable specimens (e.g. dinosaur bones or eggs). Some sites that are spatially extensive and contain an abundance of common fossils can nevertheless still be vulnerable if they also contain a few exceedingly rare fossils. Sites like these are vulnerable to irresponsible collecting and it is important that specimens are not removed from sites without the recording of crucial information including stratigraphic position (MacFadyen, 1999). And fossils can be lost from the public domain by sale to private collections or abroad (Norman, 1994; MacFadyen 2001a).

Mechanical excavators, explosives, power tools and sledgehammers have all been used to remove fossil material but have resulted in major damage to important exposures. In general, this sort of collecting leads to no scientific or educational gain and is therefore unacceptable. Instead, large amounts of rock are removed in pursuit of the rare or "perfect" specimen, resulting in loss of other, often important, fossils (Horner & Dobb, 1997; MacFadyen, 2001a). This has a major effect on our fossil heritage and is therefore to be condemned by all responsible geologists. For example, MacFadyen (1999) quotes the example of the amphibian fossil-bearing Cheese Bay "Shrimp Bed" near North Berwick in Scotland, half of which was removed illegally in a matter of hours by a collector, using a mechanical digger. Another Scottish site, Birk Knowes, known for its arthropods and early fish, attracted ruthless fossil collectors during the 1990s. Owing to the remoteness of the site, they were able to overcollect from the limited exposures causing huge damage and loss of scientific information (MacFadyen, 2001a). Sixty-five percent of the *Caloceras* ammonite beds at Doniford Bay in Somerset, England have been destroyed by irresponsible collecting (Webber, 2001). Horner and Dobb (1997) make the point that commercial collectors do not have the motivation or time to map, extract and catalogue their finds in a rigorous manner. They "cannot conduct the sophisticated analyses that enable us to identify exactly what kind of depositional environment the sediments represent... or the geological events responsible for the formation of the environments. ... when the desire to make money is paramount there's no incentive for conducting scientifically sound excavations" (Horner & Dobb, 1997, pp. 241–242).

Other examples of overcollecting of minerals or fossils that have led to the total destruction of sites can be quoted, and these can also be regarded as the equivalent of biological extinctions (see Section 7.3.3). Swart (1994, p. 321) believes that overcollecting at the world famous Ediacara Fossil Reserve in South Australia has lead to the virtual destruction of the site. Ironically, the establishment of the Reserve, intended to give the site protection, may actually have had the opposite effect by drawing attention to its importance. Collecting is only allowed at the site with permission, but its remoteness means that there is no way of monitoring visitors or activities. In Russia, some important ammonite sites have been mechanically excavated and the material shipped abroad under the label "of no scientific value" (Karis, 2002). Amateur collecting has threatened several important dinosaur egg sites in Aix en Provence, France (Gomez, 1991). Murphy (2001, p. 14) quotes the case of the Hope's Nose unique gold-bearing limestone-hosted veins that were very limited in extent. "Intensive and unconsented collecting, using heavy-duty rock saws, has removed the veins and the site is now effectively destroyed". Some mine spoil can be of geological importance, for example, for supporting rare minerals or fossils as at Writhlington in England (Jarzembowski, 1989). The indiscriminate depletion of these spoil heaps for minerals or use in track construction or fill can be a threat to this resource. A survey of fossil theft and vandalism in the US National Parks between 1995 and 1998, discovered that there had been over 1400 reported incidents, over 500 citations and 15 arrests, resulting in fines of $150,000 (Santucci, 1999).

Even research and educational activities can damage sites. Bayfield (2001) refers to an outcrop of serpentine that was entirely removed by successive visits by geology students (Speight, 1973) and this is not an isolated case. The Upper Ordovician Sholeshook Limestone Quarry in South Wales is one of many sites to have been "ruined by endless student parties" (Clarkson, 2001, p. 17). Even where sites are not destroyed, the impact of student hammering on rock outcrops can produce unsightly fresh scars. Toghill (1972, p. 514) believed that Shropshire, England "is really no place for elementary geology students" who could be taken to less sensitive coastal exposures. Impacts of palaeomagnetic coring can also be significant. Campbell and Wood (2002) describe how a valuable glacially striated rock surface in Wales first described by Sir Andrew Ramsay in 1860 has been peppered with core holes. Similar impacts have raised concerns in Tasmania so that special measures are taken to manage coring and to restore core holes (M. Pemberton, personal communication, see Box 5.4).

The concept of landscape sensitivity can therefore be extended to fossil sites. Mac-Fadyen (1999) has classified site vulnerability as follows:

- *Robust* – sites where the fossil resource is extensive and fossils are common, or the fossils are rare, difficult to collect, unspectacular and perhaps poorly preserved, or the sites are inaccessible to collectors.
- *Vulnerable* – sites where the fossil resource is substantial or of unknown extent and the specimens are generally well preserved, some of them having considerable scientific, aesthetic or commercial value.
- *Very vulnerable* – sites where there is a very small fossil-bearing resource, that could, in theory, be totally removed within hours using appropriate equipment, and/or where the fossils are of the highest scientific value and accessible to collectors.

As MacFadyen points out, site vulnerability may change over time as a result of resource re-evaluation, research excavation, commercial quarrying or discovery of new fossil material.

4.12 Climate and Sea-level Change

The most authoritative work on the progress, impacts and mitigation of climate change is the Intergovernmental Panel on Climate Change (IPCC) whose most recent report (IPCC, 2001) predicts that global temperature will warm by 1.4 to 5.8 °C by 2100 and that globally averaged sea level will rise by 0.09 to 0.88 m by 2100. The warming will, however, vary by region and there would be both increases and decreases in precipitation, changes in the variability of climate and changes in the frequency and intensity of extreme events. The impacts of these changes on physical systems are not completely known but are likely to be considerable. There are likely to be changes in the nature or rates of processes (e.g. a rise in precipitation may lead to increases in soil erosion, severe floods and limestone solution). Some processes, and their related landforms and sediments, may be reactivated while others will become fossilised.

The IPCC predicts increases in annual mean streamflow in high latitudes and southeast Asia, and decreases in central Asia, the Mediterranean, southern Africa and Australia. Rivers draining from ice masses in the Alps and the Rockies are likely to have reduced flows in summer due to disappearance of many of the glaciers in the next few decades. It is estimated that the 37 named glaciers in Glacier National Park,

USA, in the 1990s could disappear within the next 30 years (D. Fagre, personal communication) and glaciers are also shrinking rapidly in Alaska (Arendt *et al.*, 2002). Since many mountain societies rely on these glaciers for water resources, there is concern about their future disappearance (UNEP-WCMC, 2002).

One environment in which climate change is likely to have a serious impact is the periglacial environment. Although quantitative estimates of the temperature rise vary, most authors agree that global warming is likely to have the greatest impact in the world's cold climate areas. A warmer climate may not mean less snow if precipitation increases, but it may mean a shorter snow lying season. This would have serious implications for the skiing industry. Also potentially important will be melting of the significant parts of the permafrost. This would result in disruption of the delicate thermal balance of the permafrost areas and result in surface subsidence and erosion of the melted areas (thermokarst subsidence and thermal erosion). In inhabited areas, the result would be disruption to buildings and roads, which have generally been carefully engineered to preserve the permafrost but not to take into account climate change. In particular, there is a significant oil industry on the edge of the Arctic Ocean in Alaska and climate change in this area could reduce sea ice, accelerate coastal erosion and threaten onshore oil installations (Demek, 1994; Bennett & Doyle, 1997).

Rises and falls in relative sea level can have important effects by changing coastal processes, landforms or exposures. Changes in wave conditions, oceanic circulation and sea-ice cover are also likely. Coastal erosion of beaches and cliffs, flooding of low-lying coasts and islands, loss of saltmarshes and mudflats and saline water intrusion into freshwater environments are particular impacts of rising sea level in some parts of the world (French & Spencer, 2001) and these are likely to increase over the next century because of global warming. This is caused by two sets of processes:

- thermal expansion of the upper ocean;
- melting of land ice.

However, the effect of local land movements must also be taken into account since these may either accentuate or reduce the impacts of sea-level rise (Nichols & Leatherman, 1995). Although sea-level change is a natural process that has been continuous since the first oceans were formed over 4 thousand million years ago (see Section 2.3), sea-level change is currently being largely driven by human-induced global warming (IPCC, 2001).

The Bruun model has been widely used as a tool for predicting shoreline migration on soft coasts under a given sea-level rise, but has been criticised for being oversimplistic in its initial suggestion that coastal retreat will be 100 times greater than the rise in sea level. Geological and hydrodynamic factors are likely to be important factors in controlling responses to level rise. On hard coasts, coastal change will be limited and controlled by inland gradient.

The United Kingdom's MONARCH (Modelling Natural Resource Responses to Climate Change) programme has identified many impacts on UK ecosystems and geosystems on assumptions of

- temperatures 0.4 to 1.6 °C higher by the 2020s and 0.7 to 2.6 °C higher by the 2050s;
- increases in winter rainfall everywhere but a decrease in summer rainfall in southern England by >20% by the 2050s. Recent research has also shown an increase in the intensity of rainfall in the last 50 years;

- higher evaporation rates that will affect water availability, particularly in south-east England;
- sea-level rise by as much as 0.78 m by the 2050s in south-east England and southern Ireland, but less in the north-west of Britain and Ireland. Impacts on the environment may include increased flood risk, not only directly from more frequent or intense rainfall events but also indirectly through climatically induced land-use change.

Regional Climate Change Impact and Response Studies (REGIS) in East Anglia and north-west England have examined the regional implications of such changes for coastal and river flooding, agriculture, water availability and biodiversity. In particular, they have emphasised the importance of socio-economic factors on the size of climate impacts, so that "Society has an important opportunity to manage the impacts through policy choices and adaptation" (Holman *et al.*, 2002, p. 19).

Other parts of the environment may also be affected by climate change. For example, Carvalho & Anderson (2001) state that climate affects lake physics (flushing, stratification), chemistry (DOC, pH, nutrients) and biology (species diversity) and any changes in climate are therefore likely to alter many lake characteristics. Finally, soils are vulnerable to global warming, which may accelerate the decomposition of soil organic matter, potentially releasing large quantities of terrestrial carbon dioxide into the atmosphere (Puri *et al.*, 2001). In Tasmania, Australia, *Phtopthera*, a soil fungus that kills some vegetation types is now spreading rapidly probably because of higher soil temperatures (C. Sharples, personal communication).

4.13   Fire

Fire is both a natural and human-induced process. It can have advantageous effects by releasing nutrients from the ash, but there is a significant impact through the burning of peats and organic soils. In Tasmania, Australia, for example, research is being carried out on fire frequencies on the blanket bogs in the south-west of the island over the last 6000 years and the implications for organic soil accumulation and fire management.

Other impacts result from the destruction of the vegetation layer and subsequent effects. For example, fire spreading into the Mount Field National Park in Tasmania from adjacent forestry land has killed vegetation that formerly stabilised the slopes and landforms. The result has been soil erosion, landslides, burial of stream sinks, increased stream turbidity, sedimentation in limestone caves and track construction as part of salvage logging operations.

4.14   Military Activity

This can have a serious effect on many aspects of the earth science environment, though many military training areas comprise huge tracts of infrequently disturbed country and thus aid conservation. In sensitive areas, even routine operations can have damaging effects. For example, at Dungeness in England, English Nature (1998) has recognised that "the use of vehicles and explosives have caused great damage to these sensitive features", which include shingle ridges and sand dunes. Similarly, helicopter landings are doing damage to the periglacially patterned ground in Snowdonia, N. Wales. There is also compression and erosion of soils generated by such military training activities. Kiernan (1997a, p. 14) relates that "One of the more bizarre means by which Tasmania's coastal geoheritage has been diminished involved the use by naval ships of the spectacular freestanding dolerite columns on Cape Raoul, Tasman Peninsula – for target practice".

At times of war, the landscape impacts may be particularly high. The excavation of trenches and tunnels during the Great War has been the subject of much recent research (Doyle & Bennett, 2002), though this certainly increased our geological knowledge and many of the trenches themselves are now protected for their historical value. Westing and Pfeiffer (1972) calculated that the bombing of Indo-China (Vietnam, Cambodia, Laos) in the 1960s produced 26 million craters displacing 2.6 billion cubic metres of soil. This, together with landscape bulldozing to remove vegetation and soil down to the parent material destroyed at least 10% of the area's agricultural land. As the authors acknowledge (p. 28), "It has been a war against the land as much as against armies". Modern bombing methods have become even more effective, including the use of cluster bombs, and will create heavy pitting of the landscape that may do serious damage to important geological interests. Pollution of land from depleted uranium used in some modern bombs is a further issue, as are deep-penetrating bombs that can destroy cave systems. Crawford *et al.* (2003) expressed concern about the impact of the Iraq War on the 25,000 registered archaeological sites in the country, including the ancient remains of some of the world's first cities such as Ur, Babylon and Nineveh.

4.15   Lack of Information/Education

A final threat to geodiversity, and some would say the most important of all, is ignorance. A lack of survey information, documentation and designation of geoheritage has resulted in loss or degradation of sites and landscapes via inappropriate development in the past and remains a threat in many parts of both the developed and developing world. Many temporary sections from pipe laying or other excavation go unrecorded and therefore important information is lost.

As a result of this lack of information, geological conservation measures are often lacking, including integration into important land-use planning legislation, policy and practice. Joyce (1999) notes that in Australia "Problems arise because of the general lack of geologically trained staff in National Parks and similar organisations, the common lack of geological input to management plans, and the lack of a high profile for geological heritage". In England, a recent review of the Geological Conservation Strategy noted that there was "Still lots of 'wildlife-only' literature" even within English Nature, the government agency responsible for nature conservation (Prosser, 2001b). Similarly, Gray (2001) noted that a major nature conservation initiative in England (English Nature, 1998) lists 28 organisations that were consulted on the content, but not a single geological, geomorphological or geographical organisation appears in the list of consultees. Gray also noted the need for more geomorphological/Quaternary training amongst English Nature's Local Teams. In northern Ireland, a premier Jurassic fossil site at Garron Point was lost during a road reconstruction scheme in the 1990s when the Roads Department failed to consult the Environment and Heritage Department about its plans (Doughty 2002). A greater understanding of the planet's geodiversity, its value, threats and the importance of geoconservation and management are important aims of this book.

4.16 Cumulative Impacts and Sensitivity to Change

This brief review has established that there are many real threats to the geodiversity of the planet, mainly related to increasing development and environmental change. R. L. Hooke (1994) calculated that the average annual transport rate of rock and soil by humans is about 42 billion tons per year, equal to about half that transported by rivers, glaciers, oceans and wind combined, and three times that created by mountain building. The loss of cultivatable soil is now taking place at a faster rate than new soil is being created or brought into cultivation (Pimental, 1993).

An important message of this chapter is that earth surface processes are complex and often sensitive to human interference. An important essential is therefore that we understand the implications of change and development. As Boulton (2001, p. 50) puts it, "Whether we are engineering a dam or a marine wind farm, we are engineering into the environment, and we should seek to understand what the consequences will be". It also becomes important "to identify the significance or otherwise of features prior to potential damage, and to assess the likely significance of that damage, in terms of the significance of the feature itself, the extent to which it is likely to be damaged or the nature of the damage inflicted" (Kiernan, 1996, p. 15).

In some places, development pressures are high and significant cumulative losses of geodiversity have occurred. A survey of in the Peterborough area of the United Kingdom in 1989 recorded 32 important geological sites but by 1998, 26 had been lost mainly due to landfilling, vegetation growth and restoration works (Peterborough Environment City Trust, 1999). Box 4.6 lists some of the development impacts and losses of geodiversity in Tasmania, Australia, in recent years (Kiernan, 1996; Pemberton, 2001a; Sharples, 2002a).

Box 4.6 Recent Geodiversity Impact and Losses in Tasmania, Australia

- Loss of 28 Tertiary and Quaternary fossil sites destroyed or buried by quarrying, impoundments, landfill or other developments. This represents 50% of such sites identified in the last 100 years.
- Destruction of significant geological sites in road cuttings. These include an eclogite site in the designated Tasmanian Wilderness Area.
- Removal by uncontrolled collecting of valuable or rare silicified tree stumps and Thylacine subfossils from caves.
- Collection of rare or significant speleothems and minerals.
- Erosion of dunes and Aboriginal middens by off-road vehicles and cattle grazing.
- Only three out of over 50 lunette dune features are left undisturbed.
- Inundation of the globally unique landform assemblage (beach and dunes of pinkish-white quartzite sand, sand bar, spectacular sequence of sub-aqueous megaripples, smaller bedform features, lagoon system, ferromanganese concretions) by hydroelectric development at Lake Pedder (Fig. 4.13).

Continued on page 170

__ Continued from page 169 __

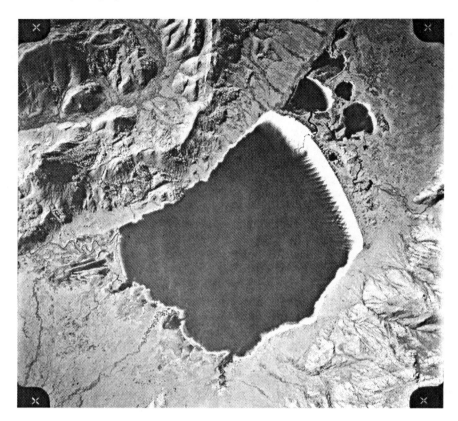

Figure 4.13 Lake Pedder, SW Tasmania, before it was flooded in the 1970s showing the beach, sand bar and megaripples on its north-eastern shore. These and the adjacent fluvioglacial landforms dunes and blanket bogs are reported to be relatively intact. It is possible that drainage and re-vegetation could reinstate this important geomorphological site (M. Pemberton, personal communication). (Photo: Reproduced by permission of Department of Primary Industries, Water and Environment (DPIWE), Tasmania)

- Damage to dunes and moraines at Lake St Clair by hydroelectric works. A road has been carved along the crest of the most prominent dune, sand has been quarried from its downvalley flank and rock rip-rap has been plastered across its up valley.
- Erosion by wake from tourist boats of significant fluvial landforms and deposits on the Gordon River.
- Removal by quarrying of a terminal moraine marking the maximum glacial ice advance in the Mersey Valley.
- Damage to the Exit Cave system from quarrying, destruction of magnesite tower karst in the mid 1980s, and degradation of spring mounds by agricultural and residential development.

__ Continued on page 171 __

___ *Continued from page 170* ___

- Infestation of coastal dunes by marram grass, altering natural processes and removing mobile sand from the system.
- Soil erosion during and following fire, including impacting on peatlands of international significance.
- Covering of important coastal sections by land reclamation and sea-walls.

In response to this history of geodiversity degradation, Tasmanian geoconservationists have constructed a 10-point scale of sensitivities indicating the types of activities that might degrade features of a given sensitivity level. In general, it is suggested that a more protective management response is required for highly sensitive features (low on the scale) while those high on the scale will require little or no active conservation management (Kiernan, 1997b; Sharples, 2002a). This scheme is shown in Table 4.4 and follows an earlier, more limited attempt in Scotland (Werritty & Brazier, 1991: See Table 4.3).

Implicit in Tables 4.3 and 4.4 is the idea that all systems are ultimately sensitive to some level of impact, but some values are robust in the face of impacts that would be highly degrading to other values. For example, the palaeontologically important, Burgess Shale site in the Canadian Rockies is very sensitive to over collecting, but this damage has little long-term impact on the landscape aesthetics of Yoho National Park.

Landscape sensitivity has been the subject of research in Scottish Mountains, particularly the Cairngorms (Werritty & Brazier, 1994; Haynes *et al.*, 1998, 2001; Gordon *et al.*, 2001; Thompson *et al.*, 2001). According to Haynes *et al.* (2001, pp. 120–121) "Sensitive landscape systems readily cross extrinsic geomorphological thresholds into new process regimes. Stable landscape systems operate far from such thresholds, and weathering often dominates over erosion so that there is a vegetation and soil cover. Robust systems are subject to constant change within intrinsic thresholds, yet retain similar forms and processes as they maintain dynamic equilibrium, with features moving across the landscape or reforming after disturbance". Werritty & Brazier (1994) illustrated the distinction between robust and sensitive behaviour (see Fig. 4.15) and applied this to the fluvial geomorphology of the River Feshie.

Figure 4.16 shows a schematic representation of the scale of impacts caused by different human-related pressures (Thompson *et al.*, 2001) and Haynes *et al.* (2001) indicate sensitivity to change from different pressures of the main topographical features on Scottish mountains.

Table 4.3   *The 5-point Scottish geosensitivity scale (after Werritty & Brazier, 1991)*

1. Activity not generally applicable
2. Activity obscures or masks the surface and stratigraphic exposures of the landform
3. Activity causes localised disruption or destruction of part of the landform surface or stratigraphy
4. Activity causes general disruption or destruction of the surface and/or stratigraphy of the landform
5. Activity causes destruction of individual or groups of landforms, permanently interrupting morphological and stratigraphic relationships within the landform assemblage

Table 4.4   *The 10-point Tasmanian geosensitivity scale (after Kiernan, 1997b; Sharples, 2002a)*

1. Values sensitive to inadvertent damage simply by diffuse, free ranging human pedestrian passage, even with care.
   Examples: fragile surfaces that may be crushed underfoot, such as calcified plant remains; calcite or gypsum hairs and straws in some karst caves that may be broken by human breath, speech or touch (e.g. see Fig. 4.14).

2. Values sensitive to effects of more focussed human pedestrian access even without deliberate disturbance.
   Examples: risk of entrenchment by pedestrian tracks; coastal dune disturbance; drainage changes associated with tracks, leading to gulleying; defacement of speleothems by touching their surfaces.

3. Values sensitive to damage by scientific or hobby collecting or sampling or by deliberate vandalism or theft.
   Examples: some fossil and mineral collecting, rock coring and speleothem sampling.

4. Values sensitive to damage by remote processes.
   Examples: degradation of geomorphic or soil processes by hydrological or water quality changes associated with clearing or disturbance within the catchment; fracture/vibration due to blasting in adjacent areas (e.g. stalactites in caves).

5. Values sensitive to damage by higher intensity linear impacts, depending upon their precise position.
   Examples: vehicle tracks, minor road construction, or excavation of ditches and trenches.

6. Values sensitive to higher intensity but shallow generalised disturbance on site, either by addition or removal of material.
   Examples: clearfelling of forests and replanting, but without stump removal or major earthworks; land degradation such as soil erosion due to bad management practices; vegetation/weathering of exposures.

7. Values sensitive to deliberate linear or generalised shallow excavation.
   Examples: minor building projects, road construction, shallow borrow pits, removal of stumps, construction of small bunds.

8. Values sensitive to major removal or addition of geomaterial.
   Examples: quarrying; dam construction, landraising.

9. Values sensitive only to very large-scale contour change.
   Examples: large opencast quarries; major reservoirs causing inundation; major river channelisation schemes.

10. Values sensitive only to catastrophic events.
    Examples: meteorite impacts; human-induced sea-level change; major landslides and tsunamis.

4.17   Conclusions

In this chapter, we have outlined the main threats to geodiversity and we have noted that these threats may apply to specific important sites or to the wider landscape. We have seen that the cumulative impacts can be considerable and that the threat will depend on the sensitivity of systems and the values placed upon them. Development is

Figure 4.14 *A fragile aragonite speleothem, Shooting Star Cave, Tasmania, Australia. The speleothem is about 300 mm wide (Photo: Reproduced by permission of Rolan Eberhard)*

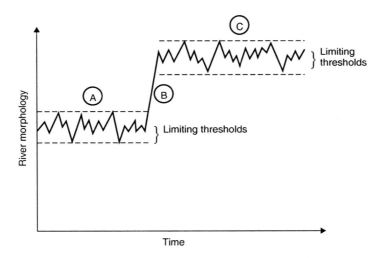

Ⓐ,Ⓒ robust behaviour–river repeatedly crossing intrinsic thresholds, but overall response stable within limiting thresholds. Negative feedback regulates change. Landforms retain stable identity as they form and reform.

Ⓑ sensitive behaviour–in response to externally imposed change river moves across extrinsic threshold to new process regime. Landforms in original regime Ⓐ destroyed and replaced by new landforms created in regime Ⓒ.

Figure 4.15 *The distinction between robust and sensitive behaviour in geomorphological river systems (After Werritty, A. & Brazier, V. (1994) Geomorphic sensitivity and the conservation of fluvial geomorphology SSSIs. In Stevens, C., Gordon, J.E., Green, C.P. & Macklin, M.G. (eds) Conserving our Landscape. English Nature, Peterborough, 100–109. by permission of English Nature)*

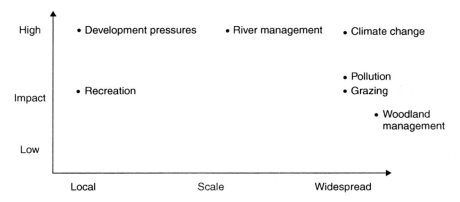

Figure 4.16   *The scale of impacts caused by different human-related pressures (After Thompson, D.B.A., Gordon, J.E. & Horsfield, D. (2001) Montane landscapes in Scotland: are these natural artefacts or complex relics? In Gordon, J.E. & Leys, K.F. (eds) Earth Science and the Natural Heritage. Stationery Office, Edinburgh, 105–119., by permission of Scottish Natural Heritage)*

of course essential to the improved prosperity of human populations and the essentials of life, but there is a need to balance these against the irreplaceable nature of much of our geoheritage. It is often possible to develop less sensitive sites or manage the land in less damaging ways. It is only by understanding the value of geodiversity and threats to it that this balance can be achieved.

5

Conserving Geodiversity: The Protected Area and Legislative Approaches

5.1 Introduction

The start of this chapter represents an important point in the book. Previous chapters have defined, described and valued geodiversity and established that there are significant threats to it. The intention has been to establish that there is a case for geoconservation. This chapter, and the subsequent two, follow directly from the preceding two, since conservation is an appropriate human response to perceived threats to features regarded as having value. The next three chapters can therefore be regarded as the most important in the book, since it is the embodiment of an environmentally sophisticated society that it acts to avoid losses and protect what is vulnerable to damage. These chapters therefore move us on to consider how geoconservation has been or can be implemented. The current chapter describes the large variety of ways in which international organisations, governments or local bodies attempt to protect areas of geological or geomorphological interest. It follows a previous review by Dixon (1996a) and a suggestion by Daly *et al.* (1994, p. 212) that research is needed on the international variation in approaches to geoconservation. As we shall see, the geoconservation systems so far established vary not only in scope but also in effectiveness, for there is more to geoconservation than simply drawing lines on a map. Chapter 6 examines several ideas for extending conservation to the wider landscape beyond protected areas. And Chapter 7 examines some important issues relating to geodiversity and biodiversity. These three chapters demonstrate that there is no shortage of implementable ideas. All that is needed is political motivation and adequate resources.

5.2 Beginnings of the Conservation Movement in North America

Conservation probably has a long history, but, as an organised activity supported by government, it began in the United States and was probably initiated as a response to overhunting, overgrazing and soil erosion (Dasmann, 1984). In the 1830s, George Catlin, an artist and naturalist, expressed his concern for the future of the buffalo and the well being of the Plains Indians who depended on it. They had therefore been careful stewards of buffalo numbers and its rangelands for centuries. Catlin was perhaps the first to propose a huge "national park" across the Great Plains where Indians

Geodiversity: valuing and conserving abiotic nature   Murray Gray
© 2004 John Wiley & Sons, Ltd   ISBNs: 0-470-84895-2 (HB); 0-470-84896-0 (PB)

and wildlife could be left in peace to pursue their traditional way of life. However, the proposal was not taken up and it was another 20 years before the issue raised its head again.

In the 1850s, Henry David Thoreau, wrote of his concern for the future of all wild nature and a hope that it could be preserved. He was also wise enough to realise that people have an important relationship with, and an obligation to, "the more sacred laws" of nature, but again few paid much attention. In 1864, George Perkins Marsh, in his book *Man and Nature*, was the first to attempt a systematic description of the human impacts on the natural environment, and formulated the view that man must live with nature. At the same time, urban dwellers came to appreciate the value of city parks and several early environmentalists like John Muir, a Scottish immigrant, argued strongly for the preservation of nature. Their arguments were successful and as a result the Yosemite Valley, California, became a protected area in 1864. As the United States had no mechanism for managing parks at a national level at that time, it was deeded to the State of California (though it was later incorporated in the Yosemite National Park). This was followed in 1872 by the world's first designated national park at Yellowstone, USA. The growth of the network was slow at first – Mackinac Island (1875), Sequioa, Yosemite and General Grant (1890), Mount Rainier (1899) and Crater Lake (1902). Canada's first national park at Banff was established in 1885. Many of the early national parks were established because of their scenic or geological values (e.g. Yellowstone (see Box 5.1), Yosemite, Mount Rainier, Crater Lake, Banff), though this fact seems to have been lost in the subsequent overwhelming emphasis on wildlife conservation, and needs to be rediscovered.

Box 5.1   Yellowstone National Park, USA

After European colonisation of the eastern part of North America, the first explorers and travellers to the western most reaches began to return with curious tales of the unearthly natural wonders of Yellowstone, particularly of its steaming ground and hot springs. But it was the Washburn–Langford–Doane expedition of 1870 that really awakened public interest in Yellowstone. There has, however, been much debate as to whether the motives were truly altruistic or, to a greater or lesser extent, commercially driven (Sellars, 1997). It was this party that, on emerging from the forest into what is now known as the Upper Geyser Basin, was met with "an immense body of sparkling water, projected suddenly and with terrific force into the air to the height of over one hundred feet. We had found a real geyser". And since it repeated its spectacular show at regular intervals, they named it "Old Faithful" (Langford, 1870).

The results of the Washburn expedition were promoted in a series of public lectures and articles by one of the participants, Nathaniel Langford. In the audience at one of these in Washington, D.C., was the Director of the US Geological Survey, Ferdinand V. Hayden, whose interest was immediately aroused and who managed to obtain congressional funding to survey the area in the summer of 1871. On his return to Washington, Hayden received the suggestion from Judge William Keeley of Philadelphia that "Congress pass a bill reserving the Great Geyser Basin as a public park forever" (Schullery, 1999). Hayden was persuaded to take up the cause and the *Yellowstone National Park Act* (1872) was signed by President Ulysses S. Grant on 1 March 1872. The park covered nearly one million hectares of public land provided for "the preservation from injury or spoilation, of all timber, mineral deposits, natural curiosities, or wonders within said park, and their retention in their natural condition". The land was "reserved and withdrawn from settlement, occupancy, or sale

Continued on page 177

___ *Continued from page 176* ___

Figure 5.1   *The northern entrance arch at Yellowstone National Park inscribed with words from the National Park Act (1872). Also note the natural geological variation in the stone blocks*

under the laws of the United States, and dedicated and set apart as a public park or pleasuring ground for the benefit and enjoyment of the people" (Fig. 5.1).

The protected area was drawn by Hayden to include all the geothermal features known at the time and the main reason for establishing the park was to protect these geological

_ *Continued on page 178* _|

__ *Continued from page 177* __

Figure 5.2   *Steam emerging from several geysers and vents in the Upper Geyser Basin, Yellowstone National Park, USA*

wonders (Schullery, 1999; L. Whittlesey, personal communication). It is estimated that there are 10,000 individual geothermal features in the park, including 200 to 250 active geysers, though with concentrations into groups (Fig. 5.2). This makes Yellowstone one of the greatest concentrations of geothermal activity on the planet. It also contains the impressive falls and canyon of the Yellowstone River and one of the world's largest calderas measuring 75 by 45 km (Fig. 5.3). However, having protected the area for its geology, this also effectively protected the wildlife within it, and over the years this became the main focus for nature conservation efforts and controversies (Pritchard, 1999; Schullery, 1999). Even today (2003), there is only one geologist employed in the park.

Since established, the park boundaries in the east have been extended and realigned to conform with the watersheds, it being recognised that rivers flowing into the park ought to be protected from exploitation and pollution. During the twentieth century, the concept of the "Greater Yellowstone Ecosystem" (GYE) also emerged as a means of co-ordinating conservation efforts by multiple agencies over a wider geographical area including Grand Teton National Park and several National Forests. Nevertheless, the greatest threat to Yellowstone's geothermal activity lies only 40 km to the west of Old Faithful where, beyond the park boundaries in southern Idaho, lies the Island Park Known Geothermal Resource Area (IPKGRA)(Fig. 5.3). It is known that the centre of igneous activity over the last 16.5 million years has shifted eastwards from southern Oregon through southern Idaho to north-west Wyoming as the American tectonic plate has drifted westwards above a volcanic hot spot at a rate of c.2.5 cm per year (Smith & Siegel, 2000). It is therefore possible that the geothermal activity related to these igneous centres is connected at depth and that any exploitation of the IPKGRA could depressurise the Yellowstone geothermal system with catastrophic results for Yellowstone's geothermal activity (H. Heasler, personal communication). If this happened "Would the very name 'Old Faithful' take on a new meaning in our culture, becoming eventually an ironic sarcasm for something sadly short of fidelity" (Schullery, 1999, p. 263). Clearly, more research is needed to understand the underground geothermal interconnectivities of the region, and to recognize that we should be referring to a "Greater Yellowstone Geoecosystem" (Smith, 2000).

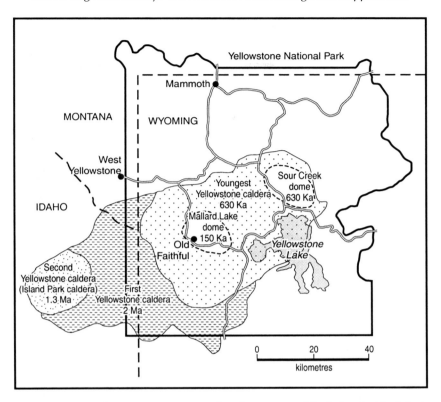

Figure 5.3 *Calderas of Yellowstone National Park, USA (Modified after Smith, R.B. & Siegel, L.J. (2000) Windows into the Earth: The Geologic Story of Yellowstone and Grand Teton National Parks. Oxford University Press, Oxford, by permission of Oxford University Press).*

In 1892, Muir founded the famous Sierra Club in California and it is still an influential NGO over a century later (Kuzmiak, 1991). But it was not until 1908 that US President Theodore Roosevelt used the name "conservation" to describe the activities and the movement (Dasmann, 1984). Roosevelt's home was in New York but he went to North Dakota in 1883 and decided to develop his interest in cattle ranching. He was brought to the west by the prospect of big game hunting, but when he arrived, the last large buffalo herds were gone as a result of overhunting and disease. Throughout his time in North Dakota, Roosevelt became more and more concerned about the disappearance of some species. Overgrazing destroyed the grasslands and with them the habitats for small mammals and songbirds. Roosevelt therefore developed a keen interest in conservation and what would nowadays be called sustainable land use. During his presidency (1901–1909), he established the US Forest Service and set aside land as national forests, signed the *Antiquities Act* (1906) under which he proclaimed 18 national monuments, and obtained congressional approval for five national parks and 55 wildlife refuges. There is now a national park in North Dakota named after this major American conservationist whose interest in politics allowed him to implement many of his ideas.

The work he started continued after his presidency had ended. On 25 August 1916, the Federal Government assumed responsibility for the ownership and management of the 40 national parks of the time through the *National Park Service Organic Act* (1916), the objective of which was to "conserve the scenery, natural and historic objects and the wildlife therein, and to provide for the enjoyment for the same in such a manner and by such means as will leave them unimpaired for the enjoyment of future generations". According to MacEwan & MacEwan (1987, p. 4), "The central idea was a democratic one. The state would keep out the freebooters of private (or public) enterprise and protect the 'natural scenery' for the enjoyment of the public".

Following the disaster of the mid-west Dust Bowl (see Box 4.5), President Franklin D. Roosevelt established the US Soil Erosion Service in 1933, later becoming the US Soil Conservation Service. Dasmann (1984, p. 9) believes that it was not until the later decades of the last century that environmental concerns amongst the general public led to more active pursuit of conservation goals and establishment of a political environmental agenda. The *Alaska Lands Bill* (1980) was a major milestone (Cutter & Renwick, 1999), but as the recent debates over further oil exploration in Alaska's National Wildlife Refuge indicate, tensions still exist in American society between economic growth and natural resource exploitation on the one hand and nature conservation on the other.

5.3   Early British Experience

In Britain in the early nineteenth century, the Romantic poets such as William Wordsworth had extolled the virtues of the Lake District's landscape and talked about it becoming "a sort of national property, in which every man has a right and interest who has an eye to perceive and a heart to enjoy" (Wordsworth, 1952, p. 127). Although there is a democratising theme here, in seeing areas of beauty as a national resource, there is also an exclusionary tone, since "if you do not have the necessary perceptual eye and emotional heart, then by implication you have no rights or interests with regard to the Lake District" (Phillips & Mighall, 2000, p. 325). Wordsworth's concerns were undoubtedly driven by a fear that the arrival of "the masses" in areas like the Lake District would ruin their "natural character".

> "A vivid perception of romantic scenery is neither inherent in mankind nor a necessary consequence of even a comprehensive education...Rocks and mountains, torrents and wild spread waters.... cannot in their finer relations to the human mind, be comprehended without opportunities of culture in some degree habitual..." (William Wordsworth, quoted in Glypteris, 1991, p. 27).

Those without the necessary insight and emotions would be better off "taking little excursions with their wives and children among the neighbouring fields within reach of their own dwellings"! Ignoring the snobbish and sexist implications of these remarks, it is interesting to note that Wordsworth's perception of the essence of the Lake District landscape was of its abiotic elements – rocks, mountains, torrents and water.

Other writers also expressed distaste for the tourist invasion. James Payne (quoted in Glypteris, 1991, p. 27) was disdainful that a steamboat "disgorges multitudes upon the pier; the excursions trains bring thousands of curious vulgar people...our hills are darkened by swarms of tourists". These fears eventually led to the establishment, in 1883, of the *Lake District Defence Society* expressly set up to oppose, successfully as it turned out, the extension of the railway line into Borrowdale. This organisation subsequently was incorporated into the *National Trust*, or to give its full title, *The*

Table 5.1 *Dates of formation of some early British nature conservation organisations (modified after Phillips & Mighall, 2000)*

Society for the Protection of Animals	1824
The Commons, Open Spaces and Footpaths Preservation Society	1865
East Riding Association for the Protection of Seabirds	1867
Association for the Protection of British Birds	1870
Thirlmere Defence Association	1878
The Lake District Defence Society	1883
The Selbourne Society for the Protection of Birds, Plants and Pleasant Places	1885
The Royal Society for the Protection of Birds	1889
The National Trust for Places of Historical Interest and Natural Beauty	1895
The Coal Smoke Abatement Society	1899
Society for the Preservation of the Wild Fauna of the Empire	1903
The British Vegetation Committee	1904
The British Association for Shooting & Conservation	1908
The Society for the Promotion of Nature Reserves	1912
The British Ecological Society	1913
The Council for the Preservation of Rural England	1926
Association for the Protection of Rural Scotland	1926
The Council for the Preservation of Rural Wales	1928
National Trust for Scotland	1931
The Standing Parks Committee	1936
Civic Trust	1957
The Council for Nature	1957
British Trust for Nature Conservation	1965
Conservation Society	1966

National Trust for Places of Historic Interest and Natural Beauty. The National Trust is one of a very large number of conservation organisations formed in the United Kingdom in the late nineteenth and the early twentieth centuries (Table 5.1), some with intriguing titles. There are also examples of early UK conservation legislation. For example, the *Seabird Protection Act* was passed in 1869 in response to threats from shooting and egg collection.

In the 1940s, Alfred Steers carried out a survey of the entire coastline of England and Wales and concluded that he could "not emphasise too strongly that if we as a nation wish to preserve one of our finest heritages for the good of the people as a whole, we must act now and act vigorously on a national scale" (Steers, 1946). In 1949, the *National Parks and Access to the Countryside Act* allowed the establishment of National Parks in England and Wales, though Scotland had to wait until 2002 for its first National Park and Northern Ireland still awaits its first. However, even these National Parks are very different from the American and international model (see Section 5.10.1). An interesting review of the history of nature conservation in the United Kingdom is given by Adams (1996) but his conclusion (p. 99) is that "Conservation in the UK has grown up without a coherent philosophy, a cultural and scientific rag-bag of passion, insight and good intentions".

5.4   The Protected Area and Legislative Approaches

McKenzie (1994, p. 127) commented that "Rarely, are Earth science features and processes effectively protected, conserved or managed successfully on a site-by-site basis without some supporting institutional framework". As the national park system (NPS) exemplifies, the traditional way in which areas for conservation have been identified and defined is by drawing boundaries on a map and then giving particular status to the land, wildlife or features within these boundaries that does not apply outside the boundaries. This "protected area" approach or boundary concept (Doyle & Bennett, 1998) is identifiable throughout the world, though it takes many different forms in different countries as applied to different types of area. Since animal life is not confined by such boundaries, a second traditional approach has been to proscribe the taking, killing or trading of named species or animal products. Thirdly, the removal or trading in historical or archaeological artefacts has been controlled by legislation and this has sometimes been extended to palaeontological objects.

These protected area and legislative approaches are explored in this chapter, beginning with the international conservation efforts and moving on to the national and sub-national scene.

5.5   International Conservation

5.5.1   The IUCN

The International Union for the Conservation of Nature and Natural Resources (IUCN) was founded in 1948 and now refers to itself as the World Conservation Union. It "brings together 78 states, 112 government agencies, 735 NGOs, 35 affiliates, and some 10,000 scientists and experts from 181 countries in a unique partnership" (IUCN web site, www.iucn.org). It has the aim "To influence, encourage and assist societies throughout the world to conserve the integrity and diversity of nature and to ensure that any use of natural resources is equitable and ecologically sustainable". In 1980, it published the *World Conservation Strategy* where the term "sustainable development" was first widely publicised (Barrow, 1999).

The IUCN has recognised that protected areas come in many different forms in different countries and has devised a list of categories (last revised in1993) to allow international comparisons (Table 5.2). At the head of this list are the Strict Nature Reserves, which are areas of land and/or sea "possessing some outstanding or representative ecosystem, geological or physiographical features and/or species, available primarily for scientific research and/or environmental monitoring", and Wilderness Areas, which are "large areas of unmodified or slightly modified land and/or sea, retaining its natural

Table 5.2   *The IUCN management categories*

I Strict Nature Reserve/Wilderness Area
Ia Strict Nature Reserve
Ib Wilderness Area
II National Park
III Natural Monument
IV Habitat/Species Management Area
V Protected Landscape/Seascape
VI Managed Resource Protected Area

character and influence without permanent or significant habitation, which is protected and managed so as to preserve its natural condition". In these areas, natural processes are allowed to operate without direct human interference, and tourism, recreation and public access are restricted. Ownership and control should be by the national or other level of government. An exception is Antarctica, which is governed by international agreement (see Box 5.6). Earlier ideas of keeping these areas inviolate proved to be a somewhat outdated concept, since, in many cases, it excluded indigenous populations with long histories of living off the land and doing so in a sustainable manner. However, attitudes have now shifted with the recognition that "conservation cannot be achieved simply by preservation, but will have to acknowledge the values, needs and aspirations of local people" (Mather & Chapman, 1995, p. 129). Thus, one of the management objectives for Wilderness Areas is "to enable indigenous human communities living at low density and in balance with the available resources to maintain their lifestyle".

According to the IUCN, by 2000, the world's 30,000 protected areas covered over 13 million square kilometres or 9% of the world (roughly equivalent to India and China combined), and it has set each nation the target of protecting 10% of their land area.

The World Commission on Protected Areas (WCPA) is an IUCN Commission that works to help governments and others to plan, manage, strengthen and enhance this world network. Sadly, it defines a protected area as "an area of land and/or sea especially dedicated to the protection and maintenance of biological diversity, and of natural and associated cultural resources" and it is criticised by Brilha (2002) for its lack of geological programmes. Although it is true that most are biological and little reference is made to geoconservation in any of its documentation, one of the networks does involve a Caves and Karst Task Force.

5.5.2   The United Nations

The United Nations is also deeply involved in international conservation mainly through its Educational, Scientific and Cultural Organisation (UNESCO) and its Environment Programme (UNEP). The latter has a World Conservation Monitoring Centre (WCMC) based in Cambridge, England, which has a Protected Areas Programme aimed at establishing a Nationally Designated Protected Areas Database. UNESCO has a World Heritage Sites (WHS) network and is also working with other agencies to support world efforts to establish international Geosite and Geopark networks.

World Heritage Sites (WHS)

The World Heritage Convention, which is concerned with "protecting the world's cultural and natural heritage", was adopted by the General Conference of UNESCO in 1972. To date, more than 150 countries have adhered to the convention, making it an important international conservation instrument. It is not intended to provide protection for all important sites in the world, but only for a select list of the most outstanding international cultural and natural areas. However, no formal limit is imposed either on the total number of sites included on the list or on the number of properties that can be submitted by any state.

The general principles to be followed in establishing the World Heritage List are given on the UNESCO web site (www.unesco.org). The World Heritage Committee invites states to nominate sites for inclusion on the list, and there must be evidence of the full commitment of the nominating government in the form of relevant legislation, staffing, funding and management plans. It is also essential that local people should participate in the nomination process to make them feel a shared responsibility with the state for the maintenance of the site. Joint nominations are encouraged in cases where outstanding areas stretch across national boundaries. Where necessary, for proper conservation, a "buffer zone" around a site where uses are restricted is encouraged in order to give an added layer of protection. A series of sites in different geographical locations may be nominated as a Single World Heritage Site provided they are related because they belong, for example, to "the same geomorphological formation", provided that it is the series as such that is of outstanding universal value and not its components taken individually. The Australian Fossil Mammal Site (Riversleigh/Naracoote) is an example of one such site (Creaser 1994b). UNESCO works with the IUCN, the International Union of Geological Sciences (IUGS) and the International Geological Correlation Programme (IGCP) to evaluate whether or not nominated sites satisfy the necessary criteria and conditions.

States are expected to monitor the conditions of sites on the list and to submit periodic reports on the condition of sites within their territory and any legislative, administrative or other actions they have taken to implement the Convention. In particular, reports are expected each time exceptional circumstances occur or when work is undertaken, which may have an effect on the state of conservation of a site on the list. Where there has been severe deterioration of a site on the list, it may be placed on the List of World Heritage in Danger. Listed examples of where this may occur include

- "severe deterioration of the natural beauty or scientific value of the property, as by human settlement, construction of reservoirs that flood important parts of the property, industrial and agricultural development including use of pesticides and fertilizers, major public works, mining, pollution, logging. . ... etc";
- "human encroachment on boundaries or in upstream areas that threaten the integrity of the property";
- "a modification of the legal protective status of the area".

Several sites have been placed on this list including two US National Parks – the Everglades National Park (threatened by urban growth, pollution from fertilisers, mercury poisoning of fish and wildlife, and fall in water levels due to flood protection measures) and Yellowstone National Park (threatened by adjacent mining operations, sewage leakage and waste contamination, illegal introduction of non-native trout, road construction, and year-round visitor pressures). Other sites placed on the List include Sangay National Park, Ecuador (unplanned road construction) and Mount Nimba Strict Nature Reserve, Guinea/Ivory Coast (iron-ore mining concession).

Sites may be deleted from the list where they have deteriorated to the extent that they have lost the characteristics that led them to be included on the list, but this would only occur after several years of monitoring. Up to 1999, no site had ever been "delisted" (Feick & Draper, 2001). The IUCN has recommended that the WHC should involve NGOs, academic institutions and local people in monitoring the conservation status of sites in order to make WHC processes more transparent and democratic (Feick & Draper, 2001).

Article 2 of the Convention states that "natural heritage" comprises

- "natural features consisting of physical and biological formations, which are of outstanding universal value from the aesthetic or scientific point of view;

- geological and physiographical formations and precisely delineated areas that constitute the habitat of threatened species of animals and plants of outstanding universal value from the point of view of science or conservation;
- natural sites or precisely delineated natural areas of outstanding universal value from the point of view of science, conservation or natural beauty".

In order to interpret the type of sites that might be suitable for inclusion on the World Heritage List as a natural feature, the *Operational Guidelines for the Implementation of the World Heritage Convention*, at paragraph 44, lists the criteria that should be met. The following are the criteria relevant to geological/geomorphological sites:

a(i)   "be outstanding examples representing major stages of earth's history, including the record of life, significant ongoing geological processes in the development of landforms, or significant geomorphic or physiographic features; or ...

a(iii) contain superlative natural phenomena or areas of exceptional natural beauty and aesthetic importance...and also fulfil the following conditions of integrity:

b(i)   The sites described in a(i) should contain all or most of the key interrelated and interdependent elements in their natural relationships; for example, an 'ice age' area should include the snow field, the glacier itself and samples of cutting patterns, deposition and colonisation (e.g. striations, moraines, pioneer stages of plant succession etc.); in the case of volcanoes, the magmatic series should be complete and all or most of the varieties of effusive rocks and types of eruptions be represented. ...

b(iii) The sites described in a(iii) should be of outstanding aesthetic value and include areas that are essential for maintaining the beauty of the site; for example, a site whose scenic values depend on a waterfall, should include adjacent catchment and downstream areas that are integrally linked to the maintenance of the aesthetic qualities of the site...

b(v)   The sites described in paragraph (a) should have a management plan...

b(vi) A site described in paragraph (a) should have adequate long-term legislative, regulatory, institutional or traditional protection. ..."

The other sections of paragraph 44 deal with biological sites.

At present (2003) there are 730 World Heritage Sites. Although UNESCO states that "efforts will be made to maintain a reasonable balance between the numbers of cultural heritage and natural heritage properties" on the list, only 144 sites are listed as natural areas, and most of these are mainly recognised for their ecological value or ecological habitats. Table 5.3 lists the 39 sites identified by Paul Dingwall of New Zealand for the IUCN as having a significant geological or geomorphological character as defined by their a(i) classification (usually denoted N(i) for natural sites).

The IUCN has undertaken a review of the geological World Heritage Sites (UNESCO, 2002) and concludes that "it appears that the current system of World Heritage sites goes a long way in representing the geological history, features and processes that support life on earth". Even ignoring this biocentric view of the role of geology, an analysis of the list would lead many geologists to reach a different conclusion. Firstly, there are some sites that are of limited geological or geomorphological importance and would not appear on any geologists' list of the world's 39 most important sites. Secondly, there are several examples of mountain landscapes, karst topography and volcanoes while

Table 5.3 The 39 World Heritage Sites designated under the N(i) category (modified after list compiled by Paul Dingwall and published with his permission)

Country	Site	Criteria	Year of inscription	Geological interest
Argentina	Ischigualasto-Talampaya	(i)	2000	Triassic palaeontology & stratigraphy
Australia	Riversleigh/Naracoote	(i), (ii)	1994	Cenozoic mammal palaeontology
	Heard & McDonald Islands	(i), (ii)	1997	Volcanic, glacial, coastal & karst topography
	Macquarie Island	(i), (iii)	1997	Oceanic crustal rocks, ophiolites
	Central Eastern Rain Forest Reserves	(i), (ii), (iv)	1986–1994	Volcanic craters and igneous rocks
	Great Barrier Reef	(i), (ii), (iii), (iv)	1981	Coral reefs
	Wet Tropics of Queensland	(i), (ii), (iii), (iv)	1988	Tablelands, escarpments, coastal topography
	Shark Bay	(i), (ii), (iii), (iv)	1991	Stromatolite analogue environments
Bulgaria	Pirin NP	(i), (ii), (iii)	1983	Mountain topography, glacial lakes
Canada	Miguasha Park	(i)	1999	Upper Devonian fish
	Dinosaur Provincial Park	(i), (iii)	1979	Cretaceous dinosaurs, badlands topography
	Gros Morne	(i), (iii)	1987	Oceanic crustal rocks, L. Palaeozoic palaeontology
	Canadian Rocky Mountain Parks	(i), (ii), (iii)	1984	Mountains & glaciers, Cambrian Burgess Shale
Costa Rica & Panama	Talamanca Ranga-La Amistad NP	(i), (ii), (iii), (iv)	1983–1990	Mountain topography
Cuba	Desembarco del Granma	(i), (iii)	1999	Karst & coastal topography
Dominica	Morne Trois Pitons NP	(i), (iv)	1997	Volcanic topography & active volcanicity
Ecuador	Galapagos Islands	(i), (ii), (iii), (iv)	1978/2001	Coastal & volcanic topography
Germany	Messel Pit Fossil Site	(i)	1995	Eocene Palaeontology
Honduras	Rio Platano Biosphere Reserve	(i), (ii), (iii), (iv)	1982	Mountain, river & coastal topography

Country	Site	Criteria	Year	Description
Hungary & Slovakia	Aggtelek & Sloval Karst	(i)	1995	Karst, caves & Quaternary speleothems
Indonesia	Lorentz NP	(i), (ii), (iv)	1999	Convergent plate margin, coastal & fluvial topography
Italy	Aeolian Islands	(i)	2000	Volcanoes
Kenya	Lake Turkana NP	(i), (iv)	1997/2001	Rift valley, saline lakes, hominid palaeontology
Malaysia	Gunung Mulu NP	(i), (ii), (iii), (iv)	2000	Karst & cave topography, limestones
New Zealand	Te Wahipounamu/SW New Zealand	(i), (ii), (iii), (iv)	1990	Mountain, glacial & fjord topography
Russian Federation	Lake Baikal	(i), (ii), (iii), (iv)	1996	Tectonic lake
	Volcanoes of Kamchatka	(i), (ii), (iii), (iv)	1996/2001	Volcanoes, coastal topography
Sweden	The High Coast	(i)	2000	Glacio-isostatic rebound
Switzerland	Jungfrau-Aletsch-Bietschorn	(i), (ii), (iii)	2001	Glaciers & glacial topography
UK	Jurassic Coast	(i)	2001	Mesozoic strat. & pal., coastal topography, landslides
	Giant's Causeway & coast	(i), (iii)	1986	Columnar basalt, lava flows, coastal topography
USA	Carlsbad Caverns	(i), (iii)	1995	Caves & speleothems in reef limestones
	Yosemite NP	(i), (ii), (iii)	1984	Glacial geomorphology
	Everglades NP	(i), (ii), (iv)	1979	Wetland & mud sedimentation
	Mammoth Cave	(i),(iii),(iv)	1981	Caves & speleothems
	Yellowstone NP	(i), (ii), (iii), (iv)	1978	Geothermal activity
	Grand Canyon	(i), (ii), (iii), (iv)	1978	Canyon & exposed stratigraphy
	Great Smoky Mountains NP	(i), (ii), (iii), (iv)	1983	Mountain topography
Venezuela	Canaima NP	(i), (ii), (iii), (iv)	1994	Upland plateau

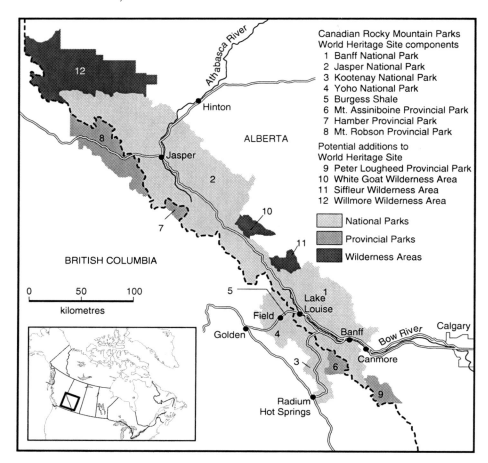

Figure 5.4 *Rocky Mountain Parks World Heritage Site in Canada comprises a number of National and Provincial Parks and further expansion is planned. The world famous Cambrian fossil site of the Burgess Shale is included in the Site (After Feick, J. & Draper, D. (2001) Valid threat or "tempest in a teapot"? An historical account of tourism development and the Canadian Rocky mountain parks world heritage site designation. Tourism Recreation Research, 26, 35–46, by permission of Center for Tourism Research and Development)*

numerous aspects of geodiversity are absent. Thus, the list by no means adequately represents global geodiversity. Thirdly, it is clear that a review of site classifications is needed since some WHS are clearly geologically important but lack an N (i) class. For example, Uluru (Ayers Rock) in Australia is listed only as a cultural site. There are also some hidden gems. The Canadian Rocky Mountain Parks WHS, for example, contains the famous Burgess Shale Cambrian fossil site (Coppold & Powell, 2000) (Fig. 5.4) and Ujung Kulon National Park WHS in Indonesia, contains the remnants of the Krakatoa volcano. Greater proactivity from UNESCO, IUGS, IUCN and others would be useful in ensuring that global geodiversity is more fully represented in the list.

Boxes 5.2, 5.3 and 5.4 give brief descriptions of World Heritage Sites with geological interests.

Box 5.2   Grand Canyon National Park, USA

This National Park in Arizona, USA, would be recognised worldwide as one of the natural wonders of the world and is richly deserving of its World Heritage Site status, which it achieved in 1979. It now covers nearly half a million hectares and is dominated by the spectacular Grand Canyon, 550 m to 30 km wide, about 1,500 m deep and 447 km long. It has formed over the last six million years by erosion of the Colorado River and its tributaries. There are over 100 named rapids along the river. The geological strata are sub-horizontal so that a trip from the canyon rim to the river takes us through over 1,000 million years of geological time from Triassic to Precambrian though there are many gaps in the record represented by unconformities (Fig. 5.5). The earliest Precambrian strata, the Vishnu Metamorphic Complex, is barren, but the late Precambrian Bass Limestone, within the Unkar Group, has early plant fossils. The Palaeozoic strata contain both marine and terrestrial fossils demonstrating alternations of submergence and emergence.

An updated management plan was completed in 1995 and involved local citizens, local Indian tribes, and public and private agencies. The park's most serious management issue is that of tourism. The annual influx of over 5 million visitors is gradually degrading the park's natural resources by, for example, footpath erosion and development, though the

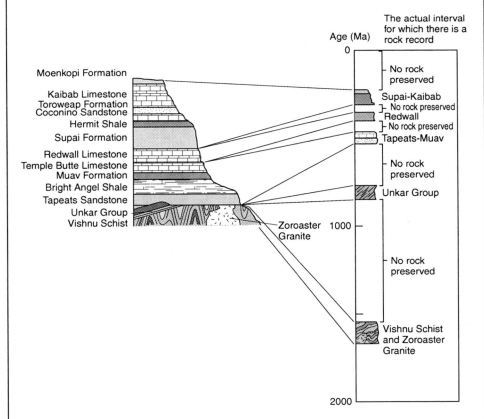

Figure 5.5   *Stratigraphy of the Grand Canyon showing the diversity of strata and occurrence of unconformities (After Marshak, S. (2001) Earth: Portrait of a Planet. W.W. Norton & Co, New York, by permission of W.W.Norton & Co.)*

Continued on page 190

__ *Continued from page 189* __

park is zoned with over 90% managed as wilderness. In 1995, the park employed 242 permanent and 159 seasonal staff but no trained geologist and this is still the case.

Box 5.3   Jurassic Coast, UK

This is a very new geological site approved by UNESCO in December 2001. The case for its inclusion on the list was made on six grounds. Firstly, the gently eastward dipping exposures along 150 km of coast (Fig. 5.6), provides a near-continuous sequence of Triassic, Jurassic and Cretaceous rocks covering 190 million years, including the internationally recognised names of Kimmeridge, Portland and Purbeck. Secondly, these rocks are highly fossiliferous and provide a very well-preserved record of the evolution of Mesozoic life and environments. Thirdly, the coastal geomorphology includes famous features such as the Chesil Beach barrier and lagoon, Lulworth Cove and the Durdle Door natural arch, and several active landslides. These demonstrate the interrelationships between coastal processes and local geology. Fourthly, the area has been a crucible of earth science investigation for over three hundred years and is associated with many of the founders of modern geology, including William Smith, Sedgwick, Murchison, de la Beche and Lyell. Fifthly, the coast's history of geological and geomorphological research as evidenced by over 5,000 referenced items, is continued to the present day and needs protection for the future. Finally, the aesthetic beauty of the coast has long been recognised, so that it remains relatively unspoilt and coastal process are allowed to operate on the underlying geology.

Box 5.4   Macquarie Island, Australia

Macquarie Island lies about 1,500 km south of Tasmania and was added to the World Heritage List in 1997 as an exposed section of a huge oceanic ridge extending upwards from a depth of 2.5 km. It started life between 30 and 11 million years ago as a small divergent margin ridge. The southern three-quarters of Macquarie Island is composed of fissure-erupted basalts, including pillow lavas. Miocene abyssal oozes are found between some of the pillows.

About 10 million years ago, the spreading stopped and instead reversed to produce an upward pressure and movement of abyssal rocks. The most important products of this upward movement occur in the north of the island where an ophiolite sequence of upper mantle rocks is exposed, comprising sea-bed lavas, a sheeted dolerite complex, massive layered gabbros and finally a mixed zone of peridotite believed to be from 6 km below the ocean floor. "No drill hole has ever penetrated these depths and these exposures provide a rare opportunity for geologists to gain an understanding of rocks from the uppermost mantle" (Pemberton, 2001b).

Other features seen on the island include fault-controlled structures and frequent earthquakes, raised shorelines, landslides, solifluction lobes and patterned ground (Viney, 2001).

Conservation objectives have been set as to

- protect World Heritage (geo)values;
- protect, maintain and monitor geodiversity;
- protect, maintain and monitor sites of geoconservation significance;

__ *Continued on page 192* __

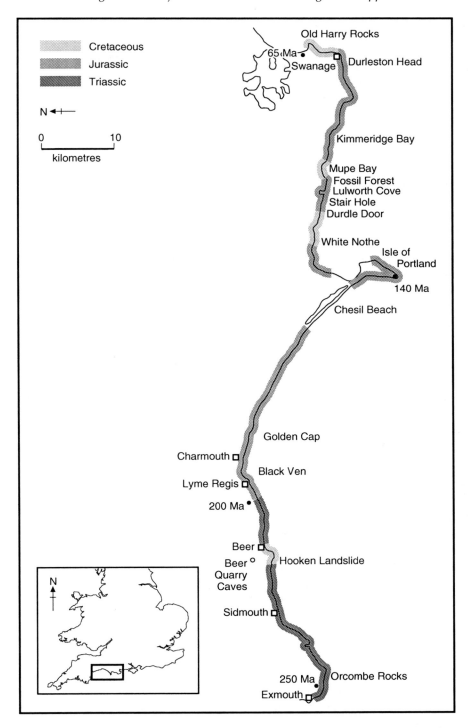

Figure 5.6   *Location map for the Jurassic Coast World Heritage Site, England (Reproduced by permission of Dorset Country Council)*

___ *Continued from page 190* ___

- maintain natural rates and magnitudes of change in earth processes;
- avoid harmful impacts on geoconservation values;
- encourage geoscientific research that is consistent with the values for which the area was nominated for World Heritage Listing;
- maintain a high standard of geointerpretation for visitors including up-to-date booklets and other interpretive material.

These conservation objectives have been framed as a series of practical policies:

- Potential adverse impacts on geodiversity and earth processes will be assessed when planning any development or action, including land rehabilitation and stabilisation.
- Management practices and development will avoid or otherwise minimise impacts on the integrity of sites of geoconservation significance.
- Scientific research will be conducted in a way that avoids impacts on geodiversity, sites of geoconservation significance or the aesthetics of significant exposures. Geoscientific research must be consistent with the values for which the area was nominated and ought to be justified in this context.
- The use of coring devices and other mechanical sampling devices for geoscientific research will not be permitted unless special permission is provided. Any approval will be strictly controlled and monitored. Similar conditions will apply to the use of explosives for geoscientific or management purposes.
- The impacts on geodiversity will be monitored, including adherence to conditions identified in scientific collecting permits.
- Sites of geoconservation significance will be identified and mapped, and recommendations will be made on their future management (Pemberton, 2001b).

Geosites

A Global Indicative List of Geological Sites (GILGES) was established in the early 1990s by UNESCO, the IUCN and IUGS. The list included hundreds of sites that were intended to be of "first-class importance to global geology... outstanding examples representing major stages of the Earth's history, significant ongoing geological processes in the development of landforms, such as volcanic eruption, erosion, sedimentation, etc., or significant geomorphic or physiographic features, for example, volcanoes, fault scarps or inselbergs..." (Cowie & Wimbledon, 1994).

Several problems arose from the establishment of this list. One was the wording of the guidelines for World Heritage Site selection which "were not particularly suited to the geosciences" (Cleal *et al.*, 2001). Another problem was the range in size of the sites from huge national parks to metre-sized fossil localities. (Cowie & Wimbledon, 1994). But the most serious problems came in judging the geological value of sites and in achieving consistency between countries (Cleal *et al.*, 2001).

Thus, in 1995, the IUGS replaced GILGES with a more rigorous and comprehensive scheme known as *Global Geosites* and this was subsequently endorsed by UNESCO. The aim is to compile a global list, with supporting documentation, of the world's most important geological sites. The work is being coordinated by the IUGS's Global

Geosites Working Group (GGWG) and the list is being stored as a computer database at the IUGS Secretariat in Trondheim, Norway. The aim is a "bottom-up approach", with geoscientists in all countries being encouraged to compile their own registers (where they do not already exist), which can then be scrutinised by the wider geological community. The end result "is not to search for token 'best sites': it is to identify natural networks of sites that represent geodiversity" (Cleal *et al.*, 2001, p. 10). An example of geosite work in Kazakhstan is given by Nusipov *et al.* (2001) and in Spain is given by Garcia-Cortés *et al.* (2001). There is multinational ("regional") collaboration (for example, for Europe, where the work is coordinated by ProGEO) with each country being asked to nominate candidate sites for the frameworks relevant to their country. When agreement is reached, the regional frameworks and candidate sites go to the GGWG for inclusion in the database (see www.iugs.org/iugs/science/sci-wgst.htm). "Potentially, an internationally important geosite, could be designated a World Heritage Site" (Cleal *et al.*, 2003), so that a mechanism exists to create a more appropriate network of World Heritage geological sites.

Geoparks

This relatively new initiative is supported by UNESCO, though it has not been accepted as a mainstream UNESCO Project. It is, however, a logical extension to the World Heritage List, but it is more than simply a second division of geological site or collection of

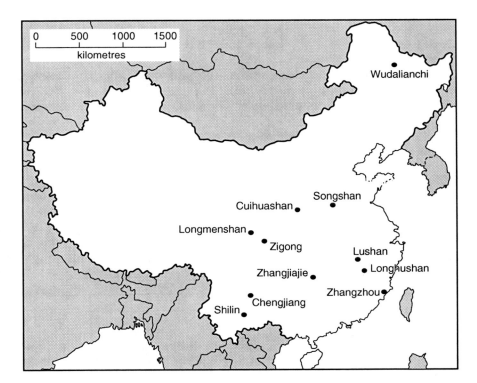

Figure 5.7 *The first 11 Chinese Geoparks (After Xun, Z. & Milly, W. (2002) National geoparks initiated in China: putting geoscience in the service of society. Episodes, 25, 33–37 by permission of International Union of Geological Sciences (IUGS))*

Table 5.4 China's 11 National Geoparks (Xun & Milly, 2002)

Name of Geopark	Area (km²)	Geological interests
Shilin Stone Forest NG	400	Limestone & dolomite karst & pillar landscape
Zhangjiajie Sandstone Peak Forest NG	3600	Quartzose sandstone pillar landscape
Songshan NG	450	Precambrian strata and unconformities
Lushan NG	500	Fault block mountains & glacial sediments
Chengjiang NG	18	Early Precambrian fossils & fault basin
Wudalianchi Volcanoes NG	720	Late Cenozoic volcanic cones and lavas
Zigong Dinosaurs NG	9	Middle Jurassic dinosaurs & other vertebrates
Zhangzhou Littoral Volcanoes NG	320	Tertiary volcanic rocks & coastal landforms
Cuihuashan Landslides NG	32	Fault–induced landslides & collapsed blocks
Longmenshan NG	1900	Nappes, thrust outliers & mountain landscapes
Mount Longhushan NG	380	Mesozoic fault basin sediments eroded into varied landforms

sites. According to Patzak and Eder (1998) and Eder (1999), the aim is to enhance the value of nationally important geological sites while creating economic development, employment and geotourism as part of an integrated programme. Parks will

- encompass one or more sites of scientific importance for geology as well as archaeology, ecology or cultural value;
- have a management plan that fosters sustainable geotourism and socio-economic development;
- provide a means of teaching geoscientific disciplines and broader environmental issues;
- be part of a global network that demonstrates best practice in Earth heritage conservation and its integration into sustainable development strategies.

It is anticipated that about 20 parks per year would be approved with an overall total of around 500 by 2025. Local involvement is seen as essential, with nomination coming from local communities and local authorities committed to developing and implementing a management plan that promotes and protects the geodiversity of the local landscape and meets local economic needs. China established 11 Geoparks in 2000 (Xun & Milly, 2002, see Fig. 5.7 and Table 5.4) and a European Geoparks Network has been established that had 15 member parks by 2003 (see Box 5.5 and Fig. 5.8).

Box 5.5   European Geoparks

The idea of a network of European Geoparks was initiated in 1997 with the triple aims of conserving geoheritage, improving public understanding of the geosciences and promoting regional economic development. Four founder members were established in 2000 (see Fig. 5.8). Haute-Provence in France (ammonites, bird footprints,

Continued on page 195

__ *Continued from page 194* __

etc.), Lesvos in Greece (petrified forest), Maestrazgo in Spain (dinosaurs) and Vulkan-eifel in Germany (volcanic craters and structures). Subsequent additions have included Astroblème Rochechouart-Chassenon in France (meteorite crater), Marble Arch Caves and Cuilcagh Mountain in Northern Ireland (Carboniferous karst) and Psiloritis on Crete (geological structures) and the network is expected to expand rapidly. Regular meetings of members of the European Geopark network help to share best practice and experiences and to develop "brand identity", geotourism marketing and geoconservation strategies.

European Geoparks must have a specific geological heritage covering a substantial area where there is an economic development strategy funded by a European Union programme. It must comprise geological sites of scientific, educational and/or aesthetic value but can also include archaeological, ecological, historical or cultural sites. One objective is to get the local population to re-evaluate their heritage and encourage them to play an active role in the economic revitalisation of their region (McKeever & Morris, 2002). The potential impact can be judged by the fact that the Haute Provence Geopark attracts 100,000 visitors to the region each year and is expected to support 400 jobs by 2006. Apart from this recognised network of European Geoparks, several independent Geoparks are being established in Europe and the number is expected to mushroom in the next few years.

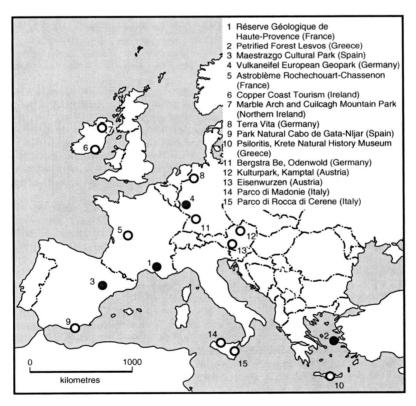

1 Réserve Géologique de
   Haute-Provence (France)
2 Petrified Forest Lesvos (Greece)
3 Maestrazgo Cultural Park (Spain)
4 Vulkaneifel European Geopark (Germany)
5 Astroblème Rochechouart-Chassenon
   (France)
6 Copper Coast Tourism (Ireland)
7 Marble Arch and Cuilcagh Mountain Park
   (Northern Ireland)
8 Terra Vita (Germany)
9 Park Natural Cabo de Gata-Nljar (Spain)
10 Psiloritis, Krete Natural History Museum
   (Greece)
11 Bergstra Be, Odenwold (Germany)
12 Kulturpark, Kamptal (Austria)
13 Eisenwurzen (Austria)
14 Parco di Madonie (Italy)
15 Parco di Rocca di Cerene (Italy)

0   1000
kilometres

Figure 5.8   *The first 15 European Geoparks (Reproduced by permission of European Geoparks Network)*

5.5.3   Other international agreements

Other international initiatives have been mainly concerned with protecting biodiversity, for example, international agreements on trade of wild animals (CITES) or whaling. The Ramsar convention for wetland conservation includes, coincidentally, many important geomorphological/hydrological sites. There is, as yet, no international Convention on Geodiversity to match the Convention on Biological Diversity (CBD). There was active interest in promoting such a convention and an international working group was established in the early 1990s (e.g. Knill, 1994; Creaser 1994c; Dixon 1996a) but this has not been sustained and perhaps now needs to be resurrected. Antarctica has its own set of treaties aimed at conserving its environment (Box 5.6).

Box 5.6   *Antarctica*

Antarctica is roughly one-and-a-half times the size of the United States. Ninety-eight per cent of it is covered by an ice sheet up to 3,700 m thick, containing about 90% of the Earth's freshwater. If melted, this would raise the sea level by about 72 m. The ice sheet reaches the sea in several places to form ice shelves and much of the coast is locked by pack ice in winter. Some areas are ice-free, particularly around the Antarctic Peninsula and on high mountain peaks of the Transantarctic Mountains. These divide the continent and the ice sheet into two parts, East and West Antarctica (Sugden, 1982).

Until recently, the main practical use of Antarctica has been for scientific research, but it has been suggested that substantial oil, gas and mineral reserves may be present on the outer continental shelf. On the Antarctic mainland, coal, oil, gas and iron ore have been found, and it has been estimated that there is sufficient iron ore in the Prince Charles Mountains to satisfy world demand for 200 years. Fortunately, there are abundant supplies of iron ore elsewhere in the world that are considerably easier to exploit. The oil and gas reserves would also be difficult and expensive to exploit but as world oil starts to run out, the pressure may fall on these resources (Sugden, 1982).

There were many disputes over sovereignty and management of Antarctica prior to 1959 when 12 governments, seven with territorial claims (Argentina, Australia, Chile, France, New Zealand, Norway and UK) and five others (Belgium, Japan, South Africa, USA and USSR) signed the Antarctic Treaty. This did not resolve the sovereignty issue but rather was "masterful in its vagueness" and "has yet to be fully tested under conditions of economic pressure" (Buck, 1998, p. 59–60). The Treaty governs the whole area south of 60°S and states that this area should be used only for peaceful purposes and the advancement of scientific knowledge. Article 1 of the Treaty explicitly prohibits any military activities and there have been various subsequent agreements dealing with environmental matters. Forty-four countries have now signed the Treaty representing over 80% of the world's population. Several further agreements have been adopted including a Protocol on Environmental Protection, which has recently banned mineral and oil exploration in Antarctica for at least 50 years (Buck, 1998), but there have been other conflicts over fishing rights and demands from non-treaty nations (Hansom & Gordon, 1998).

5.6   The European Dimension

There are as yet no specific European geoconservation directives or policies, but some of the biological conservation directives are still useful. For example, the European Habitats Directive (1992) allows the designation of Special Areas of Conservation (SACs) and the Birds Directive (1979) provides for Special Protection Areas (SPAs). Together they form a Natura network of sites that may, coincidentally, contain geological or geomorphological features. Brancucci *et al.* (2002), for example, found that at least 25% of Natura sites in Liguria, Italy contain important geological and geomorphological interests. More importantly, the conservation management of many of the sites will have to recognise the fundamental dependency of habitats on soils, geology, landforms and active processes. "For example, species of Atlantic salmon depend on the provision of adequate habitat (in the form of variable channel features such as pools, riffles and glides), which itself is a function of active processes including floods, erosion and deposition. . . . The lessons currently being learned in managing sites designated under the Habitats Directive and through the biodiversity process will be valuable in demonstrating the wider importance and relevance of Earth science understanding" (Gordon & Leys, 2001b, p. 9).

But there have been other European initiatives, including the Environmental Impact Assessment Directive (see Section 6.7), Environmental Action Programmes and establishment of the European Environment Agency. The Bern Convention of the Council of Europe was established to conserve European wildlife and habitats and has led to a network of Areas of Special Conservation Interest (ASCIs), otherwise known as the *Emerald Network*. For EU members, the Natura network serves as the Emerald Network but the latter extends into eastern Europe and north Africa and allows a more consistent approach to habitat protection. In 1999, Iceland proposed that Sites of Geological Interest (SGIs) should be included in the Emerald Network.

ProGEO is a Europe-wide organisation of geoconservationists that aims to promote the conservation of Europe's geoheritage and an integrated approach to nature conservation. It is compiling a list of European Geosites and contributing to the new Pan-European Biological and Landscape Diversity Strategy. Also relevant to geoconservation are the more recent European Water Framework Directive (2000) and European Landscape Convention (2000), the objectives of which are outlined in Chapter 6.

5.7   National Conservation Systems

As already stated, about 13 million square kilometres, or 9% of the Earth's land surface, are under national or international systems of protection. Both, the number and the area of protected sites increased rapidly during the twentieth century. However, Mather and Chapman (1995) ask us to view such figures with caution. For example, designation means little if there is no respect for the principle and no system of enforcement. This is the problem of the so-called *paper parks*. Secondly, it may be easier to designate remote locations that are not under threat and where the need for conservation is lower,

Figure 5.9   *Plaque commemorating the inauguration of the Waterton/Glacier Peace Park and World Heritage Site on the Canada/USA border*

than areas where threats from, for example, mining, logging or agriculture are greater and where conservation measures are less welcomed.

The American concept of national parks was formally adopted by the IUCN in 1969 and national parks that conform to it have been established all over the world. Most countries now define part of their territory with this name. Today, more than 100 nations contain some 1,200 national parks or equivalent preserves, though the aims differ from country to country. In some places, cross-border parks have been established, such as the Glacier/Waterton International Peace Park and World Heritage Site on the USA/Canada border (Fig. 5.9), and several in southern Africa, the largest of which is the Great Limpopo Transfrontier Park created in 2002 by combining South Africa's Kruger National Park, Mozambique's Limpopo Park and Zimbabwe's Gonarezhou National Park (Fig. 5.10).

In the sections that follow, some details of national Earth science site-protection systems are described. For more information on specific countries, government and geological web sites can often provide detailed information.

5.8   United States of America

Miller (1998, p. 388) claims that:

> "No other nation on Earth has set aside such a large portion of its land for the public's enjoyment and use. About 42% of all US land consists of public lands owned jointly by all citizens and managed for them by federal, state and local governments. Over one-third (35%) of the country's land is managed by the federal government".

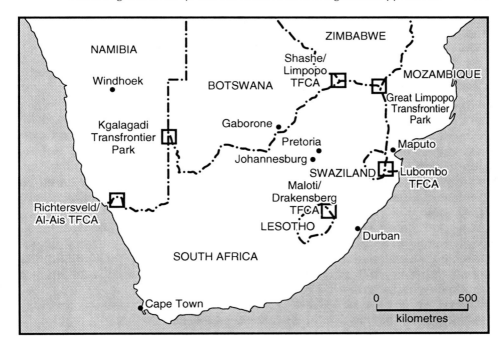

Figure 5.10 *Existing and proposed Trans-Frontier Conservation Areas & Parks in southern Africa (Modified after Magome, H. & Murombedzi, J. (2003) Sharing South Africa's national parks: community land and conservation in a democratic South Africa. In Adams, W.M. & Mulligan, M. (eds) Decolonizing Nature. Earthscan, London, 108–134, by permission of William Adams)*

However, of this 35%, 73% is in Alaska and 22% is in the western states, leaving only 5% in the central, eastern and southern states. Most of these areas are set aside, primarily for their wildlife resources rather than their geodiversity. Nonetheless, the protection of land for wildlife usually also protects the contained geology and geomorphology. Examples include the national forests, national wildlife or marine refuges, national wilderness preserves and rangeland areas. National Resource lands are mainly rangelands, administered by the Bureau of Land Management with the emphasis on providing a secure domestic supply of energy and strategically important non-energy minerals. Several categories are administered by the National Parks Service, so we begin with these.

5.8.1   The National Park System

Several types of areas/sites have been managed by the National Parks Service, on behalf of the Department of the Interior since an Executive Order in 1933 transferred 63 national monuments and military sites from the Forest Service and War Department to the National Parks Service. The *General Authorities Act* (1970) brought all areas administered by the National Parks Service into one National Parks System. Areas included in the system:

"though distinct in character, are united through their interrelated purposes and resources into one national park system as cumulative expressions of a single national heritage; that, individually and collectively, these areas derive increased national dignity and recognition of their superb environmental quality through their inclusion jointly with each other in one national park system preserved and managed for the benefit and inspiration of all people of the United States"

These were major steps in improving the integration of America's national system, which includes historical and cultural as well as scenic and scientific areas and sites. Thus, apart from National Parks, the units include National Monuments, National Rivers, National Seashores and several other categories. However, there are still anomalies in the system in that the Grand Staircase – Escalante National Monument in Utah is administered not by the National Parks Service but by the Bureau of Land Management.

Today, the United States National Park System comprises over 380 units (of all types) covering more than 35 million hectares in 49 States (none in Delaware), the District of Columbia and overseas dependencies. Some cover huge tracts of land. The largest is the Wrangell-St Elias National Park and Preserve in Alaska and Canada at over 5 million hectares, which is over 16% of the entire system.

Anyone can propose an addition to the system, but to be considered nationally significant it must meet all four of the following standards:

- It is an outstanding example of a particular type of resource.
- It possesses exceptional value or quality in illustrating or interpreting the natural or cultural themes of the nation's heritage.
- It offers superlative opportunities for recreation, for public use and enjoyment, or for scientific study.
- It retains a high degree of integrity as a true, accurate, and relatively unspoilt example of the resource.

Three other criteria are important. Firstly, it must not represent a feature already adequately represented in the system. Adequacy of representation is determined on a case-by-case basis taking account of character, quality, quantity, and so on. Secondly, it must be feasible in terms of land ownership, acquisition costs, access, threats to the resource and staff or development requirements. Thirdly, additions to the National Park System will not usually be recommended if another arrangement can provide adequate protection and opportunity for public enjoyment. This might include management by state or local governments, Indian tribes, the private sector, or other federal agency, or might include the designation of federal lands as wilderness, areas of critical environmental concern, national conservation areas, national recreation areas, marine or estuarine sanctuaries, and national wildlife refuges. Requests for boundary changes may also be made in order to address operational problems or include resources that are critical to the park's purposes.

The National Parks web site (www.nps.gov) gives a list of 10 examples of natural areas that might meet the criteria, seven of which refer to geological or geomorphological features:

- "an outstanding site that illustrates the characteristics of a landform or biotic area that is still widespread";
- "a rare remnant natural landscape or biotic area of a type that was once widespread but is now vanishing due to human settlement and development";
- "a landform or biotic area that has always been extremely uncommon in the region or nation";
- "a site that possesses exceptional diversity of ecological components... or geological features (landforms, observable manifestations of geological processes)";
- "a site that contains rare or unusually abundant fossil deposits";
- "an area that has outstanding scenic qualities such as dramatic topographic features, unusual contrasts in landforms or vegetation, spectacular vistas, or other special landscape features";
- "a site that is an invaluable ecological or geological benchmark due to an extensive and long-term record of research and scientific discovery".

This represents an impressive awareness of the geological/geomorphological heritage of the United States, with landscape, landform, processes and fossils all mentioned at least once. However, there is no mention of rocks, minerals or soils and in no sense is the system representative of the full geodiversity of the United States. Rather it conserves the most spectacular scenic areas, many geomorphological curiosities and some important fossil sites.

Areas are usually added to the National Park System by an act of Congress following an investigation by the National Parks Service, a positive recommendation to add the area, and a congressional hearing. Congress decides which of the several titles to apply to the area. Figure 5.11 shows the units in the National Park System in some of the western states and lists the main unit types, though these titles have not always been applied consistently. The title "National Park" has traditionally been reserved for the most spectacular natural areas (e.g. Fig. 5.12) with a wide variety of features, and hunting, mining and other consumptive activities such as grazing are generally prohibited in national parks. Legislation authorising national preserves, recreation areas, seashores and lakeshores sometimes allows for a wider range of activities such as oil and gas development, grazing and hunting, subject to certain limits.

Many of the units in the National Parks system have been established wholly or largely for their geological or geomorphological interest and there have been several Government Acts passed to preserve the geological heritage. The *National Park Service Organic Act* (1916) gives general protection to national park units after their adoption. However, there are often pre-existing private mineral rights, though these can, if necessary, be dealt with by land acquisition, as long as suitable compensation is paid, or regulation under the *Mining in the Parks Act* (1976). The latter requires prospective operators to submit operational plans for their mineral development, must have these approved by the National Parks Service and must deposit a bond to ensure conformity with the plan and site restoration.

Box 5.7 describes some of the sites and legislation relating to speleological sites, and Box 5.8 deals with palaeontological parks and monuments. Santucci *et al.* (2001b) have combined these interests in inventorying the palaeontological resources associated with National Parks Service caves. But several other units are well known for their

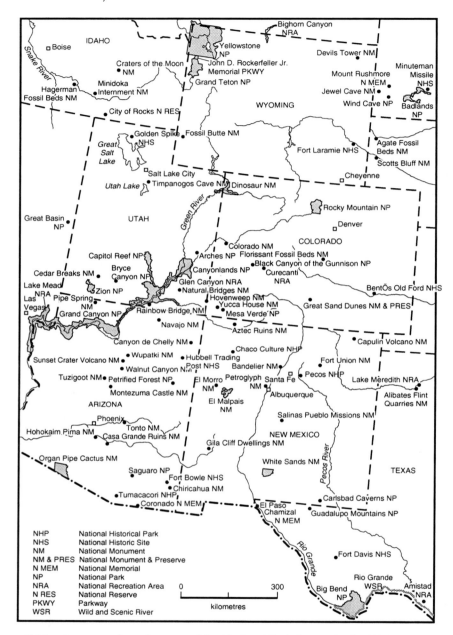

Figure 5.11 *An extract from the map of the National Park System in the USA*

geological/geomorphological interests. These include canyons (e.g. Bryce Canyon National Park), volcanoes (e.g. Hawaiian Volcanoes National Park, Hawaii; Crater Lake National Park, Oregon) geysers and hot springs (e.g. Yellowstone National Park, Wyoming; Hot Springs National Park, Arkansas), rivers (e.g. Buffalo Natural River, Arkansas; Natural Bridges, National Monument, Utah), glaciers and glacial geomorphology (e.g. Glacier Bay National Park, Alaska; Ice Age National Scenic Reserve &

Figure 5.12   *Eastern entrance to the spectacular scenery of Glacier National Park, Montana, USA*

Trail, Wisconsin) and sand dunes (Great Sand Dunes National Monument, Colorado; White Sands National Monument, New Mexico). Sprinkel *et al.* (2000) describes the geology of national park units in Utah as well as in 10 state parks (see Section 5.8.4).

Box 5.7   Caves and Karst

Caves and karst features occur in about 79 units of the National Park system, including several National Parks and National Monuments. These include Mammoth Cave in Kentucky, claimed to be the longest cave in the world with over 500 km of mapped passages in a complex system of caves and tunnels on multiple levels. The caves contain many impressive speleothems and the area around demonstrates impressive karst topography.

Similar characteristics occur at the Carlsbad Caverns National Park in New Mexico (Fig. 5.11), where the caves display an impressive array of speleothems including stalactites, stalagmites, columns (one of which, the Monarch, is one of the world's tallest at 25 m high), flowstone and other features. Unfortunately, many of the cave's smaller and more delicate formations have been damaged over the years by careless visitors and this is often irreversible, though some re-glueing has been attempted. The Carlsbad Caverns now has regulations that include the following conservation measures:

- ''Touching cave formations is prohibited. Formations are easily broken and the oil from your skin permanently discolours the rock;
- Smoking, or any use of tobacco, is not permitted. Eating and drinking are not permitted except in the Underground Lunchroom;

Continued on page 204

—— *Continued from page 203* ——

- Throwing coins, food or other objects in cave pools is forbidden. Foreign objects ruin the natural appearance of the pools and are difficult to remove. Also, the chemical reaction between foreign objects, the water and the rock can leave permanent stains;
- Photography is permitted when you tour on your own, but not when you accompany a ranger-guided tour. Photographers should not rest their tripods or other camera equipment on formations or step off the trails".

A landmark decision was the passage of the *Lechuguilla Cave Protection Act* (1993) named after a cave in the Carlsbad Caverns National Park that was threatened by oil and gas exploration on adjacent land. The Act states that the cave has "internationally significant scientific, environmental and other values" and should be protected against adverse effects of mineral exploration, tourism and development. The Act withdraws all federal lands inside the boundaries of a protected cave area from all forms of mineral and geothermal leasing. The protected area was established by an expert panel of geologists and speleologists.

Other measures have been taken to document and manage US karst and caves. In 1988, the US Congress created a major impetus for cave conservation and management by passing the *Federal Cave Resources Protection Act* (1988). Section 2 of the Act states that

- "significant caves on Federal lands are an invaluable and irreplaceable part of the Nation's natural heritage"; and
- "in some instances, these significant caves are threatened due to improper use, increased recreational demands, urban spread, and a lack of specific statutory protection".

The purposes of the act are

- "to secure, protect, and preserve significant caves on Federal lands for the perpetual use, enjoyment and benefit of all people; and
- to foster increased cooperation and exchange of information between governmental authorities and those who utilize caves located on Federal lands for scientific, education, or recreational purposes".

The Act directs the Departments of the Interior and Agriculture to record all significant caves on federal lands, to provide management of these resources and to disseminate information on them. The Act made it an offence for any person who knowingly "destroys, disturbs, defaces, mars, alters, removes or harms any significant cave" or sells, barters or exchanges any cave resource, but licences can be issued for research purposes.

This was followed up in 1998 by the passing of the *National Cave and Karst Research Institute Act* (1998) to establish a research institute within the Carlsbad Caverns National Park, with the following six aims:

- "to further the science of speleology;
- to centralise and standardise speleological information;
- to foster interdisciplinary cooperation in cave and karst research programmes;
- to promote public information;
- to promote national and international cooperation in protecting the environment for the benefit of cave and karst landforms;
- to promote and develop environmentally sound and sustainable resource management practices".

Box 5.8   Palaeontology

There is a considerable diversity in the fossil record in the National Park System. Petrified trees, leaves, wood, pollen, shells, bones, teeth, eggshells, tracks, burrows and coprolites have been found in more than 159 National Park System units (V. Santucci, personal communication). However, the palaeontological resources of the parks are still being fully assessed (see, for example, Koch *et al.*, 2002), the number of palaeontologists employed in the National Park System is very low and in many parks that have significant palaeontological resources there are no geological staff. The fossils discovered range from the Precambrian stromatolites in Glacier National Park (Montana) to the Quaternary mammal bones found throughout the Alaskan parks. Some sites in the National Park system have been established solely or mainly for their palaeontological interest, mostly in the western states (Fig. 5.11):

- Agate Fossil Beds National Monument, Nebraska, is a large fossilised Miocene waterhole where hundreds of skeletons have been discovered, but where the majority are believed to be still preserved *in situ*. Among the extinct mammal fossils discovered are *Menoceras, Morpus, Dinohyus* and *Stenomylus*.
- Dinosaur National Monument, Colorado and Utah, where many dinosaur and other fossils have been found over the last 100 years. The rocks within the Monument range from Precambrian to Quaternary and many of them are fossiliferous, but the Upper Jurassic Morrison Formation, a fluviatile sandstone, is one of the most prolific dinosaur-bearing units in the world (Santucci, 2000b). The dinosaurs include Apatosaurus (*Brontosaurus*), *Diplodocus, Stegosaurus* and *Allosaurus*.
- Florissant Fossil Beds National Monument, Colorado, where a great diversity of fossil plants and insects are exceptionally well preserved in Eocene lake mudstone. They are poorly exposed today, but since the late 1800s over 60,000 fossils have been collected for museums, universities and private collections all over the world.
- Fossil Butte National Monument, Wyoming, where 100 m of the Eocene limestones, mudstones and volcanic ash containing some of the most perfectly preserved fossil fish and other aquatic animals and plants are exposed.
- Hagerman Fossil Beds National Monument, Idaho, where over 140 vertebrate and invertebrate species have been discovered in Pliocene river sands and lake clays. Eight of the species have been found nowhere else and 44 were discovered here first, including the Hagerman horse (*Equus simplicidens*).
- John Day Fossil Beds National Monument, Oregon, where 40 million years of Cenozoic rocks (Eocene-Pliocene) contain some of the world's most diverse and rich plant and animal fossil beds, including hundreds of Eocene plant species in the Clarno nutbeds and over 100 Miocene mammals groups in the John Day Formation.
- Petrified Forest National Park, Arizona, where Late Triassic fluviatile sedimentary rocks contain silicified wood from trees such as *Araucarioxylon, Woodworthia* and *Schilderia*. These were first discovered in the 1850s, but by 1900 removal of the "petrified wood" led to calls to preserve the remaining major sites. Collecting within the park is prohibited but commercial dealers collect from the same deposits outside the park.

Fossil collecting throughout the National Parks System is prohibited by the *Antiquities Act* (1906) and the *National Park Service Organic Act* (1916), referred to above, which prevent removal of "objects of antiquity" from federal lands, though again research permits can be

Continued on page 206

___ *Continued from page 205* ___

issued. However, a large number of theft and vandalism incidents affecting fossil sites are recorded each year and detection has increased following staff training (Santucci, 1999, 2000a).

The Devil's Tower in Wyoming became the first US National Monument in 1906, having been climbed first on 4 July 1893 with the assistance of a wooden ladder. Since then, there have been thousands of bolt and piton-aided ascents that have caused significant damage and defacement to the Tower (see Section 4.10). As a result, a Climbing Management Plan has been introduced. The Tower, which is composed of phonolite porphyry, is about 180 m high and 250 m in diameter with a relatively flat top and columnar-jointed sides (Fig. 4.12). It was featured in the film *Close Encounters of the Third Kind,* and partly as a result of that publicity, it attracts over 450,000 visitors per year, 5,000 of whom have come specifically to climb the pinnacle.

The Climbing Management Plan seeks to

- monitor climbing impacts;
- assess acceptability of these and whether climbing needs to be restricted;
- educate climbers about their impacts, including a climber brochure;
- retain rock faces that are currently free of bolts;
- accept that currently bolted areas will remain as permanent impacts to the rock;
- only allow replacement of bolts and fixed pitons under certain conditions, which create little new damage;
- investigate whether holes left by removal of illegally placed bolts can be filled with a rock dust/epoxy mixture.

5.8.2   National Natural Landmarks (NNLs)

This programme was established by the Secretary of the Interior in 1962 under the *Historic Sites Act* (1935) and is administered by the National Parks Service. An NNL is a nationally significant natural area as designated by the Secretary of the Interior. To be nationally significant "a site must be one of the best examples of a type of biotic community or geologic feature in its physiographic province". Examples include rock exposures, landforms and fossils. "It is a goal of the program to identify, recognise and encourage the protection of sites containing the best remaining examples of geological and geological components of the nation's landscape" (www.nature.nps.gov/partner/nnlp). A major review of areal blocks of the United States, Puerto Rico, the Virgin Islands and the Pacific Trust Territories was recently undertaken by university scientists and others to identify sites for inclusion. About 30% of NNLs are entirely privately owned, but they are normally only designated with the agreement of the landowners. In return for their goodwill in protecting the integrity of the feature they receive a plaque and certificate. In 2001, there were 587 designated NNLs, 53 of them designated for their palaeontology. Box 5.9 gives examples of geological NNLs.

Further sites can be added following nomination by groups or individuals, often as a result of designation in state natural area programmes (see below). Whether the site is added to the programme will depend "on the primary criteria of illustrativeness and condition of the specific feature, and secondary criteria of rarity, diversity and

---

Box 5.9   *Examples of Geological National Natural Landmarks*

The following are examples of a few of the NNLs which have been designated for their geological interest.

- *Dinosaur Trackway, Connecticut,* was discovered in 1966 when a bulldozer operator discovered a slab of rock at Rocky Hill showing tracks of a three-toed creature. Originally believed to be made by the bird *Eubrontes*, it has subsequently been identified as a bipedal animal with sharp claws and a stride length of over a metre. Comparison with dinosaur bones in Arizona suggests that the tracks may have been made by *Dilophosaurus*, a carnivore about 8 m long and 2.5 m tall, though there is uncertainty about this identification.
- *Diamond Head, Hawaii,* is a prominent peak on the rim of an extinct volcano in the south-east of Oahu. It was designated as an NNL to protect its slopes from commercial development at Waikiki Beach.
- *La Brea Tar Pits, California,* is situated in downtown Los Angeles, and contains one of the richest, best-preserved and heavily researched assemblages of Quaternary vertebrates in the world. About 60 species of mammal have been found, including extinct species of native horse, camel, mammoth, mastodon, long-horned bison and sabre-toothed cat. Thousands of individuals of one sabre-toothed cat genus, *Smilodon*, have been found at La Brea, and it has been adopted as California's State fossil. Over 135 species of birds have also been found, including vultures, condors, eagles and giant, extinct, stork-like birds known as *teratorns*. Numerous mollusc, insect and plant species have also been discovered and the fossils have been dated to between 40,000 and 8,000 years BP. Many are on display at the George C. Page Museum adjacent to the tar pits.
- *Enchanted Rock, Texas,* is a large, pink granite, exfoliation dome, representing the weathered top of a Precambrian batholith, just over 1,000 million years old. It rises 125 m above general ground level.
- *Hickory Run Boulder Field, Pennsylvania,* lies within the Hickory Run State Park (see below) and comprises a large, periglacial blockfield. It covers a flat area measuring *c.*120 m by 500 m and some of the blocks are 7 to 8 m long.

---

values for science and education". For example, in 1972, it was suggested that "Lakes and Ponds" should include examples of the following: large deep lakes, large shallow lakes, lakes of complex shape, crater lakes, kettle lakes and potholes, oxbow lakes, dune lakes, sphagnum-bog lakes, lakes fed by thermal streams, tundra lakes and ponds, swamps and marshy areas, sinkhole lakes, unusually productive lakes and lakes of high productivity and high clarity.

Evaluations of sites are carried out by scientists, and landowners are notified. If accepted by the National Parks Service the sites are placed in the Federal Register for a public comment period. If confirmed by the National Parks Service, they are designated by the Secretary of the Interior and listed on the National Register of Natural Landmarks.

NNL designation does not dictate activities that can occur within the site, but the designation has been used by individuals and organisations to draw attention to development threats (Gibbons & McDonald, 2001). Additionally, the Secretary is required to provide an annual report to Congress, prepared by the National Parks Service, on damaged or threatened NNLs. If sites are deemed to have lost the values that originally led to their designation, they can be removed from the list, but to date no site has been deleted.

The scheme has the potential to create a network of sites representative of the scientific geodiversity of the United States, but currently it is a long way from doing so, and there seems little prospect of this in the near future. There is also a need for better protection of those sites already designated.

5.8.3   National Wild and Scenic Rivers

The *Wild and Scenic Rivers Act* (1968) allows rivers and stretches of rivers that have outstanding scenic, recreational, geological, wildlife, historical or cultural or other similar values to be "preserved in free-flowing condition". The legislation states that the goal is to preserve the character of the river, and generally prohibits widening, straightening, dredging, filling, damming of the rivers or mining in their vicinity. There are three classes:

- Wild – free of impoundments, generally inaccessible except by trail, watersheds or shorelines essentially primitive and waters unpolluted, "vestiges of primitive America".
- Scenic – free of impoundments, watersheds or shorelines largely primitive, shorelines largely undeveloped, accessible in places by roads.
- Recreational – readily accessible by road or railroad, may have some development along their shorelines, may have undergone some impoundment or diversion in the past.

The lengths of US rivers now designated under the Act is increasing and by 2000 had reached over 17,000 km, though much of this is in the north-western states and Alaska. Oregon, for example, has 48 designated rivers including the Klamath River. But they occur all over the United States, and include Michigan's AuSable and Pere Marquette Rivers, Connecticut's Farmington River, Louisiana's Saline Bayou, West Virginia's Bluestone River and Massachusetts' Concord, Sudbury and Assebet Rivers (www.nps.gov/rivers).

5.8.4   State Parks

All states in the United States designate state parks that are intended to be areas for public recreational activities, but many also contain features of scientific interest, including important geological sites, a situation that often leads to conflicting pressures. This is particularly true where state parks are located near urban areas. Some examples of geologically oriented state parks are outlined in Box 5.10

Box 5.10   Examples of State Parks in the United States with Geological Associations

- *Falls of the Ohio State Park* lies on the banks and bed of the Ohio River in Clarksville, Indiana. Between August and October, the low water levels reveal extensive exposures

Continued on page 209

__ *Continued from page 208* __

of Silurian – Middle Devonian limestone forming part of a patch reef system. The fossils here include corals, sponges, bryozoans, trilobites, molluscs, brachiopods, crinoids and blastoids. Fossil collecting is prohibited within the park, but there is an interpretive centre with gallery of exhibits and a video presentation, while the web site allows a virtual tour of the park (www.fallsoftheohio.org/virtualtour).

- *Sun Lakes State Park* in Washington was formerly known as Dry Falls because of the impressive 120 m high, 5 km long, dried-up waterfall that operated during the Quaternary Ice Age. Peak discharge is estimated to have been about 10 times greater than the present Niagara Falls.
- *Archbold Pothole State Park* in Pennsylvania boasts what it claims as the world's largest glacial pothole measuring about 11 m deep and 12 m in diameter. It is excavated through sandstone, shale and anthracite coal at its base. In fact, it was discovered in 1884 by a coal miner extending a mineshaft, prior to which it had been filled with hundreds of tonnes of rounded stones and other debris. It is believed that it formed as a plunge pool at the foot of a glacial moulin.

5.8.5   State Scientific and Natural Areas

Most states also designate scientific and/or natural areas. These are specifically cited for their ecological and geological interests, though the ecological sites are predominant. They are usually open to the public for nature observation and education, but are not parks or recreational areas. Consequently, there are generally no, or only limited, facilities for visitors. Most states, for example Maine, Texas and Colorado, call these localities "Natural Areas" and have passed relevant legislation (e.g. *the Maine Natural Areas Program Legislation* (1993)), but in Minnesota they are "Scientific and Natural Areas". Some examples from Minnesota are outlined in Box 5.11.

Box 5.11   Scientific and Natural Areas in Minnesota, USA

There are over 130 Scientific and Natural Areas (SNAs) in Minnesota but it is estimated that 500 are needed to adequately protect significant features. The large majority are ecological sites, but the need to protect landforms, rock outcrops and fossil remains is also recognised. The programme, begun in 1969, has the goal of ensuring that no single rare feature is lost from any region of the state. The following are some examples of geological sites and these are very well documented on the web sites, all of which show the site boundaries, access roads and parking areas (see Fig. 5.13).

- Agassiz Dunes SNA is a dune field associated with Glacial Lake Agassiz, dammed by an ice sheet located over the site of the existing Great Lakes. The Sand River flowed into the lake at this locality and formed a delta 12,000 to 9,000 years ago. Subsequently, the sand was blown into dunes, and some have been reactivated to give secondary dunes and modern blow-outs.
- Gneiss Outcrops SNA (Fig. 5.13) are Precambrian rocks dated to *c*.3.6 million years ago making them some of the oldest known rocks on earth. Other outcrops along

__ *Continued on page 210* __

— *Continued from page 209* —

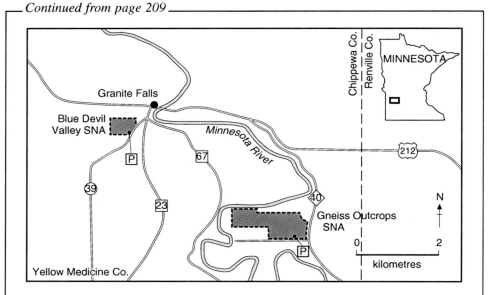

Figure 5.13   *An example of the website information for the Scientific and Natural Areas of Minnesota (Copyright 1999, State of Minnesota, Department of Natural Resources. Reprinted with permission)*

the Minnesota River have been mined and developed. The granitic gneiss is banded with light pinks, and red is most common, with grey to black hornblende-pyroxene or garnet-biotite gneiss being less common. At the nearby Blue Devil Valley SNA (Fig. 5.13) skinks can be seen sunning themselves on the granite outcrops, thus demonstrating an important geodiversity/biodiversity link.

- Grey Cloud Dunes SNA comprises two terraces of the Mississippi River. Sand dunes 3–6 m high occur on both terraces, some with impressive crescentic shapes.
- Ripley Esker SNA is a short section of a c.10 km long, steep-sided, meandering esker.
- Yellow Bank Hills SNA is a small area of kame mounds, ridges and kettle holes composed of sand and gravel.
- Sugarloaf Point SNA shows important examples of undeformed and unmetamorphosed Precambrian fluid lava flows from the North Shore Volcanic Group. Flows range from under a metre to tens of metres thick and show pahoehoe structures and pipe amygdales. The modern beach cobbles demonstrate a diversity of rock types, many of which are Canadian erratics.

5.8.6   Other legislation and protected areas

In 1987, a major review of fossil collecting in the United States was published partly in response to differences in policies between states. It made strong recommendations setting out the need for uniformity in national policy. It recommended that fossil collecting for research purposes be encouraged, but that commercial collecting should be done through permit regulated by palaeontologists. It also recommended a public education programme for collectors, amateur researchers and landowners. However, it engendered much heated argument and was not pursued within government. It was followed in 1992 by a new Bill (Known as the Baucus Bill, after Senator Max Baucus) proposing a *Vertebrate Palaeontology Resources Protection Act*. The key provisions proposed were

- recognition that vertebrate fossils on federal lands are part of the heritage of the United States and are a non-renewable resource;
- recognition of the contributions of amateur collectors and providing a means by which amateurs may lawfully collect vertebrate fossils on federal lands;
- prohibition of the sale of vertebrate fossils collected on federal lands;
- establishing that fossils collected on federal lands are the property of the nation and should be curated appropriately;
- establishing a set of financial penalties for theft, vandalism or sale of fossils from federal lands.

However, this too was not passed and instead a new Bill (HR2974) proposing a *Palaeontological Resources Preservation Act* with similar aims to the Baucus Bill but covering all palaeontological resources was introduced by Senator McGovern (Massachusetts) in 2001. Its fate is still awaited at the time of writing.

The issue was highlighted during the 1990s when the FBI seized a complete skeleton of *Tyrannosaurus rex* removed from federal land in South Dakota by commercial collectors (Norman, 1994; Horner & Dobb, 1997; Fiffer, 2000). Currently, invertebrate and plant fossils can be collected with hand tools from public lands, but vertebrate fossils found on public lands "belong to everyone. They are considered national treasures..." (Horner & Dobb, 1997, p. 238). Excavation of vertebrate fossils therefore requires a permit to ensure that proper scientific recording and excavation take place, though attempts have been made by commercial collectors to amend the legislation to allow commercial exploitation (Horner & Dobb, 1997). Palaeontological remains on private lands in the United States are generally regarded as private property, a situation that has been questioned (Sax, 2001).

The Nature Conservancy is a not-for-profit organisation dedicated to purchasing land of nature conservation value and ensuring its long-term management. Although mainly concerned with wildlife conservation, it does also purchase sites of geological value, perhaps the best example being Egg Mountain, near Choteau in Montana, famous for its Maiasaur dinosaur finds (Horner & Dobb, 1997).

5.9   Canada

5.9.1   National Parks

Canada also has a National Park system established under the *National Parks Act* and recently revised by the *Canada National Parks Act* (2000). There are currently 39 National Parks, National Park Reserves and National Marine Conservation Areas administered by Parks Canada. Many of these contain important geological and geomorphological sites or landscapes (Figs. 5.4 and 5.14), and some have been specifically designated for their geological interest, including the Miguasha fossil site in Quebec and the Gros Morne National Park in Newfoundland (Box 5.12). Four National Parks,

Figure 5.14   *Southern entrance to Jasper National Park*

including Yoho National Park and its famous Burgess Shale sites, together with three Provincial Parks make up the Rocky Mountains World Heritage Site (see Fig. 5.4).

Box 5.12   Gros Morne National Park, Newfoundland, Canada

This National Park and World Heritage Site has been established mainly for its geological interest. The Park was the site of a constructive plate margin 600 million years ago with thick basalt sequences visible at Western Brook Pond. From 570 to 420 million years ago, the area was an ocean and sedimentary strata containing a wealth of early Palaeozoic fossils were formed. However, the plate divergence was reversed about 460 million years ago, producing some unusual stratigraphic relationships. In particular, the juxtaposition of oceanic lithosphere and mantle ophiolites overlying continental crust is demonstrated in The Tablelands area of the park. Other geological points of interest include the global stratotype of the Cambro-Ordovician boundary, 40 important fossil sites including graptolites and trilobites, a massive rock sag at Bonne Bay and the famous Cow Head conglomerate. Geomorphological features include fjords, sea cliffs and intertidal platforms.

The National Parks legislation and guidance makes it clear that the primary objectives in the national parks are related to "ecological integrity" and "ecosystem management". Although "ecological integrity" is defined in the Act as including "abiotic elements" and although this is clearly understood by many park managers, there is no specific reference to geology, geomorphology or pedology in the Act or related guidance. This is clearly a weakness in Canadian geoconservation and ought to be addressed through a parallel commitment to "geological integrity". There is no national system of geosites in Canada, the "National Landmarks" programme begun in the 1970s having been abandoned at an early stage with only the Mackenzie Delta pingos having been included (D. Welch, personal communication).

All National Parks are required to have a Management Plan, which is updated every five years. Most of these include sections on geology and landforms outlining objectives and key actions for the plan period. For example, Box 5.13 and Fig. 6.15 outline and illustrate the priorities for Waterton Lakes National Park.

Box 5.13   *Geological and Landform Priorities in the Waterton Lakes National Park Management Plan*

The "strategic goal" is to allow "natural processes, including erosion and deposition, (to) shape and define the landscape and its ecosystem".

The "objectives" are "to protect and restore park landforms and associated physical processes from the impacts of development and use", and "to consider the impact of management decisions on landforms outside the park".

The "key actions" are as follows:

1. Keep disturbance of the park's landforms to a minimum. Approve activities that disturb landforms only when rehabilitation plans are finalised.
2. Whenever possible, avoid manipulating natural processes (e.g. avalanches or flooding) to protect facilities that cannot be relocated.
3. Move the Parks Canada storage area from the Blakiston Creek alluvial fan to the government compound; reclaim area.
4. Rehabilitate disturbed sites such as Red Rock Canyon.
5. Keep the need for aggregate material to a minimum.
6. Obtain construction material from suitable sources outside the park (e.g. sources that do not result in new disturbance to native vegetation, wildlife corridors, riparian areas, or scenic views).
7. Re-design the Waterton Community Golf Course's water supply to eliminate manipulation of the Blakiston Creek flood plain. Work with the golf course on a plan to use the park's treated sewage for irrigation.

All Canadian National Parks are also zoned into five categories (Special Preservation, Wilderness, Natural Environment, Outdoor Recreation and Park Services) plus "environmentally sensitive sites". Several geological/geomorphological interests are further protected by these zones. For example, in Jasper National Park, the Surprise Valley is designated as a Special Preservation Zone, being part of the Maligne karst system. This valley is drained entirely underground through limestone of the Upper Devonian Palliser Formation, and is associated with one of the largest underground river systems in North America. It also contains deep sinkholes in glacier drift, sink lakes, and some of the finest rillenkarren in North America. No new access will be provided to this area. The Yoho and Kootenay National Parks Management Plans both designate Burgess Shale sites as Special Preservation Zones and access to the classic Walcott quarry in Yoho is only allowed under licence or with a trained guide. The same is true of Rabbitkettle Hot Springs in Nahanni National Park because of the physical delicacy of the deposits.

Although Canada is divided into 39 natural regions and also has 39 national parks, this is purely coincidental since several natural regions contain more than one park. The long-term aim of Parks Canada is to establish at least one national park in each natural region so that the system as a whole is representative of the natural diversity of the country's landscapes. Several new parks are already being planned and some proposals are well advanced (D. Welch, personal communication).

5.9.2 Heritage Rivers

A particular designation of "heritage rivers" was introduced in 1986 when 110 km of the French River became the first of several Canadian Heritage Rivers, though habitat protection, cultural heritage and recreational value can be as important or more important than geomorphology in their recognition (McKenzie, 1994). In fact, the approach taken is a holistic one involving cultural, biological and geological sciences. Nonetheless, it is clear that geological processes have been and are responsible for developing these spectacular rivers, and the criteria used to decide whether a waterway can be designated include several that are geological/geomorphological. For example, a river can be designated if

- it is an outstanding example of river environments as they are affected by major stages and processes in the Earth's evolutionary history;
- it is an outstanding representation of significant ongoing fluvial, geomorphological and biological processes;
- it contains along its course, habitats with rare or outstanding examples of natural phenomena, formations or features, or areas of exceptional natural beauty.

Before designation takes place, public consultation procedures and management guidelines are drawn up, which always exclude mining as an allowable activity. After designation, monitoring of river quality, habitats and so on. takes place. Figure 5.15 shows the Heritage River System as designated in 2003 (www.chrs.ca/rivers_e.htm).

5.9.3 Provincial designations

The Canadian Provinces have considerable autonomy in designating Provincial Parks, Nature Reserves, Natural Areas, and so on and a variety of landowner agreements and private stewardship initiatives also exist (Davidson *et al.*, 2001). Almost all are owned and managed by the provinces, though some are managed jointly with local indigenous groups. For example, Nisga'a Memorial Lava Bed Provincial Park in British Columbia is managed jointly by BC Parks and the Nisga'a Tribal Council. Attempts have been made to get all provinces to list their protected areas under the IUCN categories but with limited success. Hence, a range of protected areas exists that make inter-province comparison difficult. The absence of any geological designation category in the provinces has led to some interesting treatments of geological sites. For example, in Newfoundland the global stratotype of the Precambrian – Cambrian boundary has had to be designated as an Ecological Reserve! Three provinces are discussed here – Ontario, Alberta and Quebec.

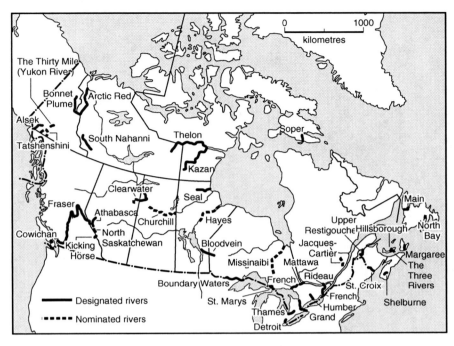

Figure 5.15 *The Canadian Heritage River System (2003) (Reproduced with permission from Canadian Heritage Rivers System)*

Ontario

Ontario's provincial park system underwent a rapid expansion during last century (Fig. 5.16) and has evolved into "one of the world's outstanding networks of protected areas" (Davidson, 2001). A major review of Provincial Park Policy took place in the 1970s on the basis of the principle that the system of protected areas should represent the diversity of biological, geological and historical features that comprise the province's landscape. This led to the designation of over 150 new parks, many at least partly based on Precambrian, Palaeozoic and Quaternary stratigraphy and geomorphology. At the same time, about 300 new geological Areas of Natural and Scientific Interest (ANSIs) on private land were identified, including key-type sections, reference sections and type areas. These are managed or overseen by the Ontario Ministry of Natural Resources (OMNR).

Private landowners are eligible for tax relief through the *Conservation Land Tax Program* if they agree to protect the geological interest of an ANSI on their property. Furthermore, under the *Ontario Heritage Act*, conservation easements can be established, the first being an Ordovician type section and fossil locality near Owen Sound. "In return for the loss of certain development rights, the landowners were provided with an income tax rebate for the full value of this gift to the province" (Davidson *et al.*, 2001, p. 230).

A further review of Ontario's parks and protected areas took place in 1996 and brought further additions, particularly of glacial landscapes, so that Ontario now has 636 protected areas covering 9.5 million hectares or nearly 9% of the province's area. Included in the system at this time were the Whitemud Conservation Reserve, a

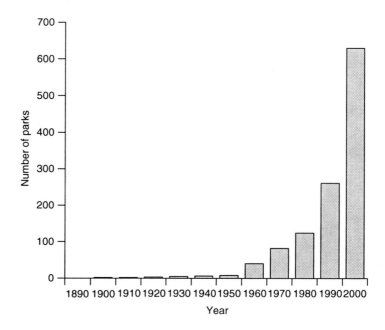

Figure 5.16   *Growth of Ontario's Provincial Park System during the 20th century (© Queen's Printer for Ontario, 1992. Reproduced with permission)*

protected area of over 17,000 ha covering moraines, kames and eskers of the Lac Seul Moraine and features of former Lake Agassiz.

There are six categories of provincial park (Wilderness Parks, Nature Reserves, Historical Parks, Natural Environment Parks, Waterway Parks and Recreation Parks), each of which is zoned with some or all of the following categories: Natural Environment Zones (NE), Development Zones (D), Wilderness Zones (WI), Nature Reserve Zones (NR), Historical Zones (HI) and Access Zones (A). The matrix of existing park classes and zone types is shown in Fig. 5.17. Management and monitoring policies are set for classes and zones, and Fig. 5.18 is an example of activity management in Nature Reserves.

Alberta

Alberta has eight existing categories of protected areas designated under the *Provincial Parks Act* (1980), *Wilmore Wilderness Park Act* (1980) and *Wilderness Areas, Ecological Reserves and Natural Areas Act* (1989). In addition, some geomorphological sites are protected under the *Historical Resources Act* (1980). The categories range from Wilderness Areas, where the only human activity allowed is foot access and which are therefore more highly protected than National Parks, to Recreation Areas, where outdoor recreation is promoted and encouraged (Fig. 5.19). There is little active enforcement in the protected areas, but among the strict management measures in existence is the public exclusion from the Plateau Mountain Ecological Reserve cave to prevent ice crystal melting.

A major review in the late 1990s has brought the protected areas together and expanded them under Alberta's "Special Places" programme. However, because of

Park class	Zone type					
	NE	D	WI	NR	HI	A
Wilderness			●	●	●	●
Nature Reserve				●	●	●
Historical	●	●		●	●	●
Natural environment	●	●	●	●	●	●
Waterway	●	●	●	●	●	●
Recreation	●	●		●	●	●

NE   Natural Environment Zones
D   Development Zones
WI   Wilderness Zones
NR   Nature Reserve Zones
HI   Historical Zones
A   Access Zones

Figure 5.17   *Ontario's Provincial Park classes and zone types (© Queen's Printer for Ontario, 1992. Reproduced with permission)*

overlaps and inconsistencies, the Alberta government is currently reviewing the legislation with the intention of

- consolidating, updating and improving the legislation under a proposed *Parks & Protected Areas Act;*
- establishing a revised system of classes of protected areas;
- clarifying the purpose and management requirements of each class;
- providing for the long-term management and protection of Alberta's expanded parks and protected areas system.

Almost half of the total of 528 protected areas in Alberta are Recreation Areas, though many of these are being downgraded or transferred to local government. Some protected areas are specifically designated for their geological/geomorphological interest or the geohabitats they provide for wildlife. Examples include sand dunes (Athabasca Dunes Ecological Reserve; Richardson River Dunes Wildland Provincial Park), glacial features (Marguerite River Wildland Provincial Park), periglacial features (Plateau Mountain Ecological Reserve), river canyons and coulees (Brazeau Canyon Wildland Provincial Park; Red Rock Coulee Natural Area), karst landscapes (Whitemud Falls Wildland Provincial Park and Ecological Reserve), erratics (Okotoks Erratic Historic Site) and fossil sites (Dinosaur Provincial Park, see Box 5.14).

In Section 6.2.4, Alberta's approach to landscape classification and its contribution to identifying new protected areas are described. Landscape classification is essentially a "top-down" approach, but Alberta has also been attempting a "bottom-up" approach by identifying special features that may be considered for inclusion in the protected areas network subsequently. Elements of special conservation concern are "those that are restricted in extent or distribution, are small in number, or are considered to be an outstanding example of that element". A study of the special features in Alberta carried out in 1998 included landform types and specific features identified from discussions

Activities and facilities Nature Reserve Parks	Zones		
	NR	HI	A
All Terrain Vehicle (ATV) travel			
Aircraft landing: water - commercial tourism			
- private			
Boating (powered) - commercial			
- private			
Campgrounds - car			
- boat-in/walk-in			
- group			
- back-country			
Canoeing/Kayaking			
Demonstration areas (eg logging exhibits etc)			
Hiking			
Historical appreciation: self-guided			
Horseback riding: trail			
Mountain biking: designated trails			
Nature appreciation: self-guided			
Orienteering			
Outfitting service			
Outpost camp (commercial			
Painting/ photography			
Picnic grounds			
Playgrounds			
Recreation programmes (organised)			
Resorts/Lodges			
Restaurants: food and beverage			
Rock climbing			
Sailing and surfboarding			
Scuba and skin diving			
Skiing (cross-country)/Snowshoeing			
Snowmobiling			
Spelunking (cave exploration)			
Sport fishing			
Sport hunting			
Swimming: facility-based			

Not compatible with this zone; if now existing, a non-conforming use to be phased out ▓

May be encouraged in this zone in certain parks where deemed appropriate ▒

Normally encouraged in this zone ☐

Nature Reserve Zones NR

Historical Zones HI

Access Zones A

Figure 5.18 *Management policies within Ontario's Nature Reserves.* © *Queen's Printer for Ontario, 1992. Reproduced with permission.*

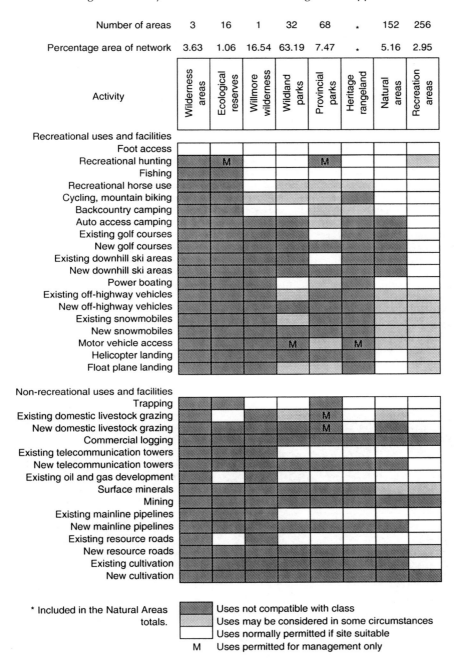

Figure 5.19 *Management policies within Alberta's protected areas. (Reproduced with permission of Archie Landals)*

with academics and literature searches. The criteria used to identify a landform element as being of special conservation status were that

---

Box 5.14   Dinosaur Provincial Park, Alberta, Canada

This 27-km-long site in the Red Deer River valley is internationally famous for its Late Cretaceous dinosaurs. Between 1979 and 1991, a total of 23,347 fossil specimens were collected, including over 300 dinosaur skeletons from at least 35 distinct species, including specimens from every known group of Cretaceous dinosaurs. No other site in the world has such an abundance of diverse dinosaur remains concentrated in such a relatively small area. Other fossils include fish, turtles, marsupials, amphibians and tropical plants.

The fossils occur in the 80-m-thick Dinosaur Park Formation (76.5–74.5 Ma) deposited by sediment-laden rivers flowing eastwards from the incipient Rockies into the shallow Bearpaw Sea that covered central North America between about 100 and 70 million years ago. Under the warm, humid and wet conditions, whole dinosaur carcasses were entombed by the huge quantities of sand and mud pouring into the system. These deposits developed into the fossiliferous claystones, siltstones, sandstones and thin coals observed today. During the last glaciation, meltwaters eroded steep-sided channels into these rocks, creating an eroded badland topography and exposing the strata for palaeontological research.

Several dinosaur fossils excavated from the site are on display in the Royal Tyrrell Museum in Drumheller, about 200 km up the Red Deer River from the Park, and others appear in the American Museum of Natural History in New York, the Canadian Museum of Nature in Ottawa, the Royal Ontario Museum in Toronto and the Natural History Museum in London (Gross, 1998). The site's international importance is recognised by its World Heritage Site status (Fig. 5.20).

Figure 5.20   *Plaque commemorating Dinosaur Provincial Park as a World Heritage Site, Alberta, Canada*

- there are five or fewer known occurrences of the landform type in the province;
- an occurrence of a landform type with more than five occurrences is considered special if it is an outstanding example as judged by geomorphological experts as noteworthy because of size, quality or representativeness in a provincial, national or international context.

This work has yet to be completed and is not currently seen as a priority in Alberta (A. Landals, personal communication).

Quebec

Quebec has about 1,100 protected sites within 17 different judicial or administrative designations. Nonetheless they cover only 2.8% of the land area of the province. To meet international obligations Quebec has agreed to increase its protected areas to 8% of the land area and has recognised that one existing deficiency is in protecting its geodiversity. A consultation exercise was carried out in 2002/03 by a working group with the aim of establishing a network of protected geosites in the province (Government of Quebec, 2002; P. Verpaelst, personal communication).

5.9.4   Other legislations

Control of fossil collecting in Canada is the responsibility of the Provinces through the passing of appropriate acts of government. The first such act (*The Historical Objects Protection Act* (1970)) was passed by Nova Scotia. All Provinces now have such acts to regulate fossil collection, which generally involves the issuing of research permits. Alberta's *Historical Resources Act* (1980) prohibits all excavation other than by permit holders, but Norman (1994) detects inconsistency from province to province in the degree to which individual fossil groups are protected and in the rigour of law enforcement. There is also a Federal Law (*the Federal Cultural Property Export and Import Act* (1975)) that controls the export of fossils from Canada.

5.10   United Kingdom

"Britain can be regarded as having more geological diversity than any other comparable area on Earth" (Trewin, 2001, p. 59) and perhaps as a result has well-developed, if complex, systems for geoconservation. As stated above, there were many early efforts at nature conservation in general and Earth science conservation, in particular, in the United Kingdom, but major steps were taken in the late 1940s. One was the establishment by Royal Charter of the Nature Conservancy as the UK government's nature conservation agency covering England, Scotland and Wales (but not Northern Ireland) and the other was the passing of *the National Parks and Access to the Countryside Act* (1949). These measures empowered the Nature Conservancy to establish National Nature Reserves and Sites of Special Scientific Interest (SSSI) on the basis of their flora, fauna, geology or "physiography" (geomorphology). The impetus was a report (Huxley, 1947) of the Wildlife Conservation Special Committee, and it is indeed fortunate that the Committee included both a geologist (A. E. Trueman) and a geomorphologist (J. A. Steers). The Nature Conservancy was later retitled the "Nature Conservancy Council", but under the *Environmental Protection Act* (1990) it was divided into three country-based agencies (English Nature (EN), Scottish Natural Heritage (SNH) and the Countryside Council for Wales (CCW)). In the case of Scotland and Wales, this was more than a simple change of name since the organisations were merged with the Countryside Commission that had responsibility for the cultural landscape. In England, English Nature and the (retitled) Countryside Agency still exist as separate entities (see Section 6.2.2), allegedly because the Thatcher government believed that a single conservation organisation for England would become too powerful. Overseeing the national agencies is the Joint Nature Conservation Committee (JNCC), which advises on nature conservation policy both nationally and internationally, undertakes research and tries to establish common standards for monitoring sites across the United Kingdom.

In the United Kingdom, a large number of site designations are used and some believe that there is an urgent need for simplification of the "existing maelstrom of countryside designations" (Adams, 1996, p. 148).

5.10.1 National Parks

British national parks are very different from those of the United States and Canada in that they are "neither national (for they are not nationally owned) nor parks (for they include large areas of farmland and private ground to which the public has no right or even custom of access), nor do they satisfy the internationally agreed criteria for national parks as being wilderness areas, which have not been materially altered by human occupation and which are owned or managed by a government primarily for the purpose of conservation" (MacEwan & MacEwan, 1987, p. 3). This "peculiarly British national park system" would be recognised internationally not as "national parks" but as "protected landscapes". "What the British call 'national parks' are intended not so much to conserve uninhabited wilderness as to protect inhabited landscapes where the land should be managed for a multiplicity of purposes – conserving their character, promoting their enjoyment and supporting human life in many diverse ways" (MacEwan & MacEwan, 1987, p. 4).

Until recently there were 10 national parks in the United Kingdom, all in England and Wales, but there has been a flurry of interest in creating new ones in recent years. First, the Norfolk & Suffolk Broads, an area of rivers and flooded peat diggings in Eastern England, which has had a status equivalent to a national park for over 10 years, has now sought permission to adopt the name – The Broads National Park. Secondly, five new national parks have been planned, two in England (New Forest and South Downs), two in Scotland (Loch Lomond & the Trossachs (created in 2002) and Cairngorms (created in 2003)) and one in Northern Ireland (Mournes) and these are shown in Fig. 5.21.

It is usually only coincidental that important geological aspects, landforms or soils are included within these areas, though their scenic beauty and often ancient geology inevitably mean that landscape and geoscientific interests often coincide. Thus, many of the British national parks, such as the Lake District, Snowdonia, Dartmoor and others, although designated for their landscape beauty, all contain important geological and/or geomorphological interests. All National Park Authorities are required to produce regularly updated management plans and many refer to geology and geomorphology. For example, the Lake District National Park Management Plan (1998, para 3.44) seeks "to protect and, where appropriate, enhance all important geological and geomorphological features in the National Park".

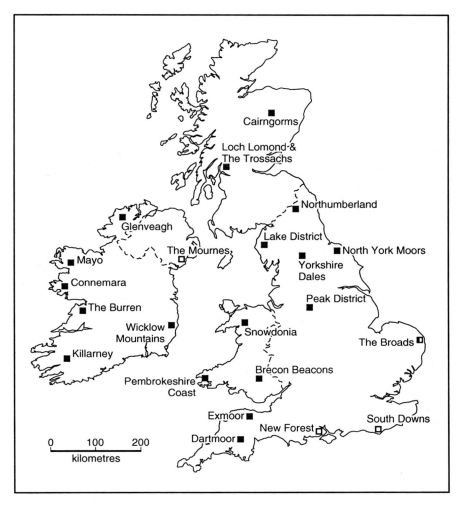

Figure 5.21 *Location of Britain and Ireland's designated (black squares) or potential (open squares) National Parks. Note that The Broads have the status of a National Park but not yet the name*

5.10.2   Nature Reserves

The Royal Charter establishing the Nature Conservancy empowered it to establish National Nature Reserves (NNRs), which are areas preserved primarily to maintain and enhance their scientific status and research potential. Marine Nature Reserves (MNRs) may also be designated. NNRs are generally owned by the State and managed by one of the three national conservation agencies (English Nature, 2000b).

A few of the NNRs and many of the MNRs contain important geological and geomorphological sites. Examples include the Wren's Nest NNR in Dudley, West Midlands designated for its ripple beds containing Silurian fossils including trilobites, Cwm Idwal in Snowdonia, Wales, which contains an important series of moraines mentioned by Charles Darwin, the Isle of Rum, Scotland, famous for its Tertiary igneous centre and the Portrush sill in Northern Ireland where there was an important historical debate between the Neptunists and Vulcanists over the origin of igneous rock (Fig. 5.22). Local authorities also have the power to designate Local Nature Reserves (LNRs) and by 2003 there were over 700 of these in England, though few, if any, have been specifically designated for their geology or geomorphology.

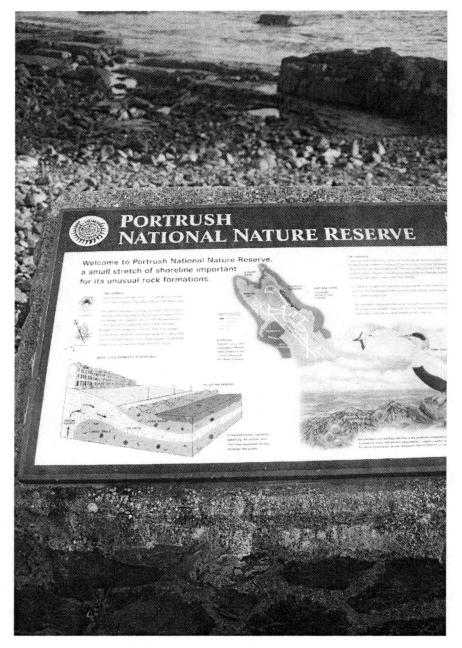

Figure 5.22 *Information panel at the Portrush National Nature Reserve, Northern Ireland*

5.10.3   Sites of Special Scientific Interest (SSSIs)

The *National Parks and Access to the Countryside Act* and subsequently the *Wildlife and Countryside Act* (1981, as amended 1985) also required the Nature Conservancy to designate sites worthy of protection purely on the basis of their scientific value. Both biological and geological SSSI were to be established, but this status did not give a site direct protection. Rather, it simply ensured that local authorities were notified of the sites and that a full consultation process took place if developments affecting SSSIs were proposed. The 1981 Act enables "potentially damaging operations" (PDOs) to be identified for each site (see Section 4.1). The *Town & Country Planning Act* (1947) and (1990) ensured that local authorities consulted the Nature Conservancy and its successors on relevant planning applications and allowed them to adopt appropriate planning policies (see Section 6.6). In Northern Ireland, Areas of Special Scientific Interest (ASSIs) give similar protection. Most SSSIs and ASSIs are on private land.

The situation has been strengthened significantly in England and Wales by the *Countryside and Rights of Way Act* (2000) (usually known as the CROW Act). This puts the emphasis on positive site management and partnership between the conservation agencies and landowners and occupiers, rather than paying them not to carry out operations that could damage sites (Prosser & Hughes, 2001). However, if agreement cannot be reached, the new Act allows it to be imposed, and this will allow action to be taken to prevent site deterioration through neglect or deliberate damage. The Act also makes it an offence for anyone, including third party fossil or mineral collectors or vandals, to knowingly or recklessly damage an SSSI. The Act also gives EN and CCW powers to introduce by-laws on SSSIs to further protect them from third party damage, and strengthens their right to enter private land to investigate offences and monitor the condition of SSSIs (Prosser & Hughes, 2001). The Act gives right of access to several categories of open land and therefore is helpful in giving access to geological exposures. Similar legislation is awaited in Scotland where SNH remains powerless to prosecute even when it has identified an irresponsible collector (MacFadyen, 2001a).

Up until 1977, SSSIs were selected on the basis of available information and consultation with the country's geoscientists, but this resulted in very inconsistent site networks (Gordon, 1994a) and a major review, the Geological Conservation Review (GCR), took place between 1977 and 1990 to establish a systematic list of important geological and geomorphological sites (Nature Conservancy Council,1990; Ellis *et al.*, 1996; Bennett & Doyle, 1997; Glasser, 2001). The site series was intended to reflect "the range and diversity of Great Britain's Earth heritage" (Ellis *et al.*, 1996, p. 45) as defined as

1. sites of importance to the international community of Earth scientists, for example, time interval or boundary stratotypes, type localities for biozones, internationally significant type localities for particular rock types, mineral or fossil species, historically important type localities or features;
2. sites that are scientifically important because they contain exceptional features;
3. sites that are representative of an Earth science feature, event or process that is fundamental to Britain's Earth history.

In practical terms, a number of operational criteria were employed (Ellis *et al.*, 1996) so that preference is given to sites

- where there is a minimum of duplication of interest between sites;
- where it is possible to conserve any proposed site in a practical sense;
- that are least vulnerable to potential threat;
- that are more accessible;
- that show an extended, or relatively complete record of the feature of interest;
- that have a long history of detailed research study and re-interpretation;
- that have potential for future study;
- that have played a significant part in the development of the Earth sciences, including former reference sites, sites where particular geological phenomena were first recognised, and sites that led to the development of new theories or concepts.

"Application of these criteria ensures that sites chosen for a particular network in the GCR have the greatest collective scientific value and can be conserved" (Ellis *et al.*, 1996, p. 75). Attempts were made to minimise the number of sites designated so that only those that are necessary to characterise "the network" were selected. Ellis *et al.* (1996, p. 76) define "the network" as those sites that "demonstrate the current understanding of the range of Earth science features in Britain. . . . ". In other words, it is a network of sites that represents the geodiversity of the country. The size of each site was also kept to a minimum, though there are exceptions in upland areas with a range of features.

About 3,000 GCR sites have been identified, but because of overlaps (e.g. a fossil site within a coastal geomorphology site), about 2,300 SSSIs are currently being designated by the three UK nature conservation agencies. This represents over one-third of United Kingdom's 6,573 SSSIs as of April 2002, the remainder being biological. GCR site descriptions are being published in a set of 42 volumes, each made up of blocks covering a particular geological period, rock or landform type and/or part of the country. The first volume on the *Quaternary of Wales* (Campbell & Bowen, 1989), was published in 1989 and an introductory volume explaining the aims, processes and products of the review was published in 1996 (Ellis *et al.*, 1996). Publication of the series is continuing and is due to be completed around 2005. Gordon & Leys (2001b, p. 7) refer to it as "the most comprehensive review undertaken by any country". In Northern Ireland, the equivalent of the GCR has identified about 300 sites, and these are being designated as Earth science ASSIs (Doughty, 2002).

SSSIs are classified into "exposure" and "integrity" sites:
Exposure sites are those whose scientific interest or educational value:

> "lies in providing exposures of a deposit that is extensive or plentiful underground. . . . and is almost certain to contain similar features to those visible at the site. . . . Such "exposure sites" are numerically the most common category of sites and include most quarries, cuttings, cliffs, outcrops and mines" (Nature Conservancy Council, 1990, p. 33).

The management priority at exposure sites is to preserve exposure of the features at the site, so that quarrying may well be welcomed as providing fresh exposures for research and teaching and allowing the three-dimensional form of sediment bodies to be revealed. In other situations, this may be controversial where coastal erosion is the process giving continuous exposure and local residents wish to stop it (see Box 6.10).

Table 5.5 *Classification of UK Earth science sites (Nature Conservancy Council, 1990)*

Exposure sites	Disused quarries, pits and cuttings	ED
	Active quarries and pits	EA
	Coastal & river cliffs	EC
	Foreshore exposures	EF
	Inland outcrops and stream sections	EO
	Mines & tunnels	EM
Integrity sites	Static geomorphological sites	IS
	Active geomorphological process sites	IA
	Caves & karst	IC
	Unique mineral, fossil or other geological sites	IM
	Mine dumps (continuum with specimen collections)	ID

Compromise solutions to conflicts between those wishing to stop erosion and geological conservation bodies wishing it to continue may be possible. For example, McKirdy (1990) described a solution of partial coastal protection that would significantly reduce the rate of coastal erosion, but retain the geological exposures.

Integrity sites are those whose scientific or educational value

> "lies in the fact that they contain finite and limited deposits or landforms that are irreplaceable if destroyed. The usual situation is that the deposit or landform is Quaternary in age, and of limited lateral extent, although many geologically older examples also fall into this category. Examples include glacial, fluvial and coastal landforms and their associated deposits, cave and karst sites, and unique mineral, fossil, stratigraphic, structural or other geological deposits and features" (Nature Conservancy Council, 1990, p. 33).

The management of these sites is quite different. Because the integrity sites are finite and irreplaceable, they are vulnerable to many of the threats described in Chapter 4, and therefore the approach taken is one of preservation and restriction of man-made changes. This category is the nearest equivalent to most biological SSSIs, which are designated because of representative, important or threatened species or disappearing habitats. There is therefore a close correspondence between biological and geomorphological site conservation, whereas most geological site conservation relies on continuous exposure of the site and removal of material.

The Nature Conservancy Council (1990) suggested a classification of both integrity and exposure sites (see Table 5.5) and Box 5.15 gives some examples of important earth science SSSIs in the United Kingdom.

Box 5.15   *Examples of UK SSSIs*

The following are some examples of the *c*.2300 earth science SSSIs currently being designated in the United Kingdom:

- Siccar Point, Berwickshire is known as Hutton's unconformity. Upturned Llandovery, Silurian greywackes and shales overlain by sub-horizontal Devonian sandstones and breccias, led James Hutton in the 1788 to recognise the principle of the angular unconformity, the essentially cyclical nature of erosion and deposition and the great age of the Earth. As such, it is one of the most important geological sites in the world (Fig. 5.23).

Continued on page 228

__ Continued from page 227 __

Figure 5.23   Hutton's unconformity at Siccar Point SSSI, Scotland

- Rhynie Chert, Aberdeenshire, is internationally famous as containing some of the world's oldest terrestrial plant fossils (Devonian), as well the earliest known wingless insect (*Rhyniella*) and a very fine micro-arthropod fauna, including mites, springtails and a small aquatic shrimp-like creature (*Lepidocaris*). The site is a silicified peat produced when the site was flooded by silica-rich hot water. This killed and preserved the fauna before the tissue decayed and so preserved a complete three-dimensional ecosystem (Ellis *et al.*, 1996; Rice *et al.*, 2002). The structure of the xylem, stomata and sporangia with spores in place can be observed.
- Blakeney Esker, Norfolk, is England's best-developed esker (Fig. 5.24). Gravel quarrying has revealed its internal structure and has demonstrated that the gravels fill a series of channels cut into the underlying marly drift, indicating that the esker was formed subglacially (Gray, 1997b). However, gravel quarrying also destroyed large parts of the esker prior to designation (Fig. 5.24).
- River Feshie, in the Scottish Highlands is one of United Kingdom's few braided, gravel-bed rivers. It regularly moves in its course and is building a fan at its confluence with the river Spey. It is therefore an important fluvial geomorphology site (Werritty & Brazier, 1994).
- Dundry Main Road South Quarry SSSI and Barns Batch Spinney SSSI in Somerset are two of the UK sites with important historical connections (Prosser & King, 1999). It was here in the 1790s that William Smith, the "Father of English Geology", whilst undertaking surveying work for a new canal, first appreciated the significance of stratigraphically organised fossils (Winchester, 2002).

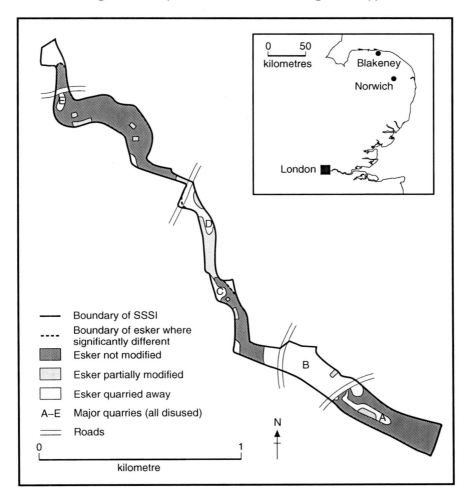

Figure 5.24 *Map of the Wiveton Downs SSSI (Blakeney Esker), England showing quarried areas*

Some site documentation is usually required before a site is designated. Normally, this will outline the site's location, importance and the potential threats to it. The 42 volumes of the GCR series give details of all the sites in terms of "Highlights", "Introduction", "Description", "Interpretation" and "Conclusions". At a more detailed level, both the CCW and SNH have produced a series of Site Documentation Reports on GCR sites in these countries. In the case of Scotland, "The primary purpose of this series of reports is to describe and explain the scientific interest of each of our GCR sites in plain English". They are written primarily for the benefit of Area Staff concerned with the conservation and enhancement of the sites, but are also used by landowners or managers. The contents are "designed to make the reader a more informed and effective advocate for earth science conservation". Over 660 of these reports have been produced (MacFadyen, 2001b). In the case of the Moss of Achnacree and Achnaba SSSI, the report runs to 85 pages including 4 maps and over 40 photographs of the site (Bentley, 1996). As well as the scientific interest of the site and annotated bibliography, the report describes the current condition of the site as a baseline for site condition monitoring programmes, geomorphological sensitivity, potential threats, site management and potential usage.

Ellis *et al.* (1996) believe that all sites need a "conservation plan" giving information about the threats to the site, how it would deteriorate naturally without intervention, and what action would be desirable, or even essential, to maintain the features(s) of interest. "This will lead naturally to a consideration of what site-specific and practical Earth heritage conservation measures will be needed to ensure that the features of special interest are not obscured, destroyed or damaged and also to indicate the recommended frequency of monitoring" (Ellis *et al.*, 1996, p. 89–90).

This type of approach has been developed in Scotland where Werritty *et al.* (1998) have developed a "Landform Attributes" system, which refers to the characteristics of an earth science site that should be taken into account for monitoring and conservation management (Table 5.6). In the case of static geological sites, attributes 2 and 5 may not be relevant. The attributes are then applied to the various landform assemblages within an SSSI and assessments of what would amount to favourable condition, the limits of acceptable change and the monitoring recommendations are made (Kirkbride *et al.*, 2001). Table 5.7 gives an example. The *Environmental Protection Act* (1990) requires "common standards throughout Great Britain for the monitoring of nature conservation" and the above approach is an example of the search for best practice in this regard.

The United Kingdom's 1990 Earth science conservation strategy (Nature Conservancy Council, 1990) contained a volume of appendices published as a handbook of earth science conservation techniques. Appendix 1 gives a series of generalised "conservation profiles" or development threats for each of the categories of site defined

Table 5.6 *Definition of terms used in the landform attributes table (Table 5.7) (after Werritty et al., 1998)*

Attribute	Definition
1. Physical attributes & morphology	The constituent materials (bedrock or sediment) and superficial form of a landform. In some cases, surface and groundwater flow is a physical attribute.
2. Processes & dynamics	The processes that interact to create landforms.
3. Composition	The range of components of a landform system and their clarity of expression.
4. Structure	The integrity, context and interrelationships between the components of a landform system.
5. Function	The status of an active landform system within its wider setting (e.g. drainage basin or coastal cell). Function may be some combination of the following: sediment source, transfer, temporary store or sink.
6. Visibility	The ease with which a landform system can be observed. The scale at which visibility is important depends on the individual landform.
7. Accessibility	The ease with which a landform can be visited and researched.

Table 5.7 *Landform attributes table for the Southern Parphe SSSI – bedrock channels (after Kirkbride et al., 2001)*

Landform attribute	Favourable condition	Limits of acceptable change	Monitoring prescription
1. Physical attributes & morphology	Natural, unregulated runoff regime maintained from upstream maintained. Bedrock features (waterfalls, pools, steps, potholes) not obscured by accumulation of sediment or growth of mosses, grasses or shrubs	Ideally, natural runoff regime should be maintained. If not possible, high flows are needed to flush out sediment that would otherwise obscure the bedrock features and potentially allow mosses, grasses and shrubs to mask features	Ground-based photographic record (every three years) of selected bedrock features to check that important components are not obscured or masked. Check flow record from gauging station: monitoring >95% flows on flow duration curve, if available, every three years
2. Processes & dynamics	Processes necessary for maintaining the clarity of the bedrock features and, in the case of potholes and plunge pools, continued evolution	Bedrock features to be maintained in as pristine a condition as possible. Potholes and plunge pools to remain operational and subject to further development	Ground-based photographic record (every three years) of representative key bedrock features (especially plunge pools and potholes) to check that they are still operational. Check inventory of specified landforms from ground-based photographic survey every three years.
3. Composition	Maintenance of individual landforms (cascades, pools, steps, ribs, potholes)	Individual landforms continue to be present and, where appropriate, subject to continued development	Check inventory of specified landforms from ground-based photographic survey every three years

(continued overleaf)

Table 5.7 (continued)

Landform attribute	Favourable condition	Limits of acceptable change	Monitoring prescription
4. Structure	Functional linkage between landforms retained	No changes in flow regime such that the functional linkage between cascades and plunge pools is damaged	Visual check from site visit every three years that functional linkages are being maintained
5. Function	Total bedrock channel system continues to operate as a highly stable reach evacuating runoff and retaining minimal amounts of bed material	Exposure of bedrock within site should remain broadly unaltered within minimal and only temporary accumulations of bed material in transit	Ground-based photographic record of selected bedrock features (every three years) to be maintained to check that important components are not obscured or masked
6. Visibility	Whole landform assemblage visible by direct inspection on the ground or via aerial survey	No obscuring of individual landform components by man-made structures. Natural vegetation regeneration acceptable except where key individual components are specified	Visual inspection and ground photography of site every three years
7. Accessibility	Access permitting all individual landform components to be observed and measured	Access by arrangement with landowner/tenants; some seasonal restrictions may be acceptable	Visual inspection of site every three years.

in the strategy (Table 5.5). Appendix 2 is a detailed guide to conservation techniques, particularly related to disused quarries and pits (see also Section 6.3.1). The techniques described in the handbook are currently being updated and revised in a project being sponsored by English Nature. In addition, Wilson (1994) and Ellis *et al.* (1996) give several examples of conservation approaches and techniques used at specific UK sites.

Maintenance and enhancement are also important elements of a comprehensive conservation plan. The UK government has set a target of having 95% of SSSIs in favourable condition by 2010. In consequence, English Nature has implemented an enhancement programme termed "Face Lift" aimed at clearing vegetation, talus or overburden from sites to re-expose the geological interest of natural or quarry sections (Jeffreys, 2000). In the 4 years up to April 2003, around a third of a million pounds was spent on enhancing 170 geological SSSIs (Murphy, 2002). Some of the work has been carried out by local amateur groups, for example, through the Ludlow series near Ludlow (Oliver & Allbutt, 2000).

Another way to conserve sites, particularly fossil and mineral ones, is by promoting collecting codes. There is a general code used in the United Kingdom, and also a number of local codes for particular areas or sites, for example, West Dorset Coast (Edmonds, 2001). Larwood & King (2001, p. 124) note that there is a "need to modify the approach to fossil collecting on a site-by-site basis, according to the extent of the resource". Larwood (2003) outlines the use of photography for monitoring site condition and visibility loss due to vegetation growth.

Even with all this protection and management in place, SSSIs can still be lost. This is particularly true where permission for a particular use predates SSSI designation. An example of this is Webster's Clay Pit in Coventry, England where landfilling in the 1990s has totally buried an important Westphalian site (Prosser, 2003; see Section 4.3). Similarly, Brighton & Hove Council in 2002 granted itself planning consent to stabilise the cliff face SSSI at Black Rock, Brighton, obscuring the Quaternary deposits and mammalian fossil assemblage (Bennett, 2003).

5.10.4   Regionally Important Geological Sites (RIGS)

The RIGS (Regionally Important Geological/geomorphological Sites) scheme was introduced in the early 1990s to meet the need for more local involvement in earth science conservation in the same way that many local wildlife groups operate. There is much local enthusiasm and goodwill amongst, principally, amateur geologists, and the aim of the Nature Conservancy Council in the early 1990s and subsequently its successor agencies has been to act as a catalyst by providing encouragement and support for county-based groups of volunteers (Nature Conservancy Council, 1990; Harley, 1994). The aim was to set up a country-wide network of sites, established and managed locally by volunteer groups. But "RIGS should not be seen as 'second-tier' SSSIs, but as sites of regional or local importance in their own right" (Nature Conservancy Council, 1990, p. 29). Potential sites are evaluated according to four sets of criteria – educational, scientific, historic and aesthetic/cultural values. There is a RIGS Handbook that gives details of the scheme, site evaluation techniques and so on (Mason & Stanley, 2001). RIGS are not designated or protected by law, but they are increasingly given some protection by local authority planning policies (see Box 6.9).

All 45 counties in England now have RIGS groups though activity is variable. In Wales, there are 4 active regional groups coordinated by an Association of Welsh RIGS Groups and it is estimated that there may eventually be 3,000 to 4,000 RIGS in Wales (Rogers, 2000). Box 5.16 shows an example of the documentation for a Regionally Important Geological Site in Wales. This gives basic details of the site such as name, location, type and date registered, followed by a "Statement of Interest", "Geological setting/context" and "References". Finally, there are sections covering such information as accessibility and safety, ownership and planning control, current and potential usage, site condition, potential threats and management options. This style of documentation is used in Wales to provide consistency of recording of RIGS sites and represents the best of existing practice.

Box 5.16   *Site Documentation for Mynydd Carreg RIGS, Wales (Edited)*

Site Name:	Mynydd Carreg
RIGS	Number: 0010
Grid Ref:	SH163292
Type of Site:	Stratigraphy (Precambrian)/Mineralogy
Local Authority:	Gwynedd Council
Site Nature:	ED: disused quarry
(Table 5.5)	EO: inland outcrop
	IM: unique mineral site
Maps:	1:50,000: Sheet 123, Lleyn Peninsula
	1:10,000: SH12NE
File Number:	GM/SM/0008
Surveyed by:	M. Wood, S. Campbell, T. Rogers
Date of Visit:	14.10.99
Date Registered:	31.7.01
Documentation authors:	M. Wood & S. Campbell
Documentation last revised:	31.7.01

RIGS Statement of Interest

Mynydd Carreg provides one of the best examples of Precambrian pillow lavas on Llyen. In addition, a disused quarry on its NW flank provides some of the finest exposures of jasper in Wales. These rocks form part of the Gwna Mélange, a disrupted sequence of various rock types, which were probably deposited along an active plate margin during late Precambrian times. The rocks exposed form a part of a large fragment that was incorporated into the mélange during the disruption event.

Jasper is normally found interstitially and in veins in the Precambrian pillow lavas of Llyen and Anglesey. At Mynydd Carreg, however, the jasper also occurs as discrete beds (>4 m thick) amongst pillow lavas, haematite mudstones and other sedimentary Precambrian rocks. Although the site has not been described in detail, there appears to be evidence for a phase of submarine volcanic activity followed by a period of relative quiescence, with associated deposition of silty iron-rich sands and mudstones, after which further pillow lavas were extruded onto the sea floor. Although pillow lavas are well developed elsewhere on Anglesey and Llyen, nowhere else shows such a remarkable juxtaposition of marine sedimentary and volcanic rocks of this age. The site provides crucial context for the environmental setting into which the pillow lavas were erupted. Moreover, the site provides probably the most accessible and easily studied rocks of this type in North Wales.

Continued on page 235

Continued from page 234

Geological Setting/Context

The late Precambrian rocks of SW Llyen are divided into two main groups, namely the Gwna Mélange and the Sarn Complex, which are separated by a steeply inclined zone of recrystallised mylonitic schists termed the Llyen Shear Zone.

On the NW side of the shear zone, the Gwna Mélange consists of a chaotic mixture of large fragments of various rock types that were deposited in a variety of settings such as deep marine trench, forearc basin, shallow shelf and even continental environments. The mélange, which is at least 3,000 m thick, includes clasts of cherty siltstone and mudstone, quartzite, limestone, mudstone, granite and basaltic volcanic rocks, all of which are contained within a matrix of mudstone, siltstone and sandstone. The Gwna Mélange has been interpreted as the product of submarine landsliding, of regional extent, which took place in response to processes directly associated with either oceanic plate subduction or plate collision during late Precambrian times.

On the SE side of the shear zone, the Sarn Complex consists of a range of poorly exposed plutonic igneous rock types including granite, tonalite, diorite and gabbro. These rocks record a phase of late Precambrian magmatic activity, which, using Rb–Sr techniques, has been dated at approximately 549 +/– 19 million years. In view of a dominantly calc-alkaline geochemical signature, this phase of magmatism was probably associated with plate subduction, the rock types seen currently representing the deeply eroded, plutonic rocks of an arc complex.

The Gwna Mélange and Sarn Complex may have been finally juxtaposed by left-lateral displacement in Lower Cambrian times.

References

Carney, J.N., Horák, J.M., Pharoah, T.C., Gibbons, W., Wilson, D., Barclay, W.J., Bevins, R.E., Cope, J.C.W. & Ford, T.D. 2000 *Precambrian rocks of England and Wales.* Geological Conservation Review Series, 20. JNCC, Peterborough

Cattermole, P.J. & Romano, M. 1981 *Llyen Peninsula.* Geologists' Association Guide, 3

Gibbons, W. & McCarroll, D. 1993 *Geology of the country around Aberdaron, including Bardsey Island.* Memoir for 1:50,000 geological sheet 133. HMSO, London

Roberts, B. 1979 *The geology of Snowdonia and Llyen: an outline and field guide.* Adam Hilger Ltd., Bristol

Practical Considerations

Access: Access is easy from a minor road at SH164289. A small unsurfaced lane, SW of Carreg Farm, leads to a National Trust car park. There are permissive paths from the car park to the summit of Mynydd Carreg where pillowed basaltic lava is exposed and there are all-round views.

Safety: The old quarry faces are not particularly high, but normal Health & Safety practices of working in disused quarries should be observed.

Conservation status: The site lies within Llyen AONB and Heritage Coast, and adjacent to Glannau Aberdaron mixed biological and geological SSSI.

Ownership and Planning Control

Owner/tenant: National Trust (Wales), Trinity Square, Llandudno, LL30 2DE

Continued on page 236

Continued from page 235

Planning authority: Gwynedd Council
Planning status/constraints: None known

Condition, Use and Management

Present use: Part of a farm and used for animal grazing
Site condition: In excellent condition with easily accessible faces in disused quarry and
    clear unobscured exposures of rock around the summit of Mynydd Carreg. Best exposures
    of jasper are below present ground level as a result of past landscaping, but adequate
    in situ rock and loose fragments can still be viewed in the old quarry.
Potential threats: Quarry infill and removal of material from summit outcrops. Invasive
    sampling of *in situ* rock should be avoided for aesthetic reasons.
Site condition: Ideal – no enhancements are envisaged.

Site Development

Potential use (general): The obvious nature of the geological features of interest, the
    potentially large number of visitors, and the accessibility of the site give it strong
    potential for an interpretative initiative.
Potential use (educational): Excellent research site and ideal for university and school visits.

The picture in Scotland is more patchy with only part of the country covered by active
groups (Lothian & the Borders, Fife, Stirling & Clackmannan) (Butcher, 1994; Leys,
2001), and in Northern Ireland the RIGS initiative is at a very early stage (Crowther,
2000). A national body (UKRIGS) was established in 1999 and has a development strategy aimed at expanding the RIGS network, strengthening work quality and partnerships
and securing significant funding for the work (www.ukrigs.org.uk). It holds an annual
conference, produces a quarterly newsletter and groups are increasingly securing significant funding for their activities. In recent years, several RIGS groups have changed
their names to reflect the fact that their aims are broader than just recording sites.
Gloucester RIGS, for example, is now Gloucestershire Geoconservation (Campbell &
McCall, 2002) and Fife RIGS is now Geoheritage Fife.

5.10.5 Limestone Pavement Orders

Limestone Pavement Orders (LPOs) are made under Section 34 of the *Wildlife & Countryside Act* (1981, as amended 1985). This is one of the few pieces of legislation to
protect a specific landform type and make it a criminal offence to damage the landforms so designated. About 100 orders have been made, and all significant limestone
pavements in England are now protected (Limestone Pavement Action Group, 1999;
Webb, 2001). Though some illegal removal still continues, the strengthening of legislation in England has meant increased removal of the Irish pavements (see Box 4.1).
Discussions between the United Kingdom and Irish governments about this situation

have taken place (Webb, 2001). Limestone pavements are also an important habitat (see Box 3.9) and are specifically identified in Annex 1 of the European Habitats Directive as a natural habitat of community interest whose conservation requires the designation of Special Areas of Conservation (SACs) (Goldie, 1994).

5.10.6   Other legislations and protected areas

Geology, and particularly geomorphology, may also be recognised in some of the other designations. For example, hill or coastal landforms and processes are often a major part of the recognition of the 41 *Areas of Outstanding Natural Beauty* (AONBs) and over 40 *Heritage Coasts* in England and Wales, the coastline of North Norfolk, for example, being recognised under both categories (Funnell, 1994; Adams, 1996). Together, these two categories cover over 12% of the land area of England and Wales and the emphasis within them is on landscape conservation. Under the *Countryside & Rights of Way Act*, local authorities are obliged to produce management plans for AONBs so there is an opportunity for these plans to include geoconservation objectives and promote the continued use of local stone for ongoing quality developments (Edmonds, 2003).

Scotland has had a chequered history of debate regarding other protected areas (Cullingford & Nadin, 1994). Until recently, it has had no National Parks (see above) but it does have about 40 National Scenic Areas (NSAs), which are equivalent to AONBs and which cover over 1 million hectares, including Ben Nevis and Glen Coe. It also has a few regional parks (including the Pentland Hills near Edinburgh) and about 35 country parks. These two categories are primarily intended to provide recreational opportunities, but they allow for conservation work and management agreements with private landowners.

A number of conservation organisations own land in the United Kingdom and manage it for conservation purposes. One example already mentioned is the National Trust and National Trust for Scotland, which together own about 350,000 ha of land or over 1% of the British land surface. Ownership includes many important mountain, moorland and coastal landscapes, including half the coastline of Cornwall and over 1,000 km of coast in total. These areas are managed to protect natural landscapes and their biodiversity, but greater attention is being given to geodiversity, for example, in relations to RIGS in SW England. Much important geoheritage is included in National Trust land holdings, for example, large parts of the glaciated landscapes and Ordovician/Silurian volcanic sequences of Snowdonia National Park in Wales.

The UK Wildlife Trusts also own and manage reserves, and although their primary aim is wildlife conservation, some also take an active interest in geodiversity and geoconservation. An example is the Scottish Wildlife Trust (www.swt.org.uk), which has a policy document on geodiversity (Scottish Wildlife Trust, 2002). In it, the Trust

- "recognises Geodiversity as an essential component of our natural heritage";
- "believes that land management practices should recognise conservation of geodiversity as a major aim and attribute high value and importance to this";
- will promote education about Geodiversity by raising awareness through interpretation on appropriate reserves and clubs for young geologists;
- will promote the conservation of Geodiversity through its works on its reserves and its support for RIGS (see above).

At least 17 of the Trust's current reserves have major geological or geomorphological interests within them and it recognises the need to "highlight this geodiversity interest" and incorporate it in management planning.

The *Wildlife & Countryside Act* (1981) provides a framework to regulate fossil collecting at designated SSSIs or NNRs, but there is still no law relating specifically to fossils. "In the light of this, prospects for the regulation of fossil collection and trading fall within the realm of aspects of Civil and Criminal Law, combined with the general understanding (legal precedent) that fossils may be regarded as minerals for the purposes of the law....in reality it puts an enormous (and unrealistic) burden upon the site owner/occupier" (Norman, 1994, p. 65). Following the proposed purchase by Staatliches Museum in Stuttgart, Germany, of "Lizzie", the earliest reptile fossil, discovered at East Kirkton Quarry, Scotland in 1988, the UK Department of Trade & Industry ruled, on legal advice, that since fossils are not manufactured or produced, they are not subject to the *Export of Goods (Control) Act* (1987). There are therefore no UK laws to control the export of fossils (Rolfe, 1990).

5.11   Ireland

Ireland is an interesting example since, apart from the geological importance of its six National Parks (Fig. 5.21) including The Burren with its important karst geomorphology and limestone pavements (Drew, 2001), "It would be fair to say that there has been no significant tradition of geological conservation (there) until relatively recently" (Parkes & Morris, 2001, p. 79). As a consequence, a major review has been undertaken and the Irish Geological Heritage Programme (IGH) instituted. It is instructive to outline the way in which a country has recently initiated a geoconservation programme more or less from scratch.

The IGH Programme is run as a partnership between the Geological Survey of Ireland (GSI), which undertakes scientific site selection, and Dúchas, the Heritage Service of Ireland, which carries out the statutory designation of Natural Heritage Areas (NHAs) under the *Wildlife (Amendment) Act* (2000) and their management. However, Parkes & Morris (2001) predict a slow process of designation, perhaps as themed blocks.

In fact, 16 themes have been identified and these are listed in Table 5.8. For each theme, expert panels are being established whose primary purpose is to "ensure inclusiveness of all acknowledged expertise and maximum scientific rigour in the selection process, as any legal challenge to the validity of a site will be based and assessed on scientific robustness" (Parkes & Morris, 2001, p. 82). Each panel then works with the GSI to refine the final choice of sites, following which consultants are appointed to undertake desk studies and site reports. Field visits to photograph sites and establish site boundaries are the final stage, with particular attention paid to personal contact with landowners.

According to Parkes & Morris (2001, p. 82) "Each theme is intended to provide a national network of NHA sites and will include all components of the theme's scientific interest". In other words, what the system is intended to establish is a representative selection of Ireland's geodiversity, but unique, exceptional and internationally important sites are also included. They also emphasise that only a minimum number of sites will be selected with minimum duplication of interest, but they recognise the potential vulnerability of this minimalist approach. The following preferences were also used in site selection:

Table 5.8 *Irish Geological Heritage programme themes*

IGH1	Karst
IGH2	Precambrian to Devonian Palaeontology
IGH3	Carboniferous to Pliocene Palaeontology
IGH4	Cambrian – Silurian
IGH5	Precambrian
IGH6	Mineralogy
IGH7	Quaternary
IGH8	Lower Carboniferous
IGH9	Upper Carboniferous & Permian
IGH10	Devonian
IGH11	Igneous intrusions
IGH12	Mesozoic & Cenozoic
IGH13	Coastal Geomorphology
IGH14	Fluvial & Lacustrine Geomorphology
IGH15	Economic Geology
IGH16	Groundwater

- Sites with an assemblage of characteristics or features or a range of interests.
- Sites with the most complete, undamaged or "natural" record.
- Sites with history of research or future research potential.
- Sites that have yielded results with a wider significance, for example, archaeological, historical or ecological interest.

Box 5.17 gives an example of an important palaeontological site.

In addition to the NHAs, a County network of non-statutory sites is being established (similar to RIGS in the United Kingdom) with some level of protection achieved through the Irish land-use planning system. Concern over damage to limestone pavements at The Burren has led to the introduction of the Burren Code, which discourages visitors from removing limestone or building cairns and dolmens from shattered limestone or field wall stones.

Box 5.17   The Valencia Island Tetrapod Trackway

A mid-Devonian tetrapod trackway comprising about 200 prints was discovered on Valencia Island, County Kerry in 1993. This makes it the most extensive and possibly the oldest of only seven comparable sites in the world, having been dated to at least 385 Ma. It represents a major evolutionary step from aquatic vertebrates to air-breathing amphibians and is therefore worthy of conservation. It is, however, threatened by unscrupulous collectors and visitor pressures and was in urgent need of protection. Since it occurs on slabs only 10 to 30 cm thick it could be removed relatively easily in sections for museum curation, but there was also a desire to conserve it *in situ* in order to retain its field context.

After negotiation, it was decided that the site would be purchased by Dúchas, which would manage it in conjunction with the GSI and local Valencia Heritage Society. The site is already within an NHA designated on ecological grounds, and day-to-day vigilance is being provided by the local population who have been made aware of the site interest and threats. Access to the site is being improved together with overlook points and interpretation panels (Parkes & Morris, 2001; Parkes, 2001).

5.12   Northern Europe

Norway's first *Nature Conservation Act* was passed in 1910 and in 1919 the first site to be protected partly for geology, was Tofteholmen island in Oslofjord. In 1923, an erratic block in southern Norway was designated and in 1931, all limestone caves in Rana/Nordland area were protected. At the time, only 5 to 8 caves were known but this is now increased to 20–30. "This protection is a good example of geological 'species protection' that was anticipated through taking care of geodiversity related to karst formation" (Johansson, 2000, p. 46). New nature protection laws were passed in 1954 and 1970 and the latter allowed national parks, landscape protection areas, nature reserves and natural objects to be protected.

Thus, during the 1970s systematic nature conservation planning was initiated, concentrating on national and regionally important areas and on features that were perceived to be in danger of becoming extinct, for example, mires, wetlands and seabird colonies. Large national parks were established and include important geological and geomorphological features within such famous areas as Jotunheimen, Jostedalsbreen and Hardangeridda National Parks. Erikstad (1994a) believes that important landscapes such as the fjords and strandflat are poorly represented because of the lack of state-owned land in these areas. The polar archipelago of Svalbard has separate legislation, but more than 50% of the land area is protected, including three national parks and two nature reserves, the largest of which, Nordaustlandet, covers over 19,000 km² (Erikstad, 1994a).

Later, work began on a regional basis on Quaternary geology, palaeontology and mineralogy. As a result, regional protection plans have been implemented for Quaternary geology of the Finnmarks and Hedermarks districts (30 sites), sedimentary rock art in the Oslo area (64 sites) and mineral localities in Sør-Norge (16 sites). About 170 geological sites are protected in Norway but there are also almost 1,000 sites of Quaternary interest registered in the national database. In addition, there is a Norwegian system of protected rivers (Huse, 1987; Smith-Meyer, 1994). There is also provision for temporary protection of threatened areas and a *Plan and Building Act,* which gives local planning authorities, in particular, the power to protect sites, though these have less status and may be more subject to periodic reviews (Erikstad, 1994a).

There have sometimes been conflicts between these designations and the desire by mineral companies to exploit the mineral resources of these areas. Erikstad (1994b) describes efforts to minimise the impact of a new Oslo airport on an important glaciofluvial complex including a distinct ice-contact zone, beyond which are two main delta formations forming a flat sandur plain more than 4 km wide. The runway was to be on a sandur surface that would survive, albeit with some channel details obscured. The main problem was anticipated as the secondary development pressures attracted by the airport location (hotels, car parks, ancillary buildings). Under Norway's *Nature Conservation Act* (1970) a 5 km² landscape protected area already existed in the south of the area and a further eight nature reserves and one natural monument were suggested within a system of three main landscape protection areas. It was hoped that this would protect "integrity" sites from airport-related developments. "In addition, special measures should be taken under area planning for protection of specified localities with aeolian sand dunes and kettle holes" (Erikstad, 1994b, p. 50).

The government body responsible for nature conservation in Norway is the Directorate for Nature Research. The Norwegian Institute for Nature Research (NINA) is also active in the field of geoconservation, although its major focus is within ecological and landscape research. The Norwegian Geological Survey is also an important institution within this field. Currently, activity within geoconservation is relatively low (Erikstad, 2000).

In Iceland, the nature conservation emphasis is to designate large wilderness areas, preferably including as many natural phenomena as possible (Thorvardardottir & Thoroddsson, 1994). Under *the Nature Conservation Act* (1971), Iceland's Nature Conservation Council can also designate nature reserves as areas considered important because of their landscape, flora or fauna. Several volcanic landscapes have been so designated, including the island of Surtsey, where there were eruptions in 1973. The Surtsey Nature Reserve is classified as IUCN Category I, a site of scientific importance, where natural processes are allowed to take place in the absence of any direct human interference and where public access is prohibited.

The Nature Conservation Council also has the power under the Act to designate unique geological formations of outstanding beauty or scientific interest. These are classified as natural monuments and include waterfalls, volcanoes, hot springs, caves, rock pillars and beds containing fossils or rare minerals. The Act stipulates that there should be a buffer zone around each monument to allow it to be appreciated. Since 1974, special attention has been given to the general protection of stalactites and stalagmites in lava caves. This is still the only general protection of geological formations in Iceland, though new legislation is currently being drafted that will improve protection of craters, coastlines, glacial landforms and other geological features from gravel extraction activities (Johansson, 2000).

Finland has 34 National Parks some of which include important geological or geomorphological features. For example, Päijänne National Park includes some of Finland's best esker systems. In fact, Finland has a conservation programme for esker protection, and also has developed management guidelines applying to all its protected areas (Metsähallitus, 2000). One of the main principles (p. 15) is "not to interfere with natural processes without good reasons related to nature conservation". Mining is prohibited within protected areas but traditional gold panning is permitted by licence.

Sweden's *Nature Conservation Law* has allowed the designation of national parks, national reserves and natural monuments incorporating landscape types, terrain forms and geology. Other legislations, policies and inventories aim to protect inland dunes, ravines, wetlands, lakes and rivers. Since 1999, the *Natural Resources Law* has allowed the selection and designation of national objects. The Nature Conservation Council and regional councils undertake this work with geoscientific input from the Swedish Geological Survey (Johansson, 2000).

In Denmark, over 200 national areas of geological interest have been identified since the 1980s and include landscapes, landforms, bedrock exposures and soils. Each of these areas/sites has a documentation including description, values, threats, references and administration. There is regular site monitoring and public interpretation of suitable sites. A survey of Denmark's coastline has been undertaken with the aim of identifying areas of geological, geomorphological or coastal dynamic interest (Johansson, 2000). The Wilhjelm Committee was appointed in 2000 to prepare a report as a basis for a government action plan on biodiversity and nature conservation. This report was completed in 2001, but it contains few references to geodiversity and geoconservation though it does promote the operation of natural processes.

Estonia has about 1,900 catalogued erratic boulders over 3 m in diameter, 226 of which are protected as natural monuments (Raudsep, 1994).

5.13   Eastern Europe

Most eastern European countries have a long history of geoconservation and detailed site networks. Cserny (1994) describes the system in Hungary, and the Czech Republic designates:

- national parks
- protected landscape areas
- national nature reserves
- national natural monuments
- nature reserves
- natural monuments,

all of which include important geological and geomorphological features. The Czech Geological Survey maintains a GIS database of over 1,200 sites of all types, including those proposed for designation and other sites. The records include geological characteristics, degree of conservation, objectives of conservation and conflicts of interest (www.geology.cz). Most sites also have a photographic record.

Poland had developed the protection of earth science features by the 1930s, but the *Nature Conservation Act* (1991) reviewed the existing series of protected areas developed piecemeal over many years. The Act established an integrated hierarchy of areas and sites and the administrative responsibilities of different levels of government (Wojciechowski, 1994). The system involves:

- nature reserves
- landscape parks
- protected landscape areas
- sites and objects
  - natural monuments
  - research stations
  - ecological sites
  - nature/landscape complexes.

The list of natural monuments is dominated by about 900 erratic boulders (Alexandrowicz, 1993). A network of 145 geosites has recently been established, 44% of which are currently unprotected (Alexandrowicz, 2002). Also limited is the enforcement of regulations and the integration with the planning system.

This is also a problem in Romania where natural monuments (including geological sites) have been established since 1973 and over 170 such sites have been recognised. But "the management of these areas is practically non-existent...due to the fragmentation of responsibility for identifying, designating and managing protected areas between central (government) and local authorities" (Grigorescu, 1994, p. 468). There is also no regulation of the collection or export of fossils (Teodorascu, 1994). Fortunately, environmental protection in Romania, including the prospects for geological site conservation, has improved greatly since 1990.

5.14   West Central Europe

Fears about inadequate geoconservation systems have also been expressed in other countries. For example, in Switzerland, Stürm (1994) and Schlüchter (1994) both expressed concern that existing procedures are too weakly implemented to protect important landscapes and deposits in the face of a booming demand for sand and gravel aggregates. Stürm's concern relates to the moraine landscape near Zurich, whereas Schlüchter mourns the loss of important Quaternary stratigraphic sections. Although the landscape described by Stürm (1994) has been added to the *Federal Register of Landscapes of National Importance* that allows the Swiss Nature Conservation Commission to intervene in the assessment of planning applications for sand and gravel extraction, the problem is that planning authorities do not always designate such landscape areas in their land-use plans. In such circumstances, the inventory "will remain limited and vague" (Stürm, 1994, p. 27). The same is true of "geotope protection areas", which are therefore similar in type to SSSIs in the United Kingdom and could influence land-use planning in a similar way (see Section 6.6).

Geotopes, defined as "distinct parts of the geosphere of outstanding geological and geomorphological interest" (Stürm, 1994) are also identified in Austria and Germany, though there are no national programmes in either country. Grube (1992) concludes that German geoconservation is in a weak condition. He proposes (Grube, 1994) that action is needed to popularise geoconservation, increase involvement by earth scientists and introduce legal designations such as "geological parks". An inventory of over 10,000 "geotopes" has been compiled by Länder, but questions of use, management and accessibility have yet to be resolved (Wiedenbein, 1994) and there is cross-Länder inconsistency.

Germany also has a Federal *Monument Protection Law* (1973) though its implementation also varies widely across the country (Norman, 1994). Baden-Württemberg uses its *Monument Protection Law* in a similar way to the Danish Museum Act, that is, to pay compensation to finders of fossils deemed to be of importance to the State. However, other Länder have no such Act. Norman (1994) believes that this inconsistency fosters conflicting attitudes to palaeontological conservation in the country and acts against trans-national consensus on the status and value of fossils.

France also has limited geoconservation programmes apart from the spectacular Haute Provence Geopark (see above) and about 12 geological nature reserves designated on palaeontological, stratigraphic, mineralogical and stratotype grounds (Martini, 1991). The Netherlands has a network of important geological sites (Fig. 5.25), but none has been formally designated (A. Kooman, personal communication).

Some countries have protection for particular features. For example, in 1928 Austria passed one of the first laws dedicated especially to the protection of a geological feature – caves (Trimmel, 1994). Criteria for establishing protected caves include scientific value, for historical research, palaeontology, geological structures, sediments and so on. The law also made it possible to create buffer zones around cave entrances. Over 10,000 caves are now documented and the protection system has been regionalised, though this has led to inconsistencies in enforcement (Trimmel, 1994).

5.15 Southern Europe

Alcalá & Morales (1994), Melendez & Soria (1994) and Alcalá (1999) describe the effectiveness of two national laws passed during the 1980s in protecting Spain's geological heritage. First, the *Law of Conservation of Natural Spaces and Wild Flora and Fauna* (1989) established four categories of protected natural area defined as "areas

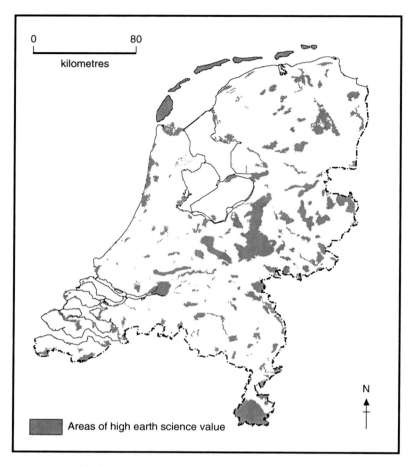

Figure 5.25 Areas of high earth science value identified but not yet generally designated in The Netherlands. (Published with permission of Dr. A. Koomen and Alterra, Green World Research, Wageningen, The Netherlands)

or natural features composed of elements of known uniqueness, rarity or beauty which merit being the object of special protection". The four categories are

National Parks
Natural Reserves
Natural Monuments
Protected Landscapes.

National Parks are designated by the Spanish Parliament and managed by both the State Administration and the Autonomous Communities in which they lie. Responsibility for designating and managing the other categories of sites lies locally. Natural monuments include geological or palaeontological sites designated for their special interest due to the unique importance of its scientific, cultural or scenic value.

The *Law of Historical Heritage* (1985) gives protection to sites of cultural interest including geological and palaeontological sites related to the history of mankind, generally treating them as subordinate to archaeological sites. These are the responsibility of the Autonomous Communities and some of the latter have also developed their own heritage laws. For example, Catalonia, Galicia, Valencia and Madrid have introduced legislation to protect their fossil heritage, but Alcalá (1999, p. 14) believes that "Spain's palaeontological heritage requires still greater recognition". Box 5.18 gives a case study from La Palma in the Canary Islands.

Box 5.18 La Palma, Canary Islands, Spain

La Palma is at the western edge of the Canary Island chain and is the most volcanically active. Despite its relatively small size (45 × 28 km) its highest point is at 2,426 m. Twenty protected areas have been designated on the island and these are shown in Fig. 5.26. Some, like P-2 and P-4 have been designated primarily for their wildlife interest, and others, like P-8 and P-10 for a combination of landscape and wildlife. However, several are of primarily geological interest and these are briefly described below:

- P-1, Caldera of Taburiente National Park was designated in 1954 as Spain's 4th National Park (Fig. 5.27). It is a huge basin 9 km in diameter and c.1500 m deep with a sharp horse-shoe rim at over 2000 m above sea level. In 1825, the German geologist Leopold von Buch applied the Spanish term "caldera" (a large, deep pot or caldron) to this feature, which he believed to be a volcanic crater. Nowadays, it is believed to be the result of large-scale volcanically induced landslides and the erosive action of 400,000 years of intensive rainfall.
- P-5, Cumbre Vieja Natural Park. This is the largest protected area on the island and comprises the volcanic spine of the southern half of the island. It contains five historically active volcanoes and many inactive cones. It has recently been predicted that further volcanic activity could result in instability of the western side of the area resulting in a landslide into the Atlantic of 500^3 km of rock. The resultant mega-tsunami would inundate coasts from Brazil to Canada with waves up to 50 m high (Ward & Day, 2001).
- P-6, Mountain of Azufre Natural Monument is an impressive volcanic cone on the eastern coast.
- P-7, Volcanoes of Aridane Natural Monument comprises four separate volcanic cones in close proximity. Santos (2000, p. 129) comments that "they have been able to avoid strong demographic and agricultural pressure, although some of them are affected by building work and aggregate extraction". This is an area of intensive banana cultivation and the setting of the four hills is certainly affected by banana fields, terraces and plastic greenhouses extending up to, into and occasionally over them (Fig. 5.28).
- P-8, Risco de la Concepción Natural Area is an igneous intrusion on the outskirts of the island's capital Santa Cruz. Santos (2000, p. 129) comments that urban pressure and, in this particular case, stone quarrying, are causing great impact on the landscape.
- P-11, Volcanoes of Teneguia Natural Monument. This protected area at the southern tip of the island contains several volcanic cones, including the most recent eruption point of Teneguia (1971) and the impressive cone of San Antonio (1677) where Fuencaliente Volcanoes Visitor Centre has recently been opened.
- P-12, Volcanic Tube of Todoque Natural Monument. This area contains a lava tube 500 m in length. It was unearthed by a mechanical digger and lies in an area of ropy

Continued on page 246

Continued from page 245

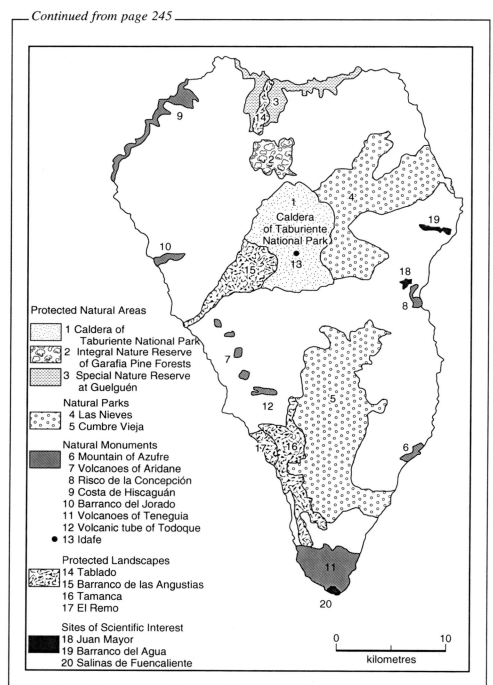

Figure 5.26 *The 20 protected areas on the Spanish Canary Island of La Palma (After Santos, J.J. (2000) La Palma: history, landscapes and customs. J.J.Santos, La Palma).*

Continued on page 247

___ *Continued from page 246* _____

Figure 5.27   *Edge of the Caldera of Taburiente on La Palma*

lava formed during the San Juan lava flow (1949). Santos (2000, p. 132) comments that "it is necessary to highlight the lack of sensitivity of the local councils and residents... All of the best ropy lava fields on La Palma were from the 1949 eruption of the San Juan volcano, but most of them have been savagely destroyed, with only small isolated areas remaining".

- P-13, Idafe Natural Monument is an upstanding dyke remnant within the Taburiente caldera and is believed to have been sacred to the indigenous population.

Melendez & Soria (1994) describe the long and bureaucratic process involved in declaring sites. Furthermore, responsibility for enforcing these laws lies with the 17 existing Autonomous Communities. Alcalá & Morales (1994, p. 57) believe that enforcement is inconsistent. In the worst cases, some "have quite simply forgotten that the State has conferred upon them this responsibility". They argue that there is a need for specific legal treatment of Spain's palaeontological heritage and believe (p. 60) that "a palaeontological heritage with a few well-protected and managed sites is preferable to an enormous, indefensible one which runs the risk of losing that which is truly exceptionable and irreplaceable".

The emphasis on geoconservation in Spain has tended to centre on landscape criteria rather than geological science stratotypes, tectonic structures or depositional systems. However, the Geological Survey of Spain has developed a National Inventory of Points of Geological Interest (PIGs), though the sites identified have yet to be given legal protection (Alcalá, 1999).

(a)

(b)

Figure 5.28 *Banana cultivation in plastic greenhouses encroaching on one of the Volcanoes of Aridane Natural Monument, La Palma: (a) the general setting of the volcanic cone and surrounding agriculture and greenhouses and (b) a greenhouse partially excavated into the cone*

Portugal has 5 natural monuments, all based on palaeontology (dinosaur footprints), and 12 natural parks, many of which contain important geological features. Attempts are being made to promote these interests (J. Brilha, personal communication). Greece has 10 national parks, 5 of which have buffer zones and geological and geomorphological interests are recognised in several of these parks. A number of the 53 "protected monuments of nature" are specifically designated for their palaeontological or geomorphological interests.

There has been an upsurge of interest in geoconservation in Italy, partly through the passing of the outline law on protected areas (L.394/91). This includes in its provisions, scope to protect geological and geomorphological features of national or international significance due to their "natural, scientific, aesthetic, cultural and recreational value...." (Gisotti & Burlando, 1998, p. 11). The categories of protected areas include parks, reserves and natural monuments. More recently, the Italian Geological Survey has been compiling an Italian Geosites database in GIS format (D'Andrea *et al.*, 2002).

5.16   Australia

5.16.1   National Parks

Australia has over 540 National Parks, most of which have been established and are managed by the States and territories under the *National Parks & Wildlife Act* (1970). The main aim has been to conserve scenery, wilderness or biological communities, and only coincidentally have geological and geomorphological sites been included. An example is Fraser Island, primarily designated for its biodiversity (Harlow, 1994) of tall rainforest growing on sand dunes, but the latter are important in their own right (see Box 5.19). Victoria contains 32 national parks, 20 of which have a significant geological interest (Markovics, 1994). Fortunately, the *Nature Conservation Act* (1992) introduced geology as an aspect to be considered in selecting subsequent national parks.

Box 5.19   Fraser Island, Australia

Fraser Island is a National Park and World Heritage Site lying off the south-east coast of Queensland, Australia. Its primary conservation interest is its globally unique ecosystem of rainforest types, but at 122 km long and 5 to 25 km wide, it is claimed to be the largest sand island in the world and the sand extends 30 to 60 m below present sea level. The dunes have been dated as a complete sequence from before the last interglacial (120,000 – 140,000 years BP) through to the early Holocene (less than 10,000 BP). The sand derives from granites, sandstones and metamorphic rocks in river catchments to the south and from the seafloor. The hydrology of the sand mass is also important with an extensive lens shaped aquifer within the porous sand mass and perched aquifers above less permeable organically bound sands. These also underlie the c.40 dune lakes, some of which contain long sediment records stretching back 300,000 years.

5.16.2   Conservation reserves and sites

In addition to national parks, there are over 2,800 other conservation reserves in Australia designated by States and territories and with a variety of names, including conservation parks, nature reserves, State parks, wilderness parks and marine reserves. However, geology is a secondary player in these reserves and indeed some prominent geological environments have been deliberately excluded because of their potential economic value.

Sites of national importance may be protected by the Australian Heritage Commission (AHC) established under the *Australian Heritage Commission Act* (1975). The Commission holds a Register of the National Estate (RNE) listing items and areas worthy of protection for future generations (Swart, 1994; Joyce, 1999). A set of criteria is used in assessing the suitability of a geological feature or site for nomination and inclusion in the RNE list. They include

- its importance in the course, or pattern, of Australia's natural. . . .history (includes processes and landscapes);
- its possession of uncommon, rare or endangered aspects of Australia's natural. . . . history (includes natural landscapes);
- its potential to yield information that will contribute to an understanding of Australia's natural. . . .places (includes research sites, teaching sites, type localities, reference or benchmark sites);
- its importance in demonstrating the principal characteristics of: (I) A class of Australia's natural. . . .places (includes landscapes).

Under this legislation, the government minister responsible must assure him- or herself that no "feasible and prudent" alternative exists if a proposed action threatens the values for which a place has been listed on the Register. However, the Commission has no legal power over state and local government or private landowners, and since "There are no provisions for financial penalties or penal clauses in the legislation, hence it merely provides moral protection" (Swart, 1994, p. 319). Several geological and geomorphological sites are included in the register, including the Franklin River, Maria Island and the Mole Creek karst system in Tasmania (Fig. 7.6).

The Geological Society of Australia (GSA) has recommended the use of the term "Significant Geological Feature" (SGF), and has carried out much work to identify "those features of special scientific or educational value which form the essential basis of geological education, research and reference" and thus are worthy of protection and preservation (Joyce, 1997). Joyce (1995, 1999) describes the methodology to be used in reviewing sites to be added to the Australian Register of the National Estate. The methodology is summarised by the acronym IDEM (Identification, Documentation, Evaluation, Management) – and to record site information, a new form called LCAN (Location, Classification, Assessment, National Estate Criteria) is to be used. The aim is to draw up a consistent classification of geological features according to physical type (section, quarry, landform, etc.), geological type (palaeontological, igneous, geomorphic, etc.), status (local to international) and use (research, education, reference). This should "provide an impetus to the development of Australia's geological heritage, and to the formal listing of further features on the Register of the National Estate" (Joyce, 1999).

However, the GSA's conservation policy is criticised by Tasmanian workers. Pemberton (2001a), for example, is critical of the GSA approach since "sites can only be significant in terms of scientific or educational use values and appears to emphasise geological features. This narrow interpretation of significance may be appropriate for the GSA, but it only encompasses part of the aims of conserving geodiversity. . . . . and does not recognise that relevant expertise may reside with non-GSA scientists". Pemberton

(2001a) also believes that "Current process sites have been seriously under represented in Australia inventories of significant sites". Similarly, Sharples (2002a) believes that "historically most geoconservation work in Australia has been focussed on a 'geological heritage' approach, in which geodiversity.... was seen as being important mainly for its value to scientific research and education". Because this approach does not address issues of intrinsic values and ecological sustainability, it "has largely been ignored as a minor issue in nature conservation programs because of its perceived lack of relevance to central issues of land management" (Sharples, 2002a). C. Sharples (personal communication) sees geoconservation as an almost essential and integral part of nature conservation, instead of it being a rather separate and almost unrelated issue, which is how the traditional "geological heritage" approach tends to be presented in Australia.

5.16.3   State activity

The Australian states have a large measure of autonomy and their jurisdiction takes precedence over the national government in conservation areas. Several states have created "site inventories" or "protected sites" of various types. For example, South Australia has created "fossil reserves" that are controlled by the South Australian Museum Board, which actively manages access and collecting. The important Ediacara Fossil Reserve, about 700 km north of Adelaide, is one example where a range of Precambrian soft-bodied animals including jellyfish was discovered in the 1950s, and is the earliest known animal assemblage in the world. Unfortunately, although collecting can only be done with permission, enforcement has not been adequate and the site is virtually destroyed (Swart, 1994, see also Section 4.11). Swart believes that enforcement and funding are major issues in geological conservation in South Australia, but notes that voluntary and community effort together with public education are probably the key to effective geological conservation (see Section 6.9). Similarly, Harlow (1994) describes several local scientific and educational initiatives in Queensland aimed at protecting valuable geological sites.

Rosengren (1994) describes Australia's Late Cenozoic volcanic province consisting of 354 volcanic eruption points. He has carefully described these features and classified them according to morphology and scientific significance. He expresses concern about the impact of quarrying on this geological and geomorphological heritage (see Section 4.2), but notes that in Victoria, the government department responsible recommends that "...the geological significance of a feature should be considered when assessing quarry applications" and there should be "...investigations...to locate alternative sources of scoria and tuff where existing pits are found to be compromising significant volcanic features..." (Guerin, 1992). Joyce (1999) describes work to identify geological sites in Victoria. The Geological Society and its sub-committees has achieved a great deal in building up site inventories for individual states/territories or parts of states/territories. Joyce (1999) also discusses the different approaches to site identification. In some states, sites are chosen as being representative of a particular geological type while in others they may be selected as being outstanding at state, national or international levels.

But the most impressive geoconservation work in Australia has undoubtedly been undertaken in Tasmania by the Department of Primary Industries, Water & Environment (DPIWE), the Parks & Wildlife Service, the Forest Practices Unit, Forestry

Figure 5.29   *Cambrian pillow lavas, City of Melbourne Bay, King Island, Tasmania. The lack of deformation and degree of preservation is unusual and as a result they have been listed on the Tasmanian Geoconservation Database as a site of international significance (Reproduced by permission of Michael Pemberton)*

Tasmania and private consultants (e.g. Kiernan, 1991, 1995, 1996, 1997a; Sharples 1993, 1995, 1997, 2002a, 2002b; Eberhard, 1994, 1996; Dixon, 1995, 1996a,b; Houshold *et al.*, 1997; Pemberton, 2001a; Jerie *et al.*, 2001). Sharples (2002a) has summarised the philosophy and practice of the approach taken in Tasmania, which sees geoconservation as an essential and integral part of nature conservation and land management. Conservation of both traditional "static" heritage sites and ongoing geomorphic and soil processes are recognised, the latter distinguishing the Tasmanian work from most of the rest of Australia. In addition, the *Regional Forest Agreement (Land Classification) Act* (1998) made "conserving geological diversity" a statutory management objective in all categories of conservation reserves in the state, while the *Mineral Resources Development Act* (1995) protects speleothems from uncontrolled destruction by collectors.

The Tasmanian Geoconservation Database (TGD) is a computerised database (with on-line access) of all existing Tasmanian geoconservation inventories and sites (e.g. see Fig. 5.29). It is actively managed by a "Reference Group" established in 1999 consisting of government department representatives, academics, industry representatives and others, and includes geologists, geomorphologists and soil scientists.

A number of Codes of Practice also exist in Tasmania. For example, the Tasmanian *Forest Practices Code* (1993) makes provision for the protection of significant landforms by management prescription or special reservation (Kiernan, 1996). The DPIWE (2001a) has issued a *Draft Reserve Management Code of Practice* to guide all activities in protected areas within the State. This contains procedures for assessing and approving new activities and developments in protected areas, and has a specific section on "Geodiversity" (Box 5.20).

Box 5.20   *Code of Practice on Geodiversity in Protected Areas in Tasmania*

The Draft Code of Practice on Tasmanian reserve management has a section on Geodiversity that includes the following measures:

- Where field assessment is required, collect information on bedrock, soil type, geomorphology and geomorphological processes operating, trends in the condition of the site and extent of disturbance, current site use, likely impact of the proposed activity and threats to identified values, condition and effectiveness of mitigation measures.
- Consult the Tasmanian Geoconservation Database and other sources for the conservation significance and sensitivity.
- Where an activity is approved at a site of geoconservation significance, measures should be implemented to avoid or minimise adverse impacts to geological, landform and soil features and processes. Specialist advice should be sought to determine these measures.
- Permits and authorities under the *National Parks & Wildlife Act* (1970) should be obtained when required.
- Threatening processes or activities should be addressed where they are likely to adversely affect a site of geoconservation significance or natural processes relevant to the integrity of a site of geoconservation significance.
- Where rehabilitation activities are undertaken, these should, as far as practicable, restore relevant natural features and processes in accordance with approved guidance.
- Modification of coastal landforms should be avoided except where there is a threat to infrastructure and the measures are consistent with *Tasmanian State Coastal Policy*.
- Retention of parts or all of artificial exposures (e.g. road cuttings, quarries, etc.) will be considered where this can contribute to maintaining the values of sites of geoconservation significance.
- Sites of geoconservation significance that are publicised or promoted should be managed to protect the values from threats arising from increased visitation. Consideration should be given to controlling public access where this is likely to result in unacceptable impacts on site values.
- Management of caves should be in accordance with the *Tasmanian Cave Classification System*. Caves will be designated "Restricted Access Caves" where unrestricted access is likely to result in unacceptable impacts to the cave environment.

Concern over the impact of coring in Tasmania has led to the adoption of a more careful approach to assessing research proposals and the restoration of cored sites (M. Pemberton, personal communication). For example:

- Proposals for coring or drilling significant outcrops (or locations in caves) are carefully considered and all other options explored, for example, use of hand specimens.
- When coring is necessary, they are taken from overhangs or similar inconspicuous locations.
- Once a core is taken, the bottom of the hole should be filled with cement and the hole plugged with the cut-off core end.
- Old drill holes can be plugged with dyed cement or preferably with rock cores taken from loose rock samples or rock dust processed from these samples and mixed with cement or resin.

Another useful concept introduced in Tasmania is the conservation status (Dixon *et al.*, 1997). Five broad categories have been proposed:

- Secure – sites, processes or systems whose geoconservation values are not degraded, and are likely to retain their integrity, because the values are robust or their protection is provided for under existing management arrangements.
- Potentially threatened – sites, processes or systems whose values are not being actively degraded, but which are sensitive and whose protection is not specifically provided for under existing management arrangements.
- Threatened – sites, processes or systems whose values have been or are subject to degrading processes, although the values remain largely intact currently.
- Endangered – sites, processes or systems whose values have been or are subject to degrading processes that have had significant impact on values.
- Destroyed – sites, processes or systems whose values have been lost owing to degrading processes.

Sharples (2002a) also suggests the development of performance indicators or targets, including data coverage indicators, condition and conservation status of sites, and indicators of the integrity of natural processes upon which significant sites and assemblages depend. A number of initiatives in this regard are discussed in Chapter 6.

5.16.4   Other legislation and protected areas

Another important geoconservation measure in Australia is the *Protection of Movable Cultural Heritage Act* (1986). Under the Act, palaeontological specimens, minerals and meteorites are included in a "Control List" of movable objects that require a permit for export. Until 1983, this applied only to fossils with a value to museums of over $1,000 (AUS), but this stipulation was removed in 1993. However, it is still the case that minerals valued at less than $10,000 (AUS) do not require an export permit (Creaser, 1994a). Applications for permits are assessed by expert examiners, but the government minister makes the decision whether to issue a permit. The maximum penalty for exporting without a permit is $100,000 (AUS) or five years' imprisonment, or both. Creaser (1994a) recounts examples of illegal fossil exporting from Australia.

Other organisations in Australia own their own reserves or lands in which conservation, including geoconservation, is practised. Each state and territory has its own National Trust and these own and protect important landscapes including coastal and upland areas. The Australian Council of National Trusts (ACNT) was formed in 1965 to represent trust interests at the federal level and increasingly coordinates the work of the constituent trusts. Another example is the Sydney Harbour Trust that produced a forward plan in 2003 in which it undertook to identify, protect, conserve and present to the community the "significant geodiversity values on trust and harbour land sites such as volcanic dykes, Pleistocene sand dunes and laterites"

5.17   New Zealand

About 20% of New Zealand has been designated as national parks or reserves since the first was established in 1887, covering the active volcanoes at Tongoriro. The 12 National Parks are principally established to protect landscapes and biological diversity but they contain some of the country's best earth science features, including, fjords, waterfalls, glaciers and volcanoes. In addition, 10 Crown Reserves have been created specifically to protect geoheritage (Hayward, 1989). Other conservation designations that include geological features include Scenic Reserves (e.g. Moeraki Boulder scenic reserve protecting large spherical septarian concretions) and Scientific Reserves (e.g. the Curio Bay Scientific Reserve protecting a shore platform exposure of a Jurassic petrified forest).

In 1983, the Geological Society of New Zealand began compiling a list of New Zealand's important and diverse geoheritage and there are now over 3,600 sites on the Geopreservation Inventory. The aim is then to ensure protection and management of these sites for future research, education and public access (Buckeridge, 1994). The inventory was compiled by relevant experts rather than by systematic field surveys. The sites have been classified into 15 categories (e.g. landform, igneous geology, fossil, mineral, etc.) and sorted by region. Each site has been given a vulnerability rating of 1(= highly vulnerable to human modification) to 4(= sites that could be improved by human activity), and 5(= sites that have been destroyed). Kenny & Hayward (1993) believed that "The overriding objective of earth science conservation in New Zealand should be to ensure the survival of the best representative examples of the broad diversity of geological features, landforms, soil sites and active physical processes".

The *Environment Act* (1986) and the *Conservation Act* (1987) both recognise the intrinsic value of natural features and systems, although geoheritage and geoconservation are not specifically mentioned in the legislation. The *Conservation Act* (1987) and *Resource Management Act* (1991) require various authorities in New Zealand to prepare Conservation Management Strategies (CMSs) and Regional Coastal Plans; some of these, including the Auckland CMS, have geoconservation objectives.

In 1987, legislation was passed to halt further drilling for geothermal steam at Rotorua, following the failure of several geysers (see Section 4.4). The situation has now stabilised with a slight increase in bore pressures. Other aspects of New Zealand's geodiversity are protected by organisations such as the Queen Elizabeth II National Trust, which protects privately owned areas of open space, usually in perpetuity, without jeopardising the rights of ownership.

5.18   The Rest of the World

In many countries in the rest of the world, geoconservation has barely begun and it is likely that significant losses of geoheritage are occurring. For example, in Central Eurasia comprising the newly independent states of Armenia, Azerbaijan, Georgia, Kazakhstan, Kyrgyzstan, Tajikistan and Turkmenistan covering 5 million km², many irreversible examples of damage can be cited (I. Fishman, personal communication). In Kazakhstan, in particular, attempts are being made to establish a Geosites network (Nusipov *et al.*, 2001). In a number of other countries, including India, China, Taiwan and South Africa, some attempts have been made to establish geoconservation systems.

India has designated about 16 geological "heritage sites" under the *National Monuments Act* (1974) (Prasad, 1994). Apart from 11 Geoparks (see above), China has over 40 national parks, several of which are geologically based World Heritage Sites. Geological interest ranks highly in the criteria used to select such areas (Jiang, 1994). China also has a law on "preservation of cultural relics" (1982), which treats vertebrate and human fossils in an identical way to cultural objects. Enforcement has been strengthened recently to prevent illegal collecting of vertebrate fossils, some of which are used in traditional Chinese medicine.

Hong Kong, prior to its reunification with China, had a well-developed system of country parks, special areas and sites of special scientific interest to protect a small territory faced with huge development pressures. The *Country Parks Ordinance* (1976) allowed 21 country parks to be established to protect wildlife, landscape and cultural sites, and also to encourage recreation and tourism. There is no reference to geological or geomorphological conservation in the Ordinance but activities such as mining are prevented in the parks. "Special areas" can be designated inside or outside country parks and several of the 14 sites are of particular geological or landscape value. SSSI designation "neither confers statutory power on the government to enforce preservation nor implies any restriction upon owners.... Designation as an SSSI is intended to ensure that due consideration is given to protecting the site when considering development proposals that may affect it" (Workman, 1994, p. 295). Of the *c.*50 SSSIs in Hong Kong, several have been designated primarily for their geological or geomorphological interest.

Taiwan has a system of national parks, nature preserves and scenic areas and a *Cultural Heritage Preservation Law* that give some protection against unsuitable land development projects (Wang *et al.*, 1994, 1999). Several national parks and three nature reserves were chosen for their geological or geomorphological interest (e.g. columnar basalt, mud volcanoes). Some activities such as mining and water abstraction predate inauguration of national parks and nature reserves and can therefore continue unchecked. A five-year Earth heritage conservation strategy was initiated in 1994 with the aim of expanding the site database, increasing the protection of sites and improving public awareness (Wang *et al.*, (1999).

About 6% of South Africa is covered by state protected areas managed by South African National Parks (SANParks). Like most of Africa's National Parks, they are primarily intended to protect wildlife, but many contain important elements of geoheritage. A particular issue in these parks has been the land claims of indigenous peoples and some settlement rights within parks are now starting to be restored, for example, the Nama at Richersfield National Park and Makuleke at Kruger National Park (Magome & Murombedzi, 2003). Another important initiative is the establishment of trans-frontier conservation areas (TFCAs) and parks. These are conceived as relatively large areas of land including one or more protected areas that straddle boundaries between two or more countries. Although they are intended to allow protection and restoration of large-scale ecosystems, including traditional migration routes, they also protect many important abiotic elements (land form, soils, active processes, etc.). Figure 5.10 shows the location of existing or proposed TFCAs in southern Africa. The Great Limpopo Transfrontier Park was formally established in 2002.

South Africa has rigorous controls on collecting and export of fossils, largely due to earlier exploitation of its rich palaeontological heritage, including the remains of hominid fossils from the Plio-Pleistocene sites in the northern Transvaal and the removal of fossil reptiles from the Permo-Triassic of the Great Karoo Basin by collectors and international expeditions. As a result, the *National Monuments Act* (1969) protects all palaeontological as well as archaeological and historical sites:

> "No person shall destroy, damage, excavate, alter, remove from its original site or export from the Republic – a. any meteorite or fossil.... palaeontological finds, materials or

object, except under the authority of and in accordance with a permit issued under this section"

5.19   International Geoconservation Revisited

A number of issues emerge for this limited review of national geoconservation systems:

- *Diversity of systems* – The first obvious point to make is that systems of protected areas and nature conservation legislation vary greatly around the world. It is clear that biodiversity dominates nature conservation efforts in most countries, but there are important geoconservation efforts being made in the United Kingdom and parts of Australia (particularly Tasmania), New Zealand, United States, Canada and Europe. The GCR programme in Britain is an outstanding effort though it is based only on scientific values. However, much more needs to be done even in these countries to raise awareness of the value and threats to geodiversity and the need for conservation efforts. In the United States, for example, there are many geological sites including world famous palaeontological sites, stratotypes and unique mineral localities that have no legal protection. Some countries have special designations for geological sites (e.g. "Geological Monuments"), but others argue that this makes geodiversity look like a special case rather than being a standard objective of nature conservation.
- *Site selection methods* – Those countries that have geoconservation site networks have different means of selecting sites. Some have used literature reviews to identify sites or panels of experts to reach consensus judgements about which sites should be included and which not. However, this may result in inventories of research undertaken rather than systematic assessments of sites and features to be protected. Ideally, systematic national surveys or inventories of the geodiversity resource should be established as the basis for selecting sites, but there are very few examples of this approach.
- *Site selection criteria* – A large number of criteria have been used to select sites for designation. Erikstad (1994c) and Gordon (1994a) have both compiled lists of criteria used in early surveys, including research value, rarity, vulnerability to threat and representativeness (Table 5.9). Clearly there is considerable overlap between many of these criteria and furthermore, site standing may be boosted where more than one criterion is present. Some countries try to select sites on the basis of significance and attempt to make the process more objective by defining sets of criteria against which to judge significance. Some use outstandingness to be the criterion, but if there are hundreds of outstanding examples of a particular type of site the impact and value of each may be diluted. Many countries have referred to the need to conserve the range of earth science interests present and in effect this means conserving the geodiversity of the country by selecting a representative sample of sites. There is, however, also the issue of the level of detail. It may not be sufficient to conserve a representative granite, drumlin or vertisol, since these themselves display a considerable geodiversity.
- *Duplication or replication?* – In selecting representative samples some countries, including the United Kingdom, try to avoid duplicating features by selecting only one example of different types of feature. While this is understandable in terms of

Table 5.9 *Criteria used by various authors in site assessment and selection (compiled from Gordon 1994a; Erikstad, 1994c)*

Educational value
Research or scientific value
Obviousness
Accessibility
Durability
Rarity
Assemblage of features
Type or reference site
Historical importance
Naturalness
Size
Combinations of morphology and process
Complexity of form and evolution
Active processes
Vulnerability to threat
Morphology
Uniqueness
Representativeness
Dateability or chronology
Usage
Diversity
Typicalness
Scientific documentation
Part of a larger unit
Potential value
Intrinsic value
Classical site or outstandingness
Key area
Importance of interpretation
Quality/beauty

trying to limit the escalation of sites, it does leave the type of feature vulnerable if the site selected becomes degraded. Replication provides some security against the complete loss of a particular aspect of geodiversity. Sharples (2002a) argues that the fewer examples of a type of feature exist, the more important it is to protect a higher proportion of the examples of the type in order "to guard against the possibility of all examples being degraded or destroyed for unforeseen reasons". He also believes that the case for replication is stronger in the case of features that are highly sensitive, support biodiversity or are poorly understood. Replication also allows nations to have representatives of particular types of feature in different parts of the country, thus supporting local (or regional) geodiversity. Further discussion of these and other issues can be found in Sharples (2002a).

• *Values* – It is also clear that sites can have value in more than one way. As outlined in Chapter 3, sites may be of intrinsic, cultural (folklore, archaeological/historical, spiritual, "sense of place"), aesthetic, economic, functional (physical, biological) or research/educational value. It is important to identify which of these values is being associated with sites. As noted above, in Britain statutory conservation and designations such as SSSIs are heavily science based, whereas the RIGS network may have wider aesthetic, educational or other values. There is an issue here related to the fact that the public finds it difficult to understand and therefore fully support a very scientific and academically led approach. For this reason, in some countries, such as Norway, educational and public-orientated criteria are integrated with scientific criteria (Daly *et al.*, 1994), and in others, such as the United States, there are integrated systems of natural and cultural protected areas (National Parks System). A related point is made by Boulton (2001) with reference to the selection of sites on scientific grounds. He points out (p. 51) that "The definition of a Site of Special Scientific Interest, for example, is not a scientific judgement, but a societal judgement about the value of science set against the value of the site for other purposes".

- *Scope* – Many systems, including the GCR programme in Britain and geological heritage programme in Australia are biased towards static sites, particularly geologically based ones, rather than dynamic processes. The New Zealand Geopreservation Inventory is one of the few to include soil sites.
- *Significance* – Sites may be graded for significance according to whether they are of international, national, regional/state or local importance.
- *Site management and enforcement* – The management regimes for protected areas vary greatly. In some cases, reliance is placed simply on the designation of the protected area and related legislation, which may include fines for violations. But laws are only as good as their enforcement and this is variable. Academic geoscientists are often very comfortable with site selection but less good at supporting practical site management or trying to influence decision-makers or a public with other agendas and priorities.

There is also the issue of low penalties for law infringement. Pemberton (2001a) quotes a recent case in Tasmania where a person was found guilty of removing quartz crystals of Devonian age from a cave, but received only a $50 fine. "That equates to a fine of about a cent for every million years. Interesting also to note that this material was stolen from the Tasmanian Wilderness World Heritage Area, an area with the highest possible level of conservation status" (Pemberton, 2001a). Likewise, Horner & Dobb (1997, p. 242) referring to vertebrate fossils in the United States believe that "Compared to what a commercial collector stands to earn from the sale of such fossils, the fines are negligible, encouraging collectors to treat them as one of the costs of doing business".

At some sites, fencing is employed to restrict all but the most persistent intruders. For example, at Yellowstone National Park, a remaining petrified tree trunk has been fenced following casual collecting, which removed the other one (Fig. 5.30). At Birk Knowes, Scotland, a restricted fossil outcrop has been fenced, collecting permits have been stopped and notice boards with anti-collecting messages in four languages have been erected (MacFadyen, 2001a). In other places, paths, boardwalks and related signage are used to try to restrict visitor access to sensitive sites (Fig. 5.31). But perhaps the largest efforts are applied to education in the hope that voluntary restraint will restrict damage to sites.

The management regime employed should reflect many of the above decisions, for example, the value and significance of sites, but many protected areas have no or little management and often suffer damage as a result. At Lake St Clair in Tasmania, Kiernan (1996, p. 207) argued that the establishment of a national park had not served to protect the glacial geoheritage. He blamed the attitude of the Hydro-Electric Commission and the low level of concern for geoheritage amongst the Parks & Wildlife Service and many of its senior officers, though the situation

Figure 5.30   *Fencing around a petrified tree to protect it from souvenir hunters who have denuded an adjacent one, Yellowstone National Park, USA*

has now improved (M. Pemberton, personal communication). Committing the energy and resources to enforcing legislation is usually as challenging as getting the legislation enacted.

- *International agreements* – There have been several calls for greater international consistency in geoconservation practice. Stürm (1994), for example, argued that an International Convention would be useful in promoting and supporting the implementation of the geotope concept within national planning systems. This would need to cover at least the following issues:
  - Definition of the relevant spatial unit (geotope).
  - Creation of international inventory of important geotopes (including criteria and procedures for selection).
  - Obligations on contracting countries to integrate geoconservation into their planning policies and legislation including the designation of geotope protection areas; take account of geoconservation at all stages of their plan-making and project approval processes at all administrative levels establish an inventory of geotopes of national and regional importance.
  - Establish in each country a statutory body which has to supervise, sustain and provide the implementation of the geotope concept.

The Geosites initiative (see above) goes at least part of the way to meeting these proposals.

Norman (1994) pointed out that fossils are a national and finite resource, but questioned their status, ownership and rights of sale on the open market. Commercial sale can remove them from the public domain into private collections or abroad. He believes that regulations should be in place to prevent contentious issues, (e.g. relating to who owns fossils, whether there is a right to sell, export and import controls)

Figure 5.31 *Signage used to discourage collecting and off track access, Craters of the Moon National Monument, Idaho, USA*

and that there should be consistency within and between nations. Specifically he recommends that

– each nation should establish an expert committee to review the status of fossils in that country as national scientific and natural heritage items;

- each nation should establish the criteria to be used in evaluating fossils objectively as heritage items;
- each nation should establish a policy for dealing with scientific, amateur and commercial fossil collecting on private and public land;
- there should be an international forum for considering transnational issues of ownership and trade of fossils.

Norman believes that this approach is essential to avoid individual countries developing policies *ad hoc* in response to individual issues or local political initiatives.

5.20   Conclusions

Despite the diversity of geoconservation systems and networks currently in use, many countries and organisations now agree that the aim of geoconservation should be to maintain the range of earth science features within their borders. Although few yet use the terminology, what this means is that they aim to conserve a representative selection of the geodiversity of their country. However, many questions remain related to the methodology for establishing geodiversity and significance, scale issues, values, replication and site management. Geoscientists must be more willing to influence the uninitiated about the value of sites, the threats to geodiversity and the need for practical management rather than discussing the merits of sites amongst themselves. There have also been many questions asked about the effectiveness and voracity of the whole protected area approach, and the following chapter discusses this issue.

6

Managing Geodiversity: New Approaches for the Wider Landscape

6.1 Sustainable Management of the Georesource

The protected area and legislative approaches outlined in Chapter 5 play an absolutely crucial role in global efforts at conserving geodiversity and providing legal support. But there has been an increasing recognition that while they are unquestionably necessary, they are not, by themselves, sufficient to enable the full and sustainable management of the world's geodiversity or geoheritage.

This chapter attempts to explain a very wide range of approaches and initiatives that have been developed in the last 15 years or so and are still being developed. They can be considered as examples of good practice for others to borrow or follow. Of course, many traditional approaches to land use have attempted to apply sustainable land management practices, for example promoting soil conservation, preventing sand dune blow-outs and protecting karst systems. But some of the new approaches and techniques described in this chapter can lead to a more holistic approach to geoconservation, thus ensuring a sound future for the physical or abiotic resources of the planet. Some of these measures have been alluded to in previous chapters, but it is worth bringing them together with some very new ideas in one management chapter.

First, it is argued in this chapter that we must go beyond the designation and management of protected areas to assess, value and protect geodiversity in the wider landscape and indeed in cities. Secondly, we must learn to work with nature rather than seeking to dominate or subjugate it. This will involve recognising local diversity and distinctiveness and the issues of landscape restoration, design of landforms and a more sustainable use of geomaterials. Thirdly, we must use any available tools of land-use policy, management and planning to safeguard the resource and we must work in partnership with other organisations and local communities, and win the hearts and minds of the general public to ensure that this is the case. Together, these approaches represent a new, sustainable approach to the management of the physical environment. The threats are many and will require constant vigilance if we are to leave future generations with a geoheritage that is not significantly depleted and degraded.

Geodiversity: valuing and conserving abiotic nature   Murray Gray
© 2004 John Wiley & Sons, Ltd   ISBNs: 0-470-84895-2 (HB); 0-470-84896-0 (PB)

6.2   Assessing the Wider Geodiversity Resource

6.2.1   Introduction

The issue of protecting sites and areas is neatly described in the following quotes from Mather & Chapman (1995). Although they are mainly directed at biological conservation, they are equally applicable to geoconservation:

> "Another problem is that a park is a discrete area, with a boundary, and therefore there is a danger that the impression is given that effective conservation can be achieved solely in such areas, and that it can be ignored elsewhere. It is now increasingly being recognised that conservation cannot successfully focus solely on protected areas (or on protected species), and that it needs to be applied to all aspects of environmental resources. The corollary is that the scientific management of protected areas and species cannot, in itself, be successful if it is divorced from the wider use of environmental resources".

> (Mather & Chapman, 1995, p. 130).

This wider perspective has certainly become widespread in the case of wildlife conservation where

> "...legislation on species protection does not ensure that ecosystem diversity is maintained.... (there is) an assumption or implication that conservation can be achieved by protecting certain areas, rather than on a countrywide basis. Over the last few years, the inadequacy of this approach has been increasingly recognised, and some progress has been made towards integrating environmental principles into production-related policies, for example in the agricultural sector in the European Union"

> (Mather & Chapman, 1995, p. 134).

This in turn reflects McNeely's (1988, 1989) view that conservation has to be integrated into the wider framework of the management of environmental resources and not viewed simply as a separate sector. He argued that conservation is too important to leave to scientists and is far more of a social challenge than a biological one. Adams (1996, p. 116) makes similar points:

> "A conservation strategy based on protected areas can make a holistic approach to conservation more difficult, encouraging the view that conservation is a sector or a land use. Their existence can also imply that the needs of conservation have been met, making it harder to achieve integrated policies that cut across economic sectors and make a difference to national policies. Protected areas tend to be treated as separate entities from surrounding land, soaking up available conservation resources. Protected area boundaries are often arbitrary lines on maps.... irrelevant to natural processes and environmental problems".

In addition, there is the danger that protected areas will lead the government and the public to "begin to think that they have a free license to do whatever they want outside of its boundaries" (Brussard *et al.*, 1992, p. 158). Adams' (1996, p. 118) conclusion is that "The traditional concern for individual protected areas must be transformed to a concern for whole landscapes". Myers (2002, p. 54) believes that the need for a wider perspective is proved by the increasing impact of humans on protected areas: "setting aside a park in the overcrowded world of the early twenty-first century is like building a sandcastle on the seashore at a time when the tide is coming in deeper, stronger and faster than ever".

Pemberton (2001a) makes the important point that in conserving geodiversity we "need to extend beyond the traditional geological monuments or significant geological features approach. . . .which has less relevance to broader issues of land management and ecological sustainability or to important links with other nature conservation values". Furthermore, Sharples (2002a) notes that a site-based approach is not appropriate to the management of fluvial or karst systems since "it is essential to protect natural geomorphic processes throughout the catchment basin, not just in the local area of the specific caves or river channels that may be thought of as significant sites". Sharples (2002a) also recognises that landforms often need to be seen as part of broader assemblages. "Thus, whilst a glacial cirque produced by past glacial processes may be of some interest in isolation, its full significance only becomes apparent when it is considered in the context of an assemblage of related features". Without the evidence provided by the assemblage, "the information that any one isolated feature can give us about past processes is limited". In other words, one of the things that we need to do in geoconservation is to think about three-dimensional geological structures, landform assemblages and system functions.

Many aspects of management of the geoheritage will be discussed in this chapter, but the aim in this section is to explore the issue of how geodiversity in the wider landscape can be assessed. Of course, in some fields of earth science, the recording of the information has a long history. Geological maps, for example, are now available for most countries in the world, and a large number have complete coverage, albeit that the quality and scale of mapping are very variable. Some countries, including the United Kingdom, produce both solid and drift maps, which distinguish between consolidated and unconsolidated materials (see Chapter 2). Geological mapping is described in many books (Barnes, 1995; Maltman, 1998; Bennison & Moseley, 2003). Soil mapping is also routinely carried out in many countries (Hodgson, 1978) and some use agricultural land classification systems. Nationally co-ordinated and systematic geomorphological mapping, apart from topographic maps, is less common, though the principles and techniques are well developed and widely applied (Cooke & Doornkamp, 1990), and include landform and process mapping. Increasingly, this information is managed within Geographical Information Systems (GIS).

In addition to these spatial recording systems, there are also well-established methods for recording vertical variability in earth materials, whether they are lithological logs, borehole logs or soil profiles. Vertical variations in a very large number of material properties can also be recorded and presented. But it is not the purpose of this chapter to describe the scientific recording of geological information. Rather, the focus here is on the less well-known assessment systems of the wider georesource intended for conservation management purposes. It is also important to note that much has been written on other aspects of geoconservation, including conservation of specimens (rock, mineral, fossil) (Collins, 1995), conservation of paper records and photographs (Ellis, 1993), and conservation of building stones (Ashurst & Dimes, 1990; Siegesmund *et al.*, 2001), but these are beyond the scope of this book.

6.2.2 European initiatives

Two recent and important European agreements apply not just to conservation of protected areas but to the environment generally and, together with other measures, are likely to have a significant and increasing impact on land and water management.

European Water Framework Directive (WFD)

The European Water Framework Directive (WFD) was adopted in 2000. Its aim is to establish a new integrated approach to the protection, improvement and sustainable use of Europe's rivers, lakes, estuaries, coastal waters and ground waters. The requirements include preventing deterioration in the status of surface water bodies, enhancing the status of aquatic ecosystems and associated wetlands, protecting, enhancing and restoring all bodies of surface water to a good status within 15 years (other than artificial and heavily modified ones), progressively reducing pollution from priority substances and ceasing or phasing out emissions, discharges and losses of priority hazardous substances. The Directive requires a series of preparatory studies before River Basin Management Plans (RBMPs) are adopted by 2009 and implemented by 2012. "Programmes of measures" must ensure that all surface waters are of "good" status by 2015. This will only be adequately achieved by partnership working amongst a great many organisations responsible, for example, for agriculture, fisheries, flood management, conservation and land-use planning. Although mainly aimed at improving water quality and aquatic ecosystems, the WFD does refer to hydromorphological systems and should lead to physical restoration of rivers and fluvial processes.

European Landscape Convention

The European Landscape Convention was adopted on 19 July 2000 and signed by 18 European countries on 20 October 2000. It requires public authorities to adopt policies and measures at local, regional, national and international level for protecting, managing and restoring of landscapes throughout Europe. It covers all landscapes, "both outstanding and ordinary", that determine the quality of the people's living environment. Article 6 requires signatory nations to increase the awareness of landscapes through education and training, identification and analysis of the landscapes within their own territories and recognition of the forces and pressures transforming them.

6.2.3   Natural Areas and Landscape Character Assessment in the United Kingdom

Natural Areas/Futures

In the United Kingdom, over the last decade or so, the sufficiency of the site-based approach to nature conservation has been reassessed. First, although designated sites represent the best of the country's biodiversity and geodiversity, they cannot by themselves preserve all that is valuable in the natural environment. Preserving only the best sites, or a representative series of sites with no duplication, is rather an elitist approach. There are many sites and areas outside the boundaries of protected areas that will be almost as valuable and many ordinary landscapes that are valued by local communities as part of their familiar and cultural environment. There are strong arguments for greater recognition and respect for the distinctiveness of regional and local geodiversity. As Jarman (1994, p. 41) remarked, "Ordinary landscapes can be rewarding too". Secondly, a site-based approach leads to a highly fragmented system of conservation and does not promote the conservation interests of the areas between sites or the wider integrity of sites in their geological/geomorphological setting.

Following the devolution of the nature conservation function in the United Kingdom (see Section 5.10), English Nature undertook a "fundamental review of its approach to wildlife and geological conservation, to analyse the strengths and weaknesses of previous activities and determine how best to take conservation forward in the 1990s" (Duff, 1994, p. 122). This review established the need not only to improve the management of designated sites (see above) but also to identify the importance of seeing nature conservation in a wider context, both geographically and in terms of countryside management. A consultation paper, issued in 1993 (English Nature, 1993), argued that English Nature must "take an integrated approach to the whole environment and not simply view sites and habitats in isolation".

The solution proposed was to base the future conservation strategy around a system of "Natural Areas" covering the whole country. English Nature has therefore divided England into 120 of these areas (97 terrestrial and 23 maritime) on the basis of similarities in landscape, landform and other natural characteristics (Fig. 6.1 and Table 6.1). "Natural Areas are tracts of land unified by their underlying geology, landforms and

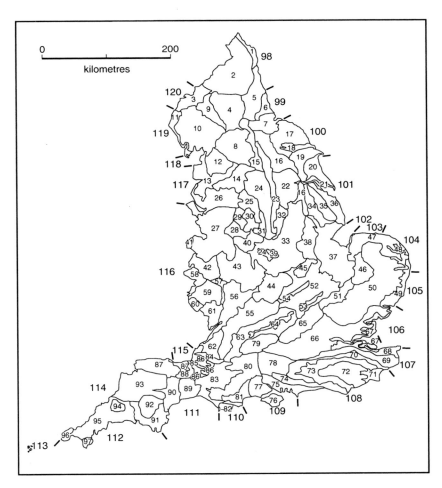

Figure 6.1 *The 120 natural areas of England as defined by English Nature (Reproduced with permission from English Nature)*

Table 6.1 *List of terrestrial and maritime Natural Areas (after English Nature, 1998)*

No.	Terrestrial Natural Area
1.	N Northumberland Coastal Plain
2.	Border Uplands
3.	Solway Basin
4.	N Pennines
5.	Northumberland Coal Measures
6.	Durham Magnesian Limestone Plateau
7.	Tees Lowlands
8.	Yorkshire Dales
9.	Eden Valley
10.	Cumbria Fells & Dales
11.	W Cumbria Coastal Plain
12.	Forest of Bowland
13.	Lancashire Plain & Valleys
14.	S Pennines
15.	Pennine Dales & Fringes
16.	Vale of York
17.	North York Moors
18.	Vale of Pickering
19.	Yorkshire Wolds
20.	Holderness
21.	Humber Estuary
22.	Humberhead Levels
23.	S Magnesian Limestone
24.	Coal Measures
25.	Dark Peak
26.	Urban Mersey Basin
27.	Meres & Mosses
28.	Potteries & Churnet Valley
29.	SW Peak
30.	White Peak
31.	Derbyshire Peak Fringe & Lower Derwent
32.	Sherwood
33.	Trent Valley
34.	N Lincolnshire Coversands & Clay Vales
35.	Lincolnshire Wolds
36.	Lincolnshire Coast & Marshes
37.	The Fens
38.	Lincolnshire & Rutland Limestones
39.	Charnwood
40.	Needwood & S Derbyshire Claylands
41.	Oswestry Hills
42.	Shropshire Hills
43.	Midlands Plateau
44.	Midland Clay Pastures
45.	Rockingham Forest
46.	Breckland
47.	North Norfolk
48.	The Broads
49.	Suffolk Coast & Heaths
50.	East Anglian Plain
51.	East Anglian Chalk
52.	West Anglian Plain
53.	Bedfordshire Greensand Ridge

Table 6.1 (*continued*)

No.	Terrestrial Natural Area
54.	Yardley-Whittlewood Ridge
55.	Cotswolds
56.	Severn & Avon Vales
57.	Malvern Hills
58.	Clun & NW Hereford Hills
59.	Central Herefordshire
60.	Black Mountains & Golden Valley
61.	Dean Plateau & Wye Valley
62.	Bristol, Avon Valleys & Ridges
63.	Thames & Avon Vales
64.	Midvale Ridge
65.	Chilterns
66.	London Basin
67.	Greater Thames Estuary
68.	North Kent Plain
69.	North Downs
70.	Wealden Greensands
71.	Romney Marshes
72.	High Weald
73.	Low Weald & Pevensey
74.	South Downs
75.	S Coast Plain & Hampshire Lowlands
76.	Ise of Wight
77.	New Forest
78.	Hampshire Downs
79.	Berkshire & Marlborough Downs
80.	South Wessex Downs
80.	South Wessex Downs
81.	Dorset Heaths
82.	Isles of Portland & Purbeck
83.	Wessex Vales
84.	Mendip Hills
85.	Somerset Levels & Moors
86.	Mid-Somerset Hills
87.	Exmoor & Quantocks
88.	Vale of Taunton & Quantock Fringes
89.	Blackdowns
90.	Devon Redlands
91.	South Devon
92.	Dartmoor
93.	The Culm
94.	Bodmin Moor
95.	Cornish Killas & Granites
96.	West Penwith
97.	The Lizard

No.	Maritime Natural Area
98.	Northumberland Coast
99.	Tyne & Tees Coast
100.	Saltburn to Bridlington
101.	Bridlington to Skegness
102.	The Wash
103.	Old Hunstanton to Sheringham

(*continued overleaf*)

Table 6.1   *(continued)*

No.	Maritime Natural Area
104.	Sheringham to Lowestoft
105.	Suffolk Coast
106.	North Kent Coast
107.	East Kent Coast
108.	Folkestone to Selsey Bill
109.	Solent & Poole Bay
110.	South Dorset Coast
111.	Lyme Bay
112.	Start Point to Land's End
113.	Isles of Scilly
114.	Land's End to Minehead
115.	Bridgwater Bay
116.	Severn Estuary
117.	Liverpool Bay
118.	Morecambe Bay
119.	Cumbrian Coast
120.	Solway Firth

soils, displaying characteristic natural vegetation types and wildlife species, and supporting broadly similar land uses and settlement patterns" (Duff, 1994, p. 121). The advantages of using these Natural Areas as the basis for developing national and local conservation strategies are seen as

- offering "a more effective framework for the planning and achievement of nature conservation objectives than do administrative boundaries" (English Nature, 1998);
- providing a framework for appraisal and evaluation to describe what is important and why, and for setting objectives to protect our characteristic biodiversity and geological heritage;
- enabling English Nature "to look at the resource in an integrated way, and not just to focus on the special sites" (Duff, 1994). In turn, this will achieve a less fragmented approach to nature conservation;
- increasing the integration of geological and biological conservation;
- acting as a framework for involving local communities in valuing the distinctiveness of their local areas and developing their sense of place in relation to nature conservation objectives.

The Natural Areas approach is therefore becoming the fundamental basis on which future nature conservation strategies in England are to be founded, though some problems have been encountered in dealing with areas that have cross-administrative boundaries.

Having adopted the principle of the Natural Areas approach and having established the boundaries of the 120 Natural Areas, English Nature has published descriptive profiles of each one compiled by about 20 local area teams based in different parts of the country. These Natural Area Profiles are available on a CD-ROM (English Nature, 1998) and other descriptions appear on the English Nature web site (www.english-nature.org.uk). Apart from describing the characteristic wildlife and geology of each area, the profiles contain details of the key conservation issues, threats and objectives in each area, lists of species and habitats that are the subject of national and international legislation and details of Local Biodiversity Action Plans (LBAPs) and Species Recovery Programmes. There are also eight regional summaries that are intended to assist in applying the principles of sustainable development to the preparation of regional economic strategies, Regional Planning Guidance (RPG) and other regional policies and strategies (English Nature, 1999).

The Natural Areas initiative is a very innovative scheme and the principle should certainly be supported. However, it has been criticised by some British geoscientists. For example, Doyle & Bennett (1998) point out that there is a danger that geology will simply be seen as the foundation for wildlife, rather than worthy of protection in its own right. "Therefore, although the Natural Area concept is clearly a step forward from the classic site-based approach to conservation, in effect it does little to promote the greater public appreciation of the fragility of Britain's earth heritage" (Doyle & Bennett, 1998, p. 56).

The present author (Gray, 2001) shares some of these views but blames the implementation of the scheme, as far as geomorphology is concerned, rather than the principle of the approach. Thus, the dominance of wildlife conservation interests, the inconsistencies in geomorphological coverage, the lack of up-to-date ideas and terminology and the weaknesses in describing the geodiversity of each area are the result of flaws in implementation rather than fundamental weaknesses in the system itself. According to Gray (2001, p. 1019):

> "A few Natural Area Profiles do start to give the essence of what is required. . . .In these descriptions we begin to see how the diversity of English landform, surficial deposits and processes – the distinctive geomorphological character of each area – can be recorded. Maps and descriptions of this type would help to crystallise what is valued in the local landscape character, the sensitivity of landscapes to geomorphological change and what needs to be done to protect the landscapes from inappropriate development".

The Countryside Agency in England has been undertaking a similar exercise, termed *Countryside Character*, but with the wider remit of including the cultural landscape in the process of sub-dividing the country. It was hoped that the two agencies would be able to agree to the same areal units, but in the end, the Countryside Agency wanted a finer sub-division of the country so that some Natural Areas comprise more than one Countryside Character Area. However, the two systems were combined into a Landscape Character Map of England.

The situation and reasoning in Scotland is quite similar, though unlike England, the countryside and nature conservation agency mergers in both Scotland and Wales have made it easier to agree to wider-based and more integrated strategies in these countries. Scottish Natural Heritage (SNH) has divided Scotland into 21 Natural Heritage Zones (now called Natural Heritage Futures (NHFs)) in very similar ways to Natural Areas in England (Gordon & McKirdy, 1997; Mitchell, 2001; Scottish Natural Heritage, 2002). However, additional roles are identified for Natural Heritage areas as

- providing a framework in which to evaluate future environmental change;
- providing a context to evaluate the significance of gaps in knowledge about the distribution and status of natural heritage resources and how best to fill the gaps.

For each of the 21 areas (Fig. 6.2), SNH has prepared a "Local Prospectus", which describes the natural heritage for the area and the processes that have produced it, discusses the changes taking place, including human impacts, and presents a series of key

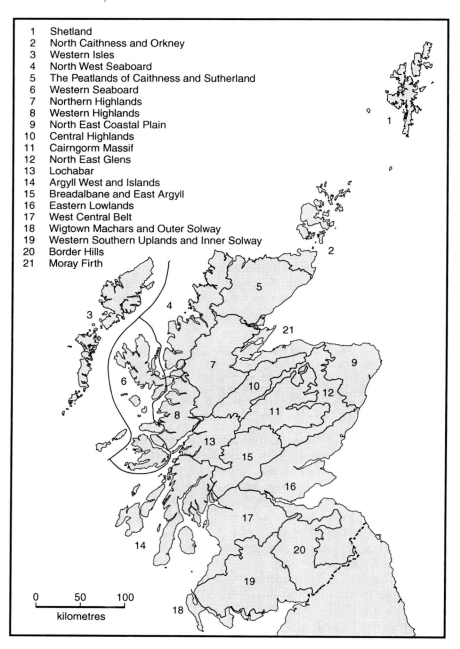

1   Shetland
2   North Caithness and Orkney
3   Western Isles
4   North West Seaboard
5   The Peatlands of Caithness and Sutherland
6   Western Seaboard
7   Northern Highlands
8   Western Highlands
9   North East Coastal Plain
10   Central Highlands
11   Cairngorm Massif
12   North East Glens
13   Lochabar
14   Argyll West and Islands
15   Breadalbane and East Argyll
16   Eastern Lowlands
17   West Central Belt
18   Wigtown Machars and Outer Solway
19   Western Southern Uplands and Inner Solway
20   Border Hills
21   Moray Firth

0   50   100
                kilometres

Figure 6.2   *The 21 Natural Heritage Zones of Scotland as defined by Scottish Natural Heritage (Reproduced with permission from Scottish Natural Heritage)*

goals and specific actions that aim to close the gap between the vision and current trends over a 25-year period (Scottish Natural Heritage, 2002). In addition, the project has also produced six thematic booklets (Hills & moors, Forests & woodland, Farmland, Fresh waters, Coasts & seas, Settlements). As these titles suggest, the geological and geomorphological content of these booklets is not strong, there is little place for geodiversity within the visions and what there is, is site-based. However, the accompanying CD-ROM has detailed descriptions of the geology, palaeontology, geomorphology, soils and landscapes of both Scotland as a whole and each of the 21 areas. There are also discussions on existing trends and pressures and the state of the georesource. Altogether, this adds up to a very significant effort to describe the geoheritage of Scotland and is a milestone that other countries could well follow (Scottish Natural Heritage, 2002). As part of this process, a pilot study has examined the feasibility of producing a geomorphological database and GIS for the north-west Seaboard (Kirbride *et al.*, 2001). This is the equivalent of the inventory approach in Australia described below.

As Mitchell (2001, p. 237) states, "Sustainable development requires working towards a long-term goal, or vision, which brings together social well-being, economic prosperity and environmental stewardship. Traditionally, planning has been carried out on a sectoral basis. In order to work towards sustainable development, various sectoral plans need to be considered over a geographical area in order to identify and resolve potential issues of conflict". The message here is therefore that geoconservation, and indeed nature conservation, in general cannot be isolated from other strategies but need to be integrated with them. This is a theme that will be returned to in Chapter 7.

The Countryside Council for Wales' Landmap Project is more closely related to a landscape character assessment approach and this is described below.

Landscape Character Assessment

Landscape Character Assessment came to prominence in the United Kingdom in the 1990s (Countryside Commission, 1993a). It is a technique that is related to the Natural Areas approach and particularly the Countryside Character initiatives described above, but at a finer scale. Landscape character is defined as "a distinct and recognisable pattern of elements that occur consistently in a particular type of landscape. Particular combinations of geology, landform, soils, vegetation, land use, field patterns and human settlement create character" (Swanwick & Land Use Consultants, 2002, p. 9). It should be noted therefore that as it stands, the technique includes the human/cultural elements of landscape, though there is no reason why it could not be restricted to natural landscape elements. In this discussion, the emphasis will therefore be placed on the physical elements of the landscape (geology, geomorphology, soils) within the context of the wider landscape character.

Characterisation is the name given to the practical steps involved in identifying two categories of landscape character:

- Landscape Character Types, which are generic in nature and which share many common combinations of geology, landform, drainage, vegetation and human influences. For example, chalk river valleys or rocky moorlands are landscape character types.
- Landscape Character Areas, by comparison, are unique and discrete geographical areas of a landscape type. For example, the Itchen Valley, Test Valley and Avon Valley in southern England are all examples of the chalk river valley type.

As Fig. 6.3 indicates, these two categories can be interlocked or nested to provide characterisation at different scales. Thus, within the Dartmoor Natural Area or Countryside Character Area, we can identify the generic landscape character types of plateau top and river valleys, and in turn these can be further subdivided. The landscape character

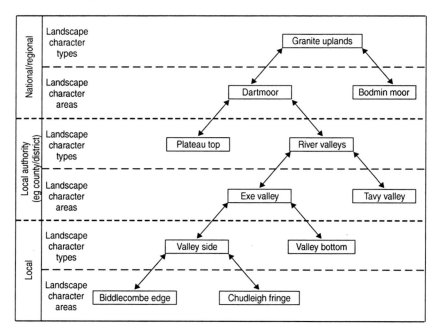

Figure 6.3   *The relationship between Landscape Character Types and Landscape Character Areas (After Swanwick, C. & Land Use Consultants, (2002) Landscape Character Assessment: Guidance for England and Scotland, prepared on behalf of The Countryside Agency and Scottish Natural Heritage by Carys Swanwick, Department of Landscape, University of Sheffield and Land Use Consultants 2002. Published by The Countryside Agency, Cheltenham (CA84))*

areas can then be mapped (Fig. 6.4) and subsequently used in landscape evaluation, the development of related land-use policies or landscape restoration and enhancement (see below).

In Scotland, SNH has co-ordinated a comprehensive national programme of Landscape Character Assessment in partnership with local authorities, and 29 separate assessments have now been published in which 366 local landscape character types and nearly 4,000 individual local character areas have been identified. A similar approach has been adopted in northern Ireland, and the Irish Republic has recently initiated a landscape character approach following some trial projects in Counties Clare and Leitrim (Heritage Council, 2002). In England, many counties and districts have also undertaken landscape character assessments for their areas, but there is some inconsistency in approach due to the fact that different consultants have worked to different briefs, and the coverage is far from complete (Swanwick & Land Use Consultants, 2002). In Wales, a national programme called *Landmap* has been developed in a similar way to landscape character assessment, but with a series of overlain layers mapping different landscape variables.

The landscape characterisation stage may be an end in itself, in which case, the landscape character map and descriptions of landscape types are value-free statements of the current appearance of the landscape. In turn, they can be used to raise awareness of the distinctiveness and diversity of the landscape and to encourage appreciation of the differences between places. However, the landscape character process will not normally be an end in itself. Rather, the process will go beyond neutral statements of

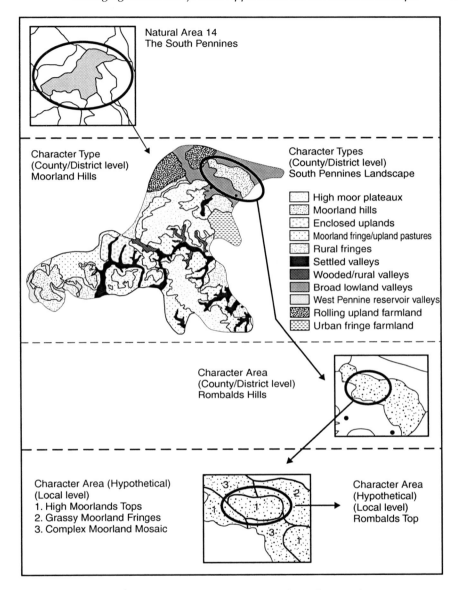

Figure 6.4 *The relationship between a Natural Area and Landscape Character Areas at various scales (After Swanwick, C. & Land Use Consultants, (2002) Landscape Character Assessment: Guidance for England and Scotland, prepared on behalf of The Countryside Agency and Scottish Natural Heritage by Carys Swanwick, Department of Landscape, University of Sheffield and Land Use Consultants 2002. Published by The Countryside Agency, Cheltenham (CA84))*

landscape character to provide an assessment of the character that can inform particular decisions (Fig. 6.5). However,

"Making judgements as part of an assessment should not concentrate only on the maintenance of existing character. This may be one part of the judgements made. The

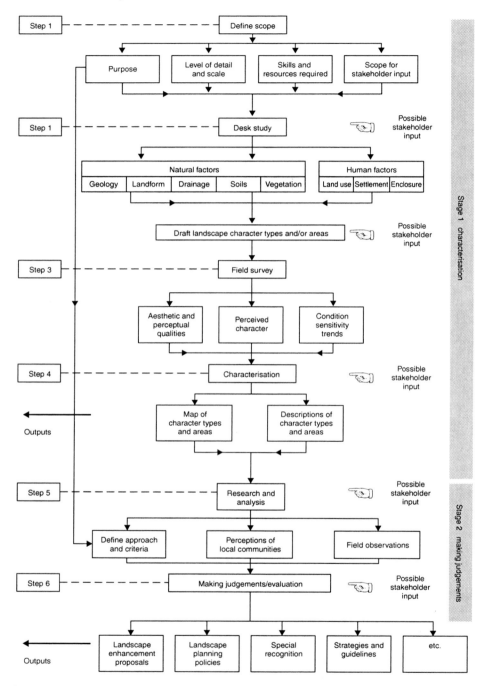

Figure 6.5 *The Landscape Character Assessment process (After Swanwick, C. & Land Use Consultants, (2002) Landscape Character Assessment: Guidance for England and Scotland, prepared on behalf of The Countryside Agency and Scottish Natural Heritage by Carys Swanwick, Department of Landscape, University of Sheffield and Land Use Consultants 2002. Published by The Countryside Agency, Cheltenham (CA84))*

focus should be on ensuring that land use change or development proposals are planned and designed to achieve an appropriate relationship (and most often a 'fit') with their surroundings, and wherever possible contribute to enhancement of the landscape, in some cases creating a new character" (Swanwick & Land Use Consultants, 2002, p. 52).

Several approaches have been taken to making judgements about landscape character, though they are often used in combination (Swanwick & Land Use Consultants, 1999). The approaches include the following:

- *Key landscape characteristics and guidelines.* These are the factors that are particularly important in creating landscape character. It follows that if the distinctive character of the landscape is to be maintained and enhanced, then the key characteristics must be protected from adverse change. Landscape guidelines can then be drawn up, which indicate the actions required to maintain and, if appropriate, enhance the landscape. Table 6.2 shows the steps recommended in developing landscape guidelines and gives examples of how this might be applied to the geoheritage.
- *Landscape quality and strategies.* The quality here refers to the physical state of repair of the landscape or parts of it, and suggests future landscape strategies. For example, if the typical character of the landscape is very apparent and the features within it are in good repair, then the strategy might reasonably be to conserve its character. Conversely, if the typical characteristics are weakly defined or in poor condition, then the strategy may be to strengthen or restore its character. This is illustrated in Fig. 6.6 (Warnock & Brown, 1998).
- *Landscape sensitivity and capacity.* Sensitivity concerns the degree to which a particular landscape character type or area can accommodate change without unacceptable adverse consequences for its character. Capacity is similar and deals with the amount

Table 6.2 *Steps in developing landscape guidelines with examples from the geoheritage (modified after Swanwick & Land Use Consultants, 1999)*

Review from field survey	
• Key characteristics of the landscape	e.g. land form, rock outcrops, active processes
• Current state of landscape – condition of elements and overall intactness	e.g. topographic remodelling, extent of quarrying, river channelisation
Identify by research & consultation	
• Trends in land use that may cause future change	e.g. increase in agricultural land, golf course demand
• Potential development pressures	e.g. urban expansion, water or mineral extraction
Predict	
• Consequences of land-use trends and pressures for the landscape	e.g. loss of geological sites, changes in soils
• Effects of predicted change on key characteristics (+ve & −ve)	e.g. creation of sections in road cuttings, loss of landforms, soil erosion
Define	
• Threats to key characteristics	e.g. loss of natural topography
• Opportunities for enhancement	e.g. river or quarry restoration,
• Guidelines on intervention	e.g. encourage use of local building stone
• Priorities for action and methods of implementation	e.g. need for change in farming practices to protect caves and karst by discussion with farmers

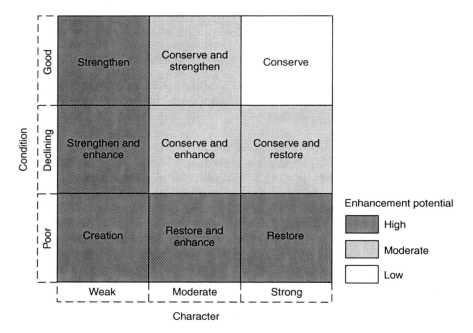

Figure 6.6   *Strategies for managing landscape character (After Warnock, S. & Brown, N. (1998) A vision for the countryside. Landscape Design, **269**, 22–26, by permission of Landscape Design)*

of change of a particular type that a landscape can accept without detrimental effects on landscape character. "Robustness" is a term used to describe both concepts. A robust landscape will have low sensitivity and a high capacity to accept change. For example, undulating landscapes may be of lower sensitivity, higher capacity and generally more robust in accepting landscape change than flat and featureless landscapes. These are familiar concepts for geomorphologists (see Section 4.16) but perhaps not for most landscape assessors.

Some practical geomorphological applications of landscape character assessment are discussed in Section 6.4.

6.2.4   Mapping and inventory approaches in Australia

In Australia, there has also been a long-standing interest in taking a wider view of the natural land resource (sometimes termed "georegionalisation"). In recent years, some of the same issues outlined above in the United Kingdom have begun to be discussed. For example, the topic of regional subdivision of Australia for geoconservation purposes was the subject of a conference in 1996 under the auspices of the Australian Heritage Commission (AHC) (Eberhard, 1997). Some examples of the mapping and inventory approaches adopted in Australia are described below.

Land systems mapping

The Land Systems Mapping approach was developed by the Commonwealth Scientific and Industrial Research Organisation (CSIRO) in Australia in the 1950s and 1960s, and it was subsequently applied in Africa, Latin America and Asia (Cooke & Doornkamp, 1990). It is based on recurring patterns of geology, landform, soils and vegetation; "thus the land system is a scientific classification of country" (Stewart & Perry, 1953, p. 55). As Cooke & Doornkamp (1990, p. 21) appreciate, the simplest criterion for mapping land system areas is landform. The system is based on a landform hierarchy in which land elements form parts of land units, which in turn combine to form land systems. The land system maps are usually supplemented by three-dimensional diagrams (Fig. 6.7) and descriptive tables (Table 6.3). The purpose of the work is to provide a basis for assessing land resources and particularly agricultural potential, rather than for conservation objectives (King, 1987). Nonetheless, it is a conceptual forerunner of the modern georegionalisation approach.

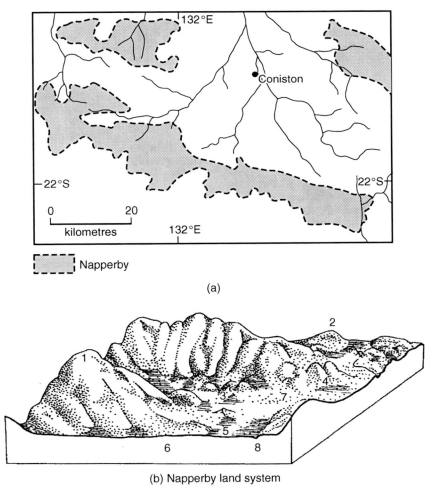

(a)

(b) Napperby land system

Figure 6.7 *(a) Location and (b) an example of a three-dimensional model of a Land System in Australia: Napperby land system near Coniston, Northern Territories (© CSIRO. Reproduced from Perry, R.A. (1962) General report on lands of the Alice Springs Area 1956-7. Land Research Series 6, CSIRO, Australia, with the kind permission of CSIRO)*

Table 6.3   *Description of the Units of the Napperby Land System, Australia (after Perry, 1962)*

Napperby Land System (2600 km^2)
Geology: massive granite and gneiss, some schist, Precambrian age
Geomorphology: Granite hills up to 150 m high and plains with branching shallow valleys,
   less extensive rugged ridges with relied up to 15 m, and a dense rectangular pattern of
   narrow, steep-sided valleys.

Unit	Area	Landform	Soil	Vegetation
1.	Large	Granite hills: tors and domes up to 150 m high; bare rock summits and rectilinear boulder covered hill slopes, 40–60%, with minor gullies; short colluvial aprons, 5–10%	Outcrop with pockets of shallow, gritty or stony soils	Sparse shrubs and low trees over sparse forbs and grasses, *Triodia spicata*, or *Plectachne pungens*(spinifex)
2.	Medium	Close-set gneiss ridges and quartz reefs: up to 15 m high; short rocky slopes, 10–35%; narrow intervening valleys		
3.	Medium	Interfluves: up to 7 m high and 0.8 km wide; flattish or convex crests, and concave marginal slopes attaining 2%	Mainly red earths, locally red clayey sands and texture-contrast soils, stony soils near hills	Sparse low trees over short grasses and forbs or *Eragrostis eriopoda* (Woollybutt)
4.	Medium	Erosional plains: up to 1.6 km in extent, slopes generally less than 1%		
5.	Small	Drainage floors: 180 – 365 m wide, longitudinal gradients about 1 in 200	Mainly texture-contrast soils, locally alluvial soils and red earths	*Eremophila* spp.- *Hakea leucoptera* over short grasses and forbs; minor *Kochia aphylla* (cotton-bush)
6.	Small	Alluvial fans: ill-defined distributory drainage; gradients above 1 in 200	Alluvial brown sands and red clayey sands	Sparse low trees over short grasses and forbs or *Aristida browniana* (kerosene grass)
7.	Small	Rounded drainage heads: up to 180 m wide and 1.5 m deep on the flanks of Unit 3	Red earths	Dense *A. aneura* (mulga) over short grasses and forbs
8.	Very small	Channels: up to 45 m wide and 1.5 m deep and braiding locally	Bed-loads, mainly coarse grit	*E. camaldulensis* (red gum), *A. estrophiolata* (ironwood) over *Chloris acicularis* (curly windmill grass)

Landform inventories

Conservation objectives are very much the theme of the systematic inventory approach
more recently developed in Australia, particularly in Tasmania. Sharples (2002a) describes three scales of geodiversity inventory as follows:

- Reconnaissance inventories, which are based largely on existing information and which have been prepared for most public land (but little private land) in Tasmania (Sharples, 1997).
- Systematic and thematic inventories, which are comprehensive comparative assessments of all features and systems in a particular region or of a given theme over a larger area (e.g. Rosengren's (1994) work on the Western Victoria lava province). A related approach is the classification-based approach in which a systematic classification of a particular aspect of geodiversity is established, an inventory of all known occurrences of each class is compiled and finally the best representative examples of each element are identified. This approach has been completed for karst, glacial and coastal features (Kiernan, 1995, 1996, 1997a). For example, in glacial geomorphology, Sugden & John (1976) and Eyles (1983) have developed classification systems for glacial landscapes, but Kiernan (1996) developed this further into a "landform communities" classification system containing 20 recognised types (Table 6.4). Kiernan argues (p. 195) that the adoption of this approach "would facilitate the advancement of geoconservation from the traditional focus on geological monuments to a proper integration with management to protect environmental diversity". Evans (2003) has recently further expanded the consideration of glacial landsystems.

On the georegional scale, Houshold *et al.* (1997) and Jerie *et al.* (2001) describe an approach to producing an inventory of stream geodiversity in Tasmania. This is specifically being done because many stream types are not included in the current reserve system and there is a need for a more strategic approach

Table 6.4 *Glacial landform communities (after Kiernan, 1996)*

Glacial erosion communities

1. Minimally eroded ice dome communities
2. Areally scoured ice dome landform communities
3. Linearly eroded outley glacier landform communities
4. Alpine valley glacier landform communities
5. Cirque glacier landform communities
6. Composite erosional landform communities

Glacial deposition communities

7. Fluted drift landform communities
8. Active ice depositional landform communities
9. Transition depositional landform communities
10. Disintegration moraine landform communities
11. End moraine landform communities

Glaciofluvial communities

12. Subglacial erosion landform communities
13. Subglacial deposition landform communities
14. Lateral margin communities
15. Terminal/proglacial communities

Other landform communities

16. Glaciolacustrine landform communities
17. Glacioaeolian landform communities
18. Fjord glaciomarine landform communities
19. Open water glaciomarine landform communities
20. Other composite communities, e.g. glaciokarst

to stream conservation in Tasmania. Jerie *et al.* (2001) are developing a GIS based Environmental Domain Analysis (EDA), which will result in a map of Tasmania showing stream regions where similar controls on stream development have acted through time. The controls include geology, topography, climate and process history. The end product will be a single map of Tasmania showing variation in system control variables across the state, and this will help improve stream conservation and management, for example, by identifying representative elements of geodiversity in a scientifically rigorous way. Other fluvial approaches adopted in Australia include the River Styles approach (Outhet *et al.*, 2001b).

- Detailed inventories, which are "detailed descriptions of particularly significant or sensitive systems at a level adequate to make specific management prescriptions for those particular systems" (Sharples, 2002a). Examples include the detailed inventories of the Exit Cave, Mole Creek and Junee–Florentine karsts in Tasmania.

Figure 6.8 shows how georegions might be identified by overlaying the significant physical factors. An example of this type of work is Dixon & Duhig's (1996) mapping of Pleistocene glacial georegions in Tasmania.

Sensitivity zoning and management

This is another approach that is being developed in Tasmania. Four broad geoconservation management options have been developed:

1. Protection – appropriate where the exclusion of artificial disturbances is necessary from sites of high significance and/or great sensitivity;
2. Special prescriptions – appropriate where features have a lower sensitivity so that their values can be preserved by special modifications of development processes to avoid degradation of the site value. Examples might include use of buffer zones, catchment management measures or reduced intensity operations.
3. General prescriptions – appropriate where the site is robust to many artificial disturbances and where general prescriptions to maintain overall environmental quality (e.g. land management codes of practice) are sufficient to maintain site value.
4. Precautionary management – appropriate where the management requirements are unknown because of poor understanding or knowledge of the site sensitivity or the response to disturbance. The precautionary principle should apply so that potentially damaging operations are avoided or reduced in intensity until further research is carried out.

This raises the possibility of mapping spatial variations in sensitivity, particularly where process–response understanding is well developed. "The ability to zone regions in this fashion provides an important planning tool which can be used to minimise conflict between conservation and development values at an early stage of planning" (Sharples, 2002a). An example of this approach is Eberhard's (1994, 1996) project to zone the Junee–Florentine Karst System into High-, Medium- and Low- Sensitivity Zones, for which he recommended Protection, Special Prescription and General Prescription respectively. A large part of this karst system lies within the State forest, and this zoning system has the potential to reduce future conflict in the area by diverting forestry operations to less sensitive locations.

Individual base maps

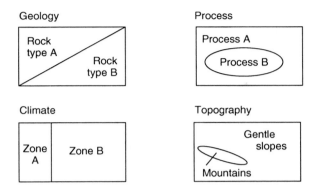

Combined georegion map (based on overlaying the above base maps)

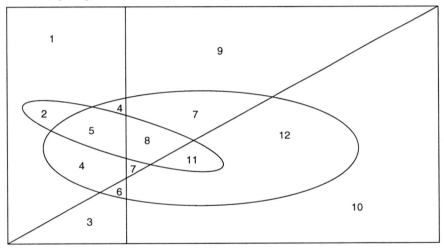

This yields 12 regions, two of which (4 and 7) comprise two separated sections.

Figure 6.8   *An illustration of how a georegional map can be constructed from individual base maps (After Sharples, C. (2002a) Concepts and Principles of Geoconservation. PDF Document, Tasmanian Parks & Wildlife Service website, by permission of Environment Australia)*

6.2.5   Natural Regions and Sub-regions in Canada

As outlined in Section 5.9.1, Parks Canada has divided the country into 39 Natural Regions and is using this scheme to try to ensure that Canada's National Park system represents the diversity of the country's natural landscapes. Some provinces have taken this further and have divided the Natural Regions within their provinces into Natural Sub-regions and are using the resulting spatial systems as a framework for managing the wider georesource. The approach is therefore similar to that being developed in the United Kingdom (see above).

One province that is developing this approach is Alberta where the six Natural Regions represented in the province have been further mapped into 20 Sub-regions (Fig. 6.9) on the basis of an amalgamation of two pre-existing schemes (Achuff, 1994). For example, the Rocky Mountain Natural Region in Alberta is divided into Alpine,

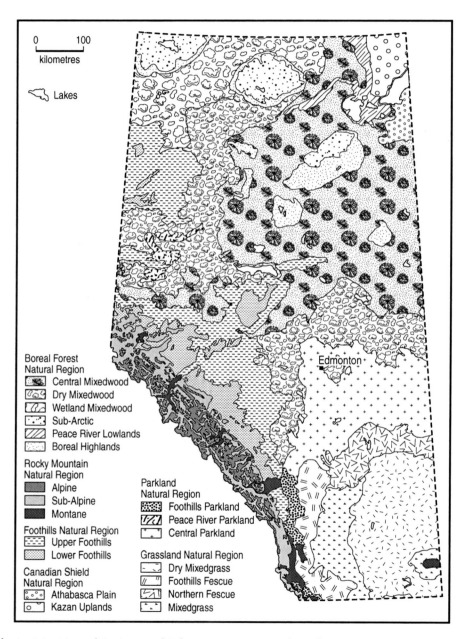

Figure 6.9 *Natural Regions and Sub-regions represented in Alberta, Canada (After Government of Alberta, (1994a) A Framework for Alberta's Special Places: Natural Regions, Report 1, Government of Alberta, Edmonton, by permission of Archie Landals)*

Subalpine and Montane Subregions, while the small section of the Canadian Shield in north-east Alberta is divided into the Athabasca Plain (underlain by Precambrian Athabasca Sandstone) and Kazan Upland (underlain by Precambrian igneous and meta-morphic rocks) Sub-regions.

A further three levels of subdivision, referred to as Natural History Themes, have then been applied.

- Level 1 Theme – a broad landscape type within a Sub-region. This is regarded as an important level for conservation and 20 Level 1 Themes have been applied across the 20 Sub-regions, resulting in 167 Level 1 Themes (not all themes are present in each Sub-region).
- Level 2 Theme – a broad habitat/vegetation type within a Level 1 Theme.
- Level 3 Theme – a specific geological feature, plant community or species within a Level 2 Theme.

Together with the Regions and Sub-regions, these three Theme Levels make up a five-level classification system (Achuff, 1994; Government of Alberta, 1994a). Figure 6.10 gives an example. This is being used in Alberta to do the following:

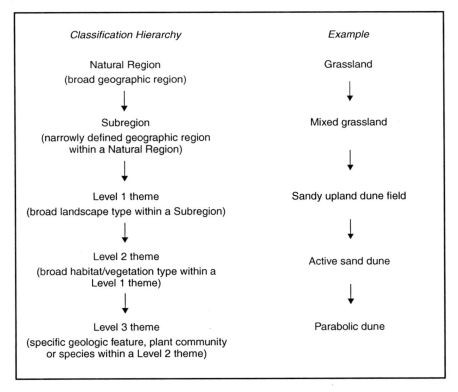

Figure 6.10   *The five-level classification scheme and examples for Alberta's landscapes (After Achuff, P.L. (1994) Natural Regions, Subregions and Natural History Themes of Alberta: A Classification for Protected Areas Management. Government of Alberta, Edmonton, by permission of Archie Landals)*

- Identify gaps and deficiencies in the protected areas system (Government of Alberta, 1994b), thus ensuring a representative conservation network. For example, a study of the Athabasca Plain and Kazan Upland Sub-regions of the Canadian Shield in 1996 (Government of Alberta, 1996) indicated a lack of protected areas and made suggestions on suitable locations for new representative protected areas.
- Act as a basis for nature conservation management, state of the environment reporting and integrated resource planning.

Other provinces have adopted similar schemes though there are some problems in reconciling boundary differences. Nevertheless, the Natural Regions landscape classification scheme is now being used for federal reporting on forestry strategies, agro-ecosystem schemes and biodiversity targets. It is therefore an important part of Canada's nature conservation strategy even if its application to geoconservation has yet to be fully developed.

6.2.6   Landscape physiographic units map of Italy

The Italian landscape physiographic units map is part of an ambitious project (Carta della Natura) introduced by national law (L.394/91) to provide a series of maps depicting the natural environment of the country, identifying the natural assets and the environmental vulnerability. The physiographic map has been compiled from satellite and aerial photographs with some field checking and has divided the country into landscape units defined as homogeneous portions of the land surface characterised by an arrangement of structural elements of the landscape and by typical patterns of morphology, lithology and land use. The system recognises both generic landscape types at the national scale and specific local areas whose local and unique features are then defined and described. For central Italy, Amadio et al. (2002) describe 21 landscape types (Table 6.5) and 285 landscape units and conclude that "geomorphology is the most important tool for the study of landscape at the regional hierarchical level to which the vegetation and the functional relationships are strictly related, to be defined" (p. 281).

The approach is similar to several of the other georegionalisation approaches described above, particularly Landscape Character Assessment. The system is organised in a GIS and will be used as a planning tool in the field of environmental quality and assessment (Amadio et al., 2002).

6.3   Georestoration

Given sufficient time, what is seen today as industrial dereliction can come to be valued as industrial archaeology. In fact, a number of mineral waste dumps are now protected as archaeological monuments (Bristow, 1994; Gray & Jarman, 2004). Furthermore, pits, quarries and other disturbed land often include very clear and important exposures of rock and sediments that need to be conserved for their scientific and educational values. There is also a great interest in mining museums as the links between geological resources and their role in local, social and cultural histories (Richards, 1996), and many are attracting significant tourist interest. Examples include the Geo-mining Historical & Natural Park of Sardinia, Italy (Arisci et al., 2002), the Kennicott Copper Mine in Alaska, USA and the Great Orme Copper Mine in North Wales.

Table 6.5   *Landscape types of central Italy (after Amadio et al., 2002)*

	Landscape type	Landscape general structure
PC	Coastal plain	Flat or sub-flat area bordered by low coast
PA	Valley floor plain	Flat or sub-flat area inside a river valley
PP	Open plain	Flat, sub-flat or undulate, wide area with variable geometry
VI	Inner valley	More or less wide valley bordered by mountains on both sides
CI	Intra-montane basin	Closed and depressed area, surrounded by relief
TC	Carbonate plateau	Flat rocky area bordered by low limestone escarpments
CV	Volcanic hills & plateaux	Conical or tabular hills of volcanic origin
CA	Clay hills	Hills with smoothed to flat tops and graded slopes, mainly clay
CT	Siliciclastic hills	Wide hills of the forechain, mainly siliciclastic
CC	Carbonate rock hills	Hills of the chain and forechain, mainly carbonate
CE	Heterogeneous hilly landscape	Hilly landscape of lithologic and morphologic diversity
PI	Areas with isolated hummocks	Group of isolated hills separated by winding valleys
CP	Terrigenous hills with rocky ridges	Hills and ridges of the chain and forechain, rising abruptly
RC	Isolated coastal relief	Isolated rocky relief, surrounded by coastal plain
RI	Isolated rocky relief	Lower area of rocky relief bordered by abrupt breaks of slope
MV	Volcanic mountains	Mainly cone shaped mountains of volcanic origin
MT	Siliciclastic mountains	Mainly siliciclastic mountains often structured in ridges
MC	Carbonate rock mountains	Mainly carbonate mountains structured in ridges and massifs
MX	Crystalline massif	Group of mountains with well-defined steep crests and valleys
IS	Small islands	Islands of less than 500 km^2 with strongly defined coastlines
AM	Metropolitan areas	Built landscapes of urban texture

However, landscape restoration is becoming increasingly important as human society comes to appreciate the need for environmental improvement and reuse of derelict sites for new buildings. In this section, we shall concentrate on the restoration of disused pits and quarries, river restoration, coastal restoration and managed retreat and contaminated land remediation. Thus, georestoration must consider restoration of both form and process.

There has, however, been considerable debate about landscape restoration. Firstly, there has been discussion about the use of terms like "restoration" and "nature" and about the extent to which a true return to natural conditions can be achieved. Secondly, there has been concern that long-term restoration proposals for new sites can be used to justify major short- and medium-term environmental impacts (e.g. Katz, 1992; Eden *et al.*, 1999), particularly where what is being proposed is not a restoration to pre-disturbance conditions. Thirdly, it is important to recognise that restoration is not about creating static places but about allowing natural environmental dynamics to operate so that the landscape continues to evolve. In Adams' words (1996, p. 169–170), "what we are doing is facilitating nature, and not making it. ... We must allow nature space to be itself, to function, to build and tear down". And fourthly, the need for integrated rehabilitation is increasingly being stressed, where geo- and bioconservation and management issues are considered together (see Chapter 7).

6.3.1   Quarry and pit restoration

Once pits and quarries have outlived their usefulness, many are simply abandoned. Over 4,000 abandoned mineral sites can be found within the US National Parks System (NPS) and thousands more occur outside the Parks. Although some are important geological or wildlife sites, others are unsightly, may be environmentally polluting and often pose safety hazards for the public. Therefore, attempts are often made to restore them. Blunden (1985, 1991) describes the principles and problems of restoration of spoil and tailings, and Gregory (2000, p. 264) asks "exactly how are the contours configured and the landscape recreated? Is advice sought from a geomorphologist or a physical geographer with an understanding of the landscape appropriate to that area, or is it undertaken by someone without such training?"

If the pits are below the water table, they may be quickly colonised by wildlife and/or water sports enthusiasts. If the pits or quarries are above the water table, they can become important in preserving the geological interest of the quarry if the important faces can be left exposed, made safe and accessible, maintained and kept clear of colonising vegetation. Low-level pit restoration by grading and returning to agriculture or woodland will result in loss of the geological exposure and may create incongruous landforms.

An alternative approach is to remodel the pit and quarry as an authentic landform, while retaining the geological interest of the quarry. An innovative attempt at this is described by Gunn (1993) and Gagen *et al.* (1993), who used restoration blasting techniques in limestone quarries in Derbyshire, England, to replicate the form of the local dry valleys. Given time, the older abandoned blackpowder-blasted quarries would develop in this way, but the scale and methods used to excavate modern quarries means that they "will continue to intrude upon the natural landscape for many centuries" (Gunn, 1993, p. 196). Restoration blasting and seeding allows the process to be accelerated so that the appearance of a mature and attractive valley can be developed very quickly. The typical dry valleys (dales) of the area contain rock buttresses, rock headwalls, scree slopes and debris flows, and by careful formulation of a drilling and blasting pattern, a predictable suite of these landforms can be created from the abandoned quarry walls. "The construction of these rock landforms together with their subsequent re-vegetation will enable quarried rock faces to be more easily harmonised with the surrounding unexcavated landscape" (Gagen *et al.*, 1993, p. 25). The restoration programme can also be designed to allow access to features of scientific interest high on the quarry walls, but would be less appropriate where all the available exposure is regarded as important or where an integrity site is involved (see Section 5.10.3).

Nonetheless, this technique represents a promising initiative, which is being adopted elsewhere. For example, a similar scheme has been promoted at Coniston in the Lake District National Park, England. Quarrying of green slate has occurred at Bursting Stone Quarry for many years, but it occupies a prominent position half way up the east flank of the hill called Coniston Old Man where the slate waste tips are artificially terraced. A planning application was submitted in 1997, which proposed re-grading the slate waste to more natural profiles, seeding the slopes created and blasting the quarry face to create rock buttress features (Stephens Stephenson, 1998).

This is also the approach being used in the United States where the National Parks Service has an *Abandoned Mineral Lands Program*, which focuses on re-establishing landscapes and environments that mimic the surrounding undisturbed lands. Volunteer labour is often used in this work, which applies not only to the mines but also to the access roads. Good examples of such successful restoration occur at Redwood National Park and Joshua Tree National Park, California. At Lingerbay, in the Scottish Hebrides, the restoration aimed to create a flooded glacial corrie (Box 6.1).

Box 6.1   *Lingerbay Superquarry, Scotland*

In March 1991, Redlands Aggregates Ltd applied for permission to develop a large coastal superquarry in the Precambrian anorthosite outcrop at Lingerbay on the Isle of Harris in the Scottish Outer Hebrides. The plan was to remove 550 million tonnes of anorthosite over a 60-year period for use as general aggregate and armourstone removed by ship to south-east England, continental Europe and perhaps America (Owens and Cowell, 1996; McIntosh, 2001). A large proportion of Mount Roineabhal would have been removed (459 ha) and a substantial sea loch created (McKirdy, 1993; Bayfield, 2001). Figure 6.11 shows the restoration scheme, which aimed to restore the quarry basin progressively as a flooded coastal corrie. However, objectors questioned whether the resultant landform could be given a natural appearance in view of the difficulty in mimicking natural corrie back walls and the need to comply with slope stability criteria (Owens & Cowell, 1996).

The main concerns of SNH and local people were over the visual impact of the operations over a long period of time and the loss of a valued local landscape (McIntosh, 2001; Warren, 2002). The applicants' case was that these impacts had to be balanced against the economic benefits to the island in terms of local employment.

The application was "called in" by the Secretary of State in 1994 and an 85-day public inquiry was held in 1995. The Inquiry Inspector concluded that the proposals would "completely change the landscape characteristics of Lingerbay by changing the scale and character of the coastline and its hinterland. . . . The quarry would create an area of massive disturbance", but her overall conclusion was that there was a justified need for the aggregate that would make an essential contribution to national prosperity and was therefore in the national interest. However, this was not accepted by the environment minister of the newly devolved Scottish government who refused the application in November 2000, nine years after the application was made, on the grounds of landscape impact.

This has not ended the issue since further legal challenges have been made by the applicants and the latest legal advice to Scottish ministers is that the 2000 decision letter was legally flawed (*Planning*, 30 August 2002, p. 7).

An alternative means of restoration of pits and quarries above the water table is by filling them with waste (after suitable lining systems have been installed), capping with suitable sediments and soil and returning them to an appropriate after-use, (e.g. grassland, scrub, woodland, recreation). Carefully done, and given time, this can return the landscape to a similar or enhanced condition compared to the pre-excavation landscape (Box 6.2). However, if the quarry or pit contains important rock or sediment geological exposures, full land filling will result in loss of the exposure. According

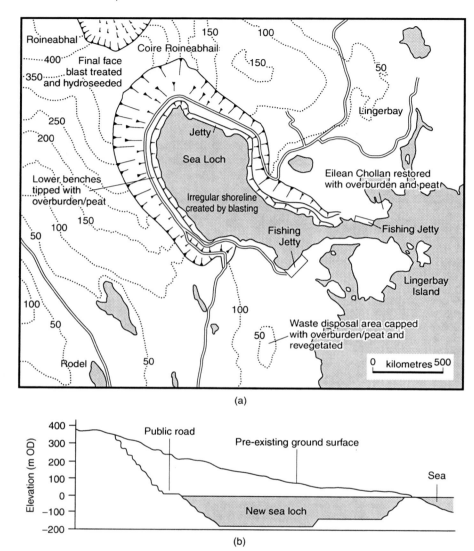

(a)

(b)

Figure 6.11 *Restoration proposals for the Lingerbay superquarry as a flooded coastal cirque, (a) plan view (Modified after McKirdy, A. (1993) Coastal superquarries: a blot on the Scottish landscape. Earth Science Conservation, **33**, 18–19) by permission of Earth Heritage), (b) long profile (After Bayfield, N. (2001) Mountain resources and conservation. In Warren, A. & French, J.R. (eds) Habitat Conservation: Managing the Physical Environment. Wiley, Chichester, 7–38)*

to Bennett (1994), 31% of all Quaternary Sites of Special Scientific Interest in England are located in disused pits and quarries. In the case of large quarries, it may be possible to engineer the site to leave a conservation face (see Fig. 6.12). This becomes impractical and uneconomic in the case of small quarries unless the important strata are towards the top of a quarry face. Further information on restoration of pits and quarries is given by Gordon (1996).

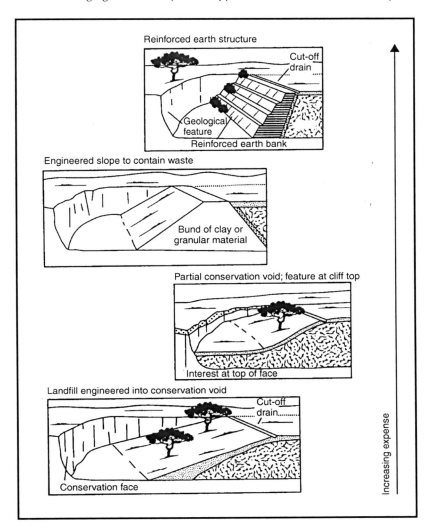

Figure 6.12 *Some restoration schemes for waste disposal in quarries while retaining the geological interest of a face (Reproduced from Bennett, M.R. & Doyle, P. (1997) Environmental Geology. Wiley, Chichester, by permission of Wiley)*

6.3.2   River restoration

In his forward to Brookes and Shields' (1996a) book on *River Channel Restoration,* Gore (1996, p. xiii) comments that "the degradation of riverine systems over the entire planet is dismaying. . .. In the 'Lower Forty-Eight', in the United States, the Yellowstone River remains the only medium to large river which is unimpounded and does not suffer the impacts of regulated flows". It is estimated that 98% of Danish streams and 96% of those in lowland England have been modified in some way (Iverson *et al.,* 1993; Brookes & Shields,1996b). Allan (1995) estimates that although freshwater environments cover less than 1% of the Earth's surface, they hold 12% of the world's species. River engineering therefore impacts significantly on biodiversity (Soulsby & Boon, 2001).

Box 6.2   *Restoration of Coal Mining Areas in the Lower Rhine, Germany*

The largest mining area in Germany is found in the Lower Rhine area around Aachen, Cologne and Monchengladbach where lignite seams between 10 and 100 m thick are quarried in open pits up to 300 m deep. Mining has occurred here since the eighteenth century. The pits currently cover an area of about 90 km² with a further 150 km² having been previously worked. Most of this has now been restored by waste infilling and rehabilitation to forestry, agriculture, horticulture or industrial uses, but the changes to the landscape, soils and hydrology of the area have been significant. Nonetheless, the older restored areas south of Cologne have matured into attractive lake and woodland landscapes used for recreational purposes. In the more recently quarried areas, the overburden is very carefully used to provide different soil types and different land uses. For example, thick mixtures of gravels, sand and loess are used on slopes where forestry is planned, but loess and loess loam are spread on flatter ground meaning that extensive areas of land have been returned to agriculture (Aust & Sustrac, 1992).

Fortunately, over the last 15 years or so, there have been attempts to reverse this process and there is now a large body of theory and practice on river restoration, only a brief summary of which can be attempted here. Readers requiring more information are referred to Brookes and Shields (1996a), Graf (1996), Rosgen (1996), de Waal *et al.* (1998) and several papers in Rutherfurd *et al.* (2001).

There has been some debate in the literature about the use of terms. Cairns (1991) defined restoration as "the complete structural and functional return to a pre-disturbance state", but the pre-disturbance state is not always well recorded and may be impractical to achieve, given ground and cost constraints. Instead, terms such as "rehabilitation", "enhancement", "creation" and "naturalization" have been used (Brookes & Shields, 1996a; Gregory, 2000) with definitions as shown in Table 6.6. Graf (1996, p. 443) believes that "geomorphic and ecologic changes related to the dams are not completely reversible. The issue of what is natural, and how closely restored systems can approximate natural conditions downstream from dams are challenging policy and scientific questions for fluvial geomorphology". His conclusion (p. 469) is that the best we may be able to do is to "make them more natural than they are at present by selective removal of dams and alteration of operating rules for the remaining structures". There have also been significant attempts to restore water quality through tighter regulation and reduced sewage and industrial effluent discharges into rivers.

Table 6.6   *Some terms used in river restoration (modified after Sear, 1994)*

Recovery	The act of restoration of a river to an improved/former condition
Re-establishment	To make a river secure in a former condition
Enhancement	Any improvement of a structural or functional attribute
Rehabilitation	Partial return to a pre-disturbance structure or function
Reinstatement	To restore a river to a former condition
Restoration	The act of restoring a river to a former or original structural or functional condition
Creation	Development of a morphological and/or ecological resource that did not previously exist
Naturalisation	To return a system to a condition sustained by natural processes

In the early stages of river restoration work, designs were often based on trial and error, guided by what those involved regarded as "natural" designs. However, the aims were often ecologically driven and constructed by engineers with the result that the reaches created were not always geomorphologically authentic. The involvement of geomorphologists has strengthened in the last decade because of the realisation that an understanding of river process, landforms and sediments as well as flow regimes is crucial to the success of river restoration schemes.

Figure 6.13 shows the main steps in restoration appraisal and design (Brookes & Sear, 1996). The first priority is to decide on the aims of the project. This will involve decisions on which type of restoration in Table 6.6 is required. Is the aim to improve the aesthetics of the river, improve habitats, allow public access and recreation and so on? Is intervention necessary or should the river be allowed to recover naturally? The next steps are to collect geomorphological and other data on the river, which will include historical information, hydrological data, bed sedimentology, catchment area processes, land-use data, land ownership, and so on. The final steps include drawing up and evaluating restoration options and choosing a final design.

Among the issues to be resolved at the design stage are

- design of the river planform, which will depend on functional constraints, floodplain constraints, and so on. It will also be necessary to decide what level of future channel instability will be acceptable and how additional unwanted instability should be constrained (Gippel *et al.*, 2001);
- design of long profile (pools, riffles, bars, etc. See for example Wilkinson *et al.*, 2001; Outhet *et al.*, 2001a);
- design of cross-sectional channel shapes through the reach, which will depend on planform design, substrate and bank materials, desired low-flow width, and so on.

Over recent years, a more integrated approach has been taken, not only in reconnecting rivers to their floodplain systems (Brookes *et al.*, 1996; Richards *et al.*, 2002) and viewing them in the context of their catchment areas (Kondolf & Downs, 1996; Abernethy & Wansborough, 2001) but also in linking geomorphological, ecological, water quality, recreational and other aims into integrated river management strategies (Brookes & Shields,1996c; Outhet *et al.*, 2001b) that cover longer timescales (Brierley & Fryirs (2001). Thus, Lake (2001) sees three basic ways of stream restoration – restoration of particular reaches, restoration of longitudinal and lateral connectivity and restoration of drainage basins involving co-ordinated activities at the stream catchment level. Hart & Poff (2002) discuss the "emerging science of dam removal", while Pizzuto (2002) examines the effects of dam removal on river form and process.

There has also been an increasing interest in project appraisal carried out at various stages from conception through several years of post-completion monitoring (Bruce-Burgess, 2001).

In terms of case studies, reviews of restoration projects in the United States (Haltinder *et al.*, 1996; Rhodes & Herricks, 1996; FISRWG, 1998), Europe (Brookes, 1996; Nielsen, 1996; Iversen *et al.*, 2000), Australia and New Zealand (Brizga & Finlayson, 2000; Schofield *et al.*, 2000; Pigram, 2000; Rutherfurd *et al.*, 2001), Canada (Karr *et al.*, 2000) and Japan (Waley, 2000) have been published. Box 6.3 describes probably the world's most ambitious river restoration project on the Kissimmee River in

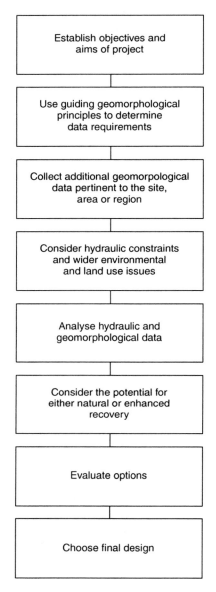

Figure 6.13 *Main steps in river restoration appraisal and design (Reproduced from Brookes, A. & Sear, D.A. (1996) Geomorphological principles for restoring channels. In Brookes, A. & Shields, F.D. (eds) (1996a) River Channel Restoration. Wiley, Chichester, 75–101, by permission of Wiley)*

Florida, USA. Powell (2002) describes attempts to restore more natural flows and sediment transport regimes on the Colorado River, which has been drastically altered since the construction of the Glen Canyon Dam. In the United Kingdom, the River Restoration Centre (formerly Project) is an independent, non-profit organization whose aim is to encourage and co-ordinate good practice in river restoration (www.therrc.co.uk), and Fig. 6.14 is an example of a UK scheme supported by them.

Box 6.3   *River Restoration on the Kissimmee River, Florida, USA*

The Kissimmee River lies in central Florida and drains into Lake Okeechobee. It has a 7800-km^2 drainage basin, and a 90 km length of flood plain varying in width between 1.5 and 3 km (Toth, 1996). Prior to 1962, the river meandered for 166 km over this floodplain, which was a rich wetland ecosystem supporting over 300 fish and wildlife species including resident and overwintering waterfowl and wading birds. "The diversity and persistence of these biological resources were linked to dynamic river and floodplain habitat characteristics provided by basin hydrology and channel geomorphology" (Toth, 1996, p. 369).

However, the river was totally channelised between 1962 and 1971 to provide drained farmland for the expanding agricultural economy of central Florida. The major features of the scheme were a 9-m-deep rectilinear canal cut through the floodplain and divided into five level reaches by water-control dams and separated from the floodplain by levees. The 26 headwater lakes were connected by canals and regulated as flood-storage reservoirs. The scheme was successful in draining two-thirds of the floodplain but at the cost of a dramatic loss of habitat and wildlife.

This impact soon led to calls from local communities and environmental groups for the river to be restored, and in the 1980s and early 1990s a number of plans and feasibility studies were carried out, including a demonstration project between 1984 and 1989 on a 19-km reach (Pool B). In this scheme, water from the canal was diverted back through sections of the original channel, leading to improved river aesthetics and wildlife habitats.

Lessons learnt from these studies led to a state-federal partnership plan for the dechannelization of the river, including backfilling of 35 km of canal with original spoil, which had been spread on the floodplain, removal of two dams and associated levees and re-excavation of 14 km of former river channel. Restoration work began in 1999 and is due to be completed by 2010 at a cost of over $400 million at 1997 prices. Flood protection is maintained in the residential stretches and peak discharges are reduced by increasing flood-storage capacities in the headwater lakes (Toth, 1996).

Despite these and many other river restoration initiatives, it is clear that there are still major constraints to river restoration, including lack of legislation, lack of funding, lack of organisational structures to disseminate good practice and land-use and land-ownership issues (Brookes, 1996). In some countries, legislation has aided river restoration. This is certainly the case in Denmark where the *Watercourses Act* (1982) restricts maintenance practices in order to safeguard local stream ecology. It also incorporated special provisions for stream restoration and it's funding, and has led to several major restoration projects (Madsen, 1995). As in many cases, geomorphology has benefited from predominantly ecologically driven measures.

6.3.3   Coastal restoration and managed realignment

In many parts of the world, coastal erosion is resulting in loss of land and property. An understandable reaction to this is to try to prevent it from occurring by erecting sea defences, which may take the form of sea walls, revetments, groynes, embankments, and so on. Apart from preventing coastal erosion, these constructions may also prevent coastal flooding and allow marshland inside the defences to be drained and cultivated. However, this engineering approach has been questioned on a number of grounds over the last 20 years or so (Bush *et al.*, 1996; Hooke, 1998, 1999; French & Reed, 2001).

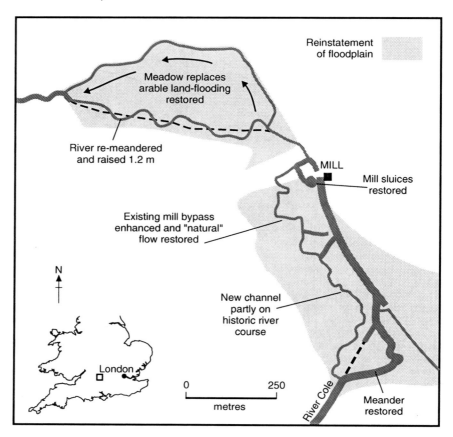

Figure 6.14   *The river restoration scheme for the River Cole, England (Reproduced by permission of Robert Burns)*

Firstly, sea defences are expensive to install and maintain. Wave attack usually means that, even with regular maintenance, a sea wall will have only a limited life before a replacement is needed, often to a larger design. Secondly, preventing coastal erosion and long-shore drift through the construction of sea walls and groynes starves the coastline of sediment in a down-drift direction and may accelerate coastal erosion there. In other words, human intervention is preventing the operation of natural coastal processes and disrupting natural systems. The evidence is that the traditional methods of coastal defence are not just ineffective but actually exacerbate erosion problems. Thirdly, sea defences such as concrete sea walls or giant armour blocks often obscure coastal geological exposures and are not a very aesthetically pleasing addition to the coastline (see Section 4.5). In other words, they inevitably reduce natural coastal geodiversity.

The alternative approach that has been adopted in many areas over the last 20 years or more is what is often referred to as a *soft engineering approach*. The principle of this approach involves working with nature rather than against it, firstly in understanding the operation of local coastal processes, and then using this understanding to achieve appropriate coastal management solutions. These will usually involve the use of natural coastal defence systems such as beaches, storm ridges and sand dunes and techniques such as beach replenishment/recharge from offshore sources, ridge reinforcement and dune stabilisation. They may be aided by use of artificial reefs, breakwaters or other structures, though these also tend to be visually intrusive. At coastal towns and seaports, it is anticipated that artificial sea defences will still be required, but in less-developed areas, the restoration of a more naturally operating coastline is now being encouraged in several parts of the world. Scottish Natural Heritage (2000c, p. ii), for example, "advocates approaches to erosion management which retain the natural coastal habitats, processes and landscapes and which enable Scotland's coastlines to evolve naturally with minimal human intervention". However, as an approach, it is often opposed by those who will lose land and property through its implementation, unless there are suitable compensation payment systems in place. Box 6.4 examines the managed re-alignment approach in the United Kingdom.

Box 6.4   Managed Realignment and Shoreline Management Plans in the United Kingdom

Shoreline Management Plans are being drawn up for the whole coastline of England and Wales according to government guidelines (MAFF, 1995) and are also being considered in Scotland where only a few have been prepared so far (Hansom *et al.*, 2000). Each plan is based on a set of sediment cells or sub-cells, and each presents a strategy for coastal defence for that stretch of coast, taking into account coastal processes, human influences and other relevant factors. Given the complex administrative and legislative framework of coastal management responsibilities, the aim is to improve the integration of coastal management both within individual coastal areas and between them, thus bringing a more strategic approach (Hansom *et al.*, 2000).

A Shoreline Management Plan is "a document which sets out a strategy for coastal defence of a specified length of coast, taking account of natural coastal processes and human and other environmental influences and needs" (MAFF, 1995). They should be heavily geomorphology-based, though geomorphologists are not always involved in the work (Hooke, 1999). The end product is usually a map of a coastal cell showing zones where particular shoreline management strategies are recommended. "The choice of strategy is between four management options: hold the line, retreat the line, advance the line, or do nothing" (Hooke, 1999, p. 380–381). Variations on these themes may include installing limited defence schemes, which slow down coastal erosion in order to retain important coastal exposures but prolong the lifetime of properties threatened by coastal erosion. (McKirdy, 1990; Brampton, 1998; Barton, 1998).

Many of the estuaries and marshlands of south-east England have been enclosed by embankments and drained over the last two centuries to provide flood defence and additional grazing or arable land. This has resulted in coastal squeeze, which occurs when rising relative sea levels raise the low water mark, while the high water mark is held in place by the embankment. The width of the intertidal zone is therefore reduced with significant loss of geo- and biodiversity and buffering protection afforded to the embankment. These then become open to attack, with resultant need for expensive maintenance (Burd, 1995). Managed realignment involves setting back the line of actively maintained sea defences to a new line inland of the original and encouraging the operation of natural intertidal processes (saltmarsh or mudflat) in the area between the old and new defences. The original front defence line can either be allowed to degrade naturally or may be wholly or partially removed. Since estuaries are naturally wider at the seaward end, managed retreat should be widest here and narrow towards the upper part of the estuary (Pethick, 1994).

Continued on page 298

__ Continued from page 297 __

An experimental site at Tollesbury in Essex was established in 1995 when the old sea wall was breached allowing the sea to flood low-lying agricultural farmland for the first time in over 150 years. Since then, the site has been studied to discover the biotic and abiotic changes that are occurring as a direct result of sea-water inundation and the effect on ebb and flow rates within the existing saltmarsh creek systems. The conclusion was that managed realignment has the potential to alleviate some of the problems of rising sea level in this area and produces benefits for biodiversity. In 2002, two full-scale schemes were established in England, allowing arable land to return to saltmarsh at Freiston in Lincolnshire and at Abbott's Hall in Essex (Fig. 6.15).

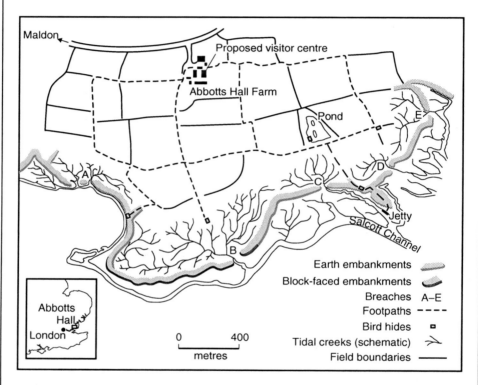

Figure 6.15   *The coastal realignment scheme at Abbott's Hall, England (Source: WWF-UK)*

There are also situations where coastlines have been used to dump mine and quarry waste and where restoration has been carried out or could be carried out (Saiu & McManus, 1998). An example is the "Turning the Tide" project in County Durham, England, where, for much of the last century, five coal mines had dumped at least 100 million tonnes of colliery waste over the coastal cliffs and pipelines pumped black liquid sludge into the sea. Following closure of the last mine in 1993, the project began to restore the coastline by removing the waste from the cliffs and beaches and restoring their profiles and materials, as well as returning the pithead landscapes to grassland. A coastal path has been created and the public encouraged to return to a restored coastline from which they had been excluded for so long (www.turning-the-tide.org.uk).

6.3.4   Contaminated land rehabilitation

"Contaminated land is land which, because of the substances contained within it, is causing significant harm, or has the potential to do so; or affects controlled waters or has the potential to do so" (UK *Environmental Protection Act* (1990) and *Environment Act* (1995)). The United Kingdom has adopted a risk management approach to contaminated land that uses a source – path – receptor methodology. It aims to restore sites for specific purposes and is therefore based on a "fitness for purpose" methodology (Nathanail, 1997). Techniques of risk assessment have been very useful in determining locations where contamination has significant potential to cause harm and has greatly assisted in the management and remediation of such land (Cairney, 1995). Some other European countries use fixed national or supranational numerical standards, but the UK approach is gaining support.

Several techniques for remediation of contaminated land may be employed, but the three main approaches are (Nathanail, 1997):

- *Remove/neutralise the source* – This may include removal of the soil to a suitable landfill site (may require licence and payment of landfill tax) or bioremediation in which particular bacterial species, fungii or plants draw up the soil contaminants;
- *Divert/block pathway* – For example, by capping of the site with concrete or textile, or installation of a permeable reactive barrier (PRB) that contains reactive material to remediate contaminated water as it passes through the treatment zone;
- *Remove target* – For example, change use of site to a less sensitive one, restrict access to site or alter the layout of the proposed development.

Responsibility for remediation should fall on the polluter, but it is not always easy to identify those involved and in the case of historical incidents, the company involved may no longer be trading. For more details on the assessment and remediation of contaminated land with particular reference to the United Kingdom, see Hester & Harrison (2001).

6.3.5   Other land restoration projects

Community efforts at environmental restoration are increasingly important. One successful example is Landcare in Australia that now involves over 3000 groups and which has spawned other initiatives such as Coastcare and Rivercare (Sutherland & Scarsbrick, 2001). Eberhard & Houshold (2001) describe community efforts to restore degraded karstic features in the Mole Creek karst area of Tasmania, Australia. This includes removing sediment and rubbish from cave floors and entrances and cleaning sediment from delicate calcite formations.

In many places in National Parks and other areas in North America, attempts are being made to restore landscapes, for example, by removing and landscaping redundant forestry tracks, restoring pits and garbage dumps and reinstating wetlands. For example, the objectives and key actions in the Management Plan for the Waterton Lakes National

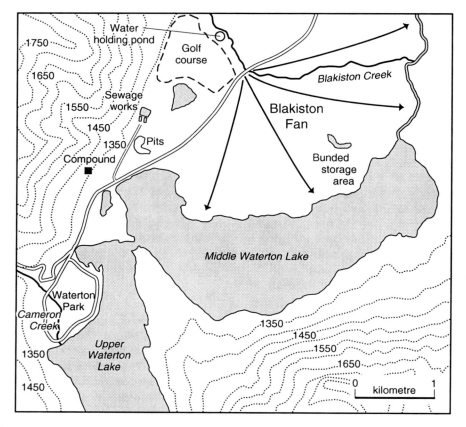

Figure 6.16   *The landform restoration scheme at Waterton Lakes, Canada (see Box 5.13)*

Park focus on land restoration and the impacts of management decisions inside the park on landforms beyond its boundaries (see Box 5.13 and Fig. 6.16).

In Hampshire, England, the construction of a motorway cutting (M3) through chalk downland (Twyford Down), enabled the closure and restoration of 1.5 km of the route of the old A33 at St Catherine's Hill. The old road was broken up and chalk spoil from the cutting was used to recontour the over-steepened hillside and recreate the pre-A33 slope (Eden *et al.*, 1999). It was recognized, however, that this was not a return to an entirely natural state. Similarly, the National Trust of Scotland has restored high-altitude vehicle tracks on Beinn a' Bhúird in the Cairngorms.

A number of projects have involved restoration of drained bogs and wetlands by blocking channels and ditches in order to raise water levels and restore the bog to a more natural condition. Projects of this type have been undertaken in, for example, eastern Canada (Rochefort & Campeau, 1997), England (Johnson, 1997) and Northern Ireland (Gunn, 1995). Projects are also being undertaken in the drained fenlands south of the Wash in England to restore them back to wetland from arable.

Finally, in Box 5.20 we saw how old research core holes or climbers bolt holes could be restored by plugging them with dyed cement or with rock cores or rock dust taken from loose stones of the same rock type.

6.4   Landform Design

Section 1.1 contained a quote from Marsh (1997) about the need to design new developments to respect the diversity of local landscapes. One of Marsh's concerns is the scale of developments, which may be dictated by national policies, rules or standards that are applied irrespective of landscape character. "From place to place the environment is different in fundamental ways; unless these differences are made part of the information base for decision making, we will continue to build missized and unsustainable infrastructures and land-use systems" (Marsh, 1997, p. 5). He cites the case of Franconia Notch, a narrow valley in the White Mountains of New Hampshire, USA, where federal and state transportation planners had designed a standard four-lane interstate highway squeezed into the centre of the valley with huge entry and exit ramps. Fortunately, the local people objected and after a decade of argument, including court action, the highway planners agreed to modify their standard approach and adopted a smaller design better suited to the topography, drainage and scale of the landscape. This has resonances of McHarg's (1995) famous plea to *Design with Nature*.

But it is not just the design of infrastructure and buildings that needs to be considered. New landforms are also being created by excavation and construction and examples are given in Table 6.7. This section will examine some of the impacts in more detail, paying particular attention to conservation of sensitive physical landscapes and design of authentic landforms. Several of these issues are discussed in more detail in Jarman (1994), Marsh (1997) and Gray (1997a), while Gregory (2000) has discussed the role of geomorphologists in environmental design as part of a "cultural physical geography". Haigh (2002) asks "what greater challenges and what greater vindication can there be for a discipline than to create a new landscape or to recycle land that has been sacrificed to human well-being".

Table 6.7   Some anthropogenic landforms (modified after Haigh, 1978; Goudie, 2000)

Excavational	
Digging	e.g. drainage ditches
Cutting	e.g. road & rail cuttings
Quarrying	e.g. pits & quarries
Cratering	e.g. bomb craters
Constructional	
Tipping	e.g. tailings heaps, landraising, coastal reclamation
Infilling	e.g. infilling of hollows
Mounding & bunding	e.g. for visual & noise screening
Embanking	e.g. river and coastal flood defence, road embankments
Excavation and construction	
Terracing	e.g. vineyards & rice terraces on slopes
Ridge & furrow	e.g. agriculture
Water features	e.g. agriculture reservoirs, canals, ponds, moats
Remodelling	e.g. golf courses, regrading of slopes
Other landform effects	
River engineering	e.g. channelisation, straightening, dredging, damming
Coastal engineering	e.g. sea walls, armouring
Subsidence	e.g. due to extraction of fluids and minerals underground
Slope failure	e.g. due to loading, lubrication, undercutting

Many parts of western Europe have been densely populated for centuries and have lost most of their natural vegetation. It is often said that natural landscapes no longer exist. While this may be true of biological landscape elements, the same is not true of the geological and geomorphological landscape elements. Although there has been much engineering work done to alter the form of the European land surface since the Industrial Revolution (Hoskins, 1955), not to mention the smoothing out of irregularities achieved by centuries of ploughing, the shape of the land surface over large parts of rural Europe has remained substantially unaltered since the major changes of the Pleistocene and modification by Holocene and modern processes. In contrast with the natural vegetation, the natural land form generally remains intact. Note the use of the term "land form" rather than "landforms" here, since in the wider countryside the topographic form of the land may not have formal landform names and yet may still be natural (Gray, 2002b). Similar views are expressed by Kiernan (1997a, p. 9) in arguing that "Landforms are defined by their contours. Hence any unnatural changes to the contours of a landform by definition damages the natural geomorphology.... The geoconservation significance of the damage is what is important...." It is the aim of this section to argue that society ought to pay greater attention to respecting, conserving and designing with this natural geomorphological heritage, which is a very important element in landscape character and local distinctiveness (see above).

Firstly, the loss of natural landform character often accompanies development. Jarman (1994, p. 42) notes that "Human agency has long been ironing out the surface irregularities of the land. Ploughing smooths over breaks of slope; farmers fill wet hollows; hill paths are surfaced with fill from glacial hummocks.... Wherever development takes place, there is an inevitable tendency to eliminate rather than incorporate local landform character".

Secondly, the design of new landforms may be even more damaging, particularly where engineering standards or traditions dictate regularity. Jarman (1994, p. 42) decries the "evenly graded road embankments, rectilinear cut-and-fill shapes, level earth dams, improbably convex screen bunds. The classic incongruity is the square, flat-top, standard-batter grass covered water tank or military installation perched on a rugged moorland". Many other examples of incongruous landform creation can be cited.

6.4.1 Golf courses

There was a time when golf course design utilised existing landforms. An example is the British Open links courses at St Andrews, Troon and Turnberry, where raised beach sands have been blown into dunes, stabilised by vegetation growth and turned into golf courses with a minimum of remodelling. Price (1989, 2002) was even able to classify Scotland's golf courses by their geomorphological characteristics. However, the construction of modern golf courses often involves huge re-engineering works to remodel the topography. The creation of raised tees, bunkers, water features, mounds and slopes can radically alter the topographic landscape. This can result in a landform that is out of keeping with the local landscape. An example occurs at Dunston Hall near Norwich, England, where the natural medium-scale rolling agricultural landscape has been carved into a much finer-scale landscape of hummocks, fairway channels and ponds (see Fig. 6.17). Apart from the landform impact, the topographic changes decrease the reversibility of the area to agricultural land (Jones, 1996).

Figure 6.17 *Landform change created by golf course remodelling, Dunston Hall, England. On the right is the natural landform of the agricultural landscape with its long low-angle slopes. On the left, this landform has been remodelled to create fairways, mounds, etc.*

Yet, in the United Kingdom at least, there have been signs of the need for regulation of golf course design through planning and policy guidance. Planning Policy Guidance Note 17 (PPG17), for example, refers to the need to locate and design golf courses that are in harmony with the landscape, and to take special care with proposals in designated areas. A Countryside Commission (now Agency) publication on *Golf Courses in the Countryside* (Countryside Commission, 1993b) has a chapter on topographical change. This chapter concludes that while all new golf courses will require some earth movement during their construction, "large-scale remodelling is not essential to the quality of a golf course and can be highly inappropriate. In particular, topographical changes should reflect the local topographical character so that the final landform is indistinguishable from the surrounding landform" (Countryside Commission, 1993b, p. 25). It argues that of all the features of a new golf course, "mounding is often the most alien to the landscape setting". It notes that "a flat landscape can accept very little by way of grading and almost certainly no mounding", whereas on steeply sloping or hilly sites, significant earthworks can be justified as long as care is taken to avoid "slopes that are unacceptably severe because of the risk of appearing wholly unnatural; they may also be liable to erosion" (Countryside Commission, 1993b, p. 25).

An example of the latter problem occurs at Fraser Hill in Malaysia, where massive cut-and-fill slopes have been created by the construction of a golf course in a tropical mountain rainforest. In spite of immediate hydroseeding aimed at stabilising the slopes and surfaces, not surprisingly in this environment, some slope failures have occurred (Bayfield, 2001).

A recently produced guidance booklet for Scottish golf course development (Scottish Natural Heritage & Scottish Golf Course Wildlife Group, 2000) has very little to say on topographical change. However, the organisations did achieve revisions to the planned extension to Dunkeld Golf Course, including the elimination of bunkers on some holes due to the impact on one of Scotland's National Scenic Areas.

6.4.2   Landraising

Another example of incongruous landform creation is in the aboveground dumping of waste, a practice known as *landraising*. The hills of waste are created to allow for settlement of waste as it decomposes and to provide land drainage rather than water infiltration, but a further reason is simply to get more waste in, particularly where there is a shortage of landfill void space (Gray, 1998b, 2002a).

Some spectacular landforms have been created in this way. For example, the Packington site near Birmingham, England, rises 50 m above the surrounding landscape and intrudes 12 m into Birmingham Airport's airspace! However, others have been refused permission partly because of the incongruous landforms proposed. This is particularly true of hills of waste in areas of flat topography such as till plains, river terraces or coastal flats. For example, at Rivenhall Airfield in Essex, England, a planning application was made in 1993 to extract 10 million tonnes of sand and gravel from a 100-ha area and overfill the void with household and commercial waste to create two waste hills rising to 18 m and 12 m above the surrounding till plain and separated by a small valley (Fig. 6.18). According to the applicants, the object was to replace the current "flat and featureless" airfield site with a more undulating and interesting topography. Fortunately, the scheme was refused permission, partly because "the proposed landform would not relate well visually to any existing feature and when completed would not appear to be an authentic part of the local scene".

But, even when attempts have been made to design landraised sites in keeping with the local topography, the results have not always been very successful. For example, at Fleetwood in Lancashire, England, a waste hill was designed to resemble the local drumlins around the Wyre estuary, but an analysis of the hill proposed demonstrates that it falls short of an authentic local drumlin in terms of size, shape and orientation (Gray & Jarman, 2004). The applicants claimed that this was a "carefully designed landform (which) will greatly assist in integrating the landfill into the surrounding landscape", but Gray & Jarman (2004) concluded that this had been a rather weak attempt to design a drumlin in keeping with the local morphology. One suspects that what we have here is a *post hoc* justification for a landform that was designed with little reference to the natural landforms of the area. Yet, a simple landform redesign could have achieved an authentic drumlin morphology whilst retaining a similar waste capacity.

6.4.3   Bunding

One method of screening unsightly developments such as landfill sites, gravel pits or new roads is by constructing soil mounds or ridges to obscure them from view and reduce noise. As the UK's Landscape Insititute (1995, p. 59) states "major works in themselves may create adverse landscape and visual impact, and care should be taken to ensure that a new landform looks natural and appears as an integral part

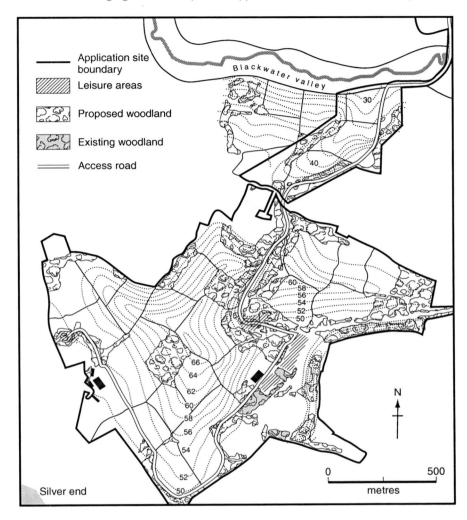

Figure 6.18 *The land-raising scheme proposed at Rivenhall Airfield, England*

of the landscape". This is particularly good advice in low-relief areas where steep-sided, rectilinear ridges and even low mounds can appear completely out of place in their landscape context. Approved planting schemes in the countryside now generally encourage the use of native species, and the same approach needs to be taken to geomorphological landscaping. In many areas, the linear bund can be regarded as the geomorphological equivalent of the *leylandii* hedge.

A detailed study of bunding was undertaken in South Norfolk, England, by Gray (1997a). The area has a predominantly flat topography as part of the East Anglian till plain and is therefore very sensitive to topographical change such as bunding. The study demonstrated a growing and worrying tendency to include bunding and mounding as part of landscaping schemes (Fig. 6.19). Part of the reason is the cost of removing subsoil and topsoil from the sites, including payment of Landfill Tax (see below). It is usually cheaper and easier to simply mound this material on site.

Figure 6.19   *Roadside bunding up to 4 m high, Brockdish bypass, Norfolk, England*

6.4.4   Pond and reservoir formation

The digging of ponds and agricultural reservoirs raises at least three geodiversity issues. Firstly, pond shape may not be in keeping with the local character. Secondly, the spoil may be heaped into incongruous bunds and mounds. And thirdly, topsoil may be buried by the spoil rather than being stripped, stored and replaced over the mounds.

Two examples of all these issues may be cited from the South Norfolk study cited above (Gray, 1997a). Figure 6.20 shows the design of a garden pond and agricultural reservoir in this area. The natural shape of ponds in Norfolk is circular or oval as in the Breckland meres or Diss Mere. The garden pond, however, is highly intricate in shape. On the other hand, the agricultural reservoir is rectilinear. Neither has an authentic shape, and this is compounded by the nature of the associated bunding, the burial of topsoil and the absence of a topsoil cover on the bunds. Agricultural reservoirs allowing winter storage of water are being encouraged in the east of England where climate change may lead to wetter and drier summers, but their design needs to be more carefully considered.

6.4.5   Urban landform

As well as design of rural landform as described above, towns also have a geomorphology that can be eroded or buried under new developments. As Turner (1998, p. 37) states:

> "Modern man (sic) has tended to conceal or destroy the landform of cities. We have found it easy to adjust land-shape for various purposes and, in making a city, enormous volumes of earth are moved from site to site. It is usually done without a comprehensive plan. ... Yet the old landform often remains as a feature to be exploited in urban renewal. Landscape planners should seek out landform, just as Michelangelo looked at a block of marble and saw a statue concealed within".

He suggests (p. 38) that "every city needs a landform plan". Birmingham, England, is an example of a city where the landform has been concealed by buildings, whereas

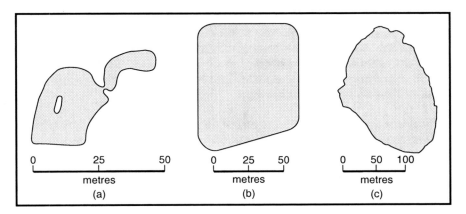

Figure 6.20 *Examples of pond design in Norfolk, England: (a) new garden pond at Fundenhall, (b) new agricultural reservoir at Rushall and (c) the more natural shape of South Norfolk ponds–Diss Mere*

Robert Adam's designs for Georgian Edinburgh, Scotland, utilised the landform variation of the New Town site (Gray, 1997a).

6.4.6 Conclusions

Human alteration of the form of the land has a long and not very distinguished history. Landform may be incongruous in its local area due to its size, height, gradient, shape or detailed variation. This should not be taken to imply that all geomorphological change is unacceptable. There are situations, such as construction of flood defence levées, where landform changes may be essential or desirable. It is also true that the natural landscape contains many incongruous features and that some man-made features eventually become an accepted and even valued part of the landscape, for instance, archaeological earthworks. But what is needed is an intelligent and aware approach to what we are changing and why we are changing it. Geomorphological design ought to be important elements in landscape character assessment (see above). Key characteristics of landform and process character need to be identified and included in character guidelines, geomorphological sensitivity to development should be recognised and opportunities for restoration or enhancement should be taken.

It is also possible that government environmental initiatives can have unforeseen effects. The UK Landfill Tax (see below) imposed a charge on the deposition of waste materials, but landscaping was granted an exemption. As a result, significant quantities of inert waste were dumped and remodelled on golf courses and bunds in order to avoid payment of the tax.

6.5 Geomaterials

Prosser (2001c) argues that a number of pressures have impacted on the minerals industry and have made it more receptive to geoconservation issues. These include corporate social responsibility and sustainability pressures that mean that businesses are acting more openly and responsibly in the way they work. It includes a social responsibility to respect the rights of local populations and an environmental responsibility to ensure that environments are not totally devastated by mineral extraction. In turn, it is leading to increased dialogue, agreements and partnerships with the minerals industry, including Statements of Intent and Memoranda of Understanding between the minerals industry and nature conservation agencies (Prosser, 2001c). English Nature (2003) has published guidance on how the minerals industry can contribute to geodiversity conservation. This includes the following:

- Careful planning of new quarries by identifying as early as possible any features of geoheritage value on the site and taking appropriate steps (recording, protecting, etc).
- Operating quarries so that new discoveries are brought to geologists' attention, allowing research and recovery to take place and facilitating long-term management.
- Restoring quarries to retain features of interest and provide safe access to these (see above).
- Taking a corporate perspective by developing Company Geodiversity Action Plans (see below) and company targets, and reporting on successes.

This section explains in more detail some of the initiatives being used to ensure a more sustainable approach to mineral extraction.

6.5.1 Sustainable mining

The end products of mining are essential to modern societies, and mining activities should bring significant direct and indirect employment opportunities to local populations. Yet, as explained in Chapter 4, mining and quarrying activities have often had a devastating effect on local environments, landscapes and even economic and social systems (Humphreys, 2002). Owing to its poor reputation in environmental management and other areas, the Global Mining Initiative was launched in 1998 by nine international mining companies in order to redefine the role of the global mining industry in relation to sustainable development.

The initiative has involved three elements. Firstly, a Mining Minerals and Sustainable Development Project (MMSD) was carried out to analyse the challenges facing the industry and ways in which the issues might be resolved. The final report (MMSD, 2002) recognises that "simply meeting market demand for mineral commodities falls far short of meeting society's expectations of industry". Instead, it proposes an "agenda for change" and a "vision for the mineral sector" to create a picture of what it would look like if it were to maximise its contribution to sustainable development.

Secondly, the International Council on Mining and Metals (ICMM) was founded (replacing its predecessor, the International Council on Metals and the Environment), with the aim of being "the clear and authoritative global voice of the world's mining and metals industries, developing and articulating their sustainable development case, discovering and promoting best practice on sustainable development issues within the industries and acting as the principal point of engagement with the industries for stakeholders at the global level". Among the issues being confronted in the ICMM Charter (www.icmm.com) are mineral economics, environmental stewardship, social and cultural responsibility, human health and safety, product stewardship, stakeholder engagement and innovation in technology (see Box 6.5).

Thirdly, an international conference was held in Toronto, Canada in May 2002 to debate the ideas emerging from the MMSD Project. At this conference, 20 chief executives or chairmen from mining companies around the world together with other industry leaders engaged directly with 25 government officials responsible for mining in their countries including state ministers, together with industry associations, academics and representatives of 74 non-governmental organisations (NGOs).

Box 6.5   ICMM's "Fundamental Principles of Sustainable Development"

In order to meet society's requirements for minerals and metals, while contributing to sustainable development and enhancing shareholder value, as ICMM members we:

- recognise that sustainable development is a corporate priority;
- recognise the importance of integrating environmental, social and economic aspects into the decision-making process;
- acknowledge that consultation and participation are integral to balancing the interests of local communities and other affected organizations and to achieving common objectives;
- support the use of sound scientific, technical and socio-economic analysis in developing policies related to the mining and metal industries;
- employ risk management strategies, based on sound science and valid data in the design, operation and decommissioning of mining and metal processing operations, including the handling and disposal of hazardous materials and waste;
- implement effective environmental management systems, conduct regular environmental assessments and act on the results;
- acknowledge that neither our operations nor our products should present unacceptable risks to employees, communities, customers, the general public or the environment;
- encourage product design, technologies and uses that promote recyclability as well as the economic collection and recovery of metals;
- contribute to biodiversity conservation and protection;
- contribute to and participate in the social, economic and institutional development of the communities where our operations are located;
- adhere to ethical business practices and, in so doing, contribute to the elimination of corruption and bribery, to increased transparency in government–business relations, and to the promotion of respect for human rights internationally; and
- provide public reports on progress relating to economic, environmental and social performance.

6.5.2   Over exploitation of minerals

One of the reasons for the interest in sustainable mining is the looming shortage of some minerals, in some cases, related to potential political instability. Many earth materials "are so much a part of everyday living that. . . .they are generally taken for granted and little heed is taken of the impact of their use on the longer-term reserves" (Kelk, 1992, p. 34). But, as Prentice (1990, p. x) remarks, as raw materials become scarcer, "The proper use of our mineral resources can only be achieved if we use all our geological skills to ensure that they are used to the best advantage".

Bulk minerals are unlikely to be exhausted for a very long time. "There is, however, a clear possibility that there will be local shortages, and because the cost of transport is so large a factor, these shortages may be difficult to remedy by bringing in material from further away" (Prentice, 1990, p. 5). Prentice quotes the example of aggregates in south-east England where the traditional river terrace and valley floor sources have largely run out. Other reserves have been sterilised by unplanned extension of housing across their surface. He believes that the region will be obliged to seek its aggregate from sea-dredged materials, and from crushed rock imported from as far away as Scotland, with the inevitable increase in cost that such long hauls will bring. He sees a similar example in north-east United States, where the nearest hard rock sources are being increasingly populated or used for recreation, with the result that distant coastal superquarries are being opened to transport crushed aggregate by boat (see Chapter 3). An example of this is the Glensanda superquarry in Scotland, which has supplied aggregate to highways in Texas and autobahns in Germany (McKirdy, 1993).

Fossil fuel shortages are also looming. We are living in what Marshak (2001) calls the "Age of Oil" which began in the late nineteenth century. A number of authors have attempted to predict future production levels based on several approaches. Hubbert predicted in 1956 that American oil production (lower 48 states) would peak in 1969 and decline thereafter (Hubbert, 1956). Although castigated at that time, American oil production in fact peaked in 1970, and has declined since then and is likely to dwindle to insignificant levels by 2010 (Cutter & Renwick, 1999). Using similar methods, a number of authors have warned of a looming oil crisis with most believing that peak global oil production will occur before 2020 (Fig. 6.21) and possibly as early as 2004 to 2008 (e.g. Hatfield, 1997; Kerr, 1998; Campbell & Laherrère, 1998; Deffeyes 2001; Rifkin, 2002). According to Campbell & Laherrère (1998), new finds, improved technology and use of substitutes such as heavy oil or tar sands are unlikely to have much impact, and oil demand is rising by about 2% per year. Furthermore, as at least 50% of the remaining reserves are in the Middle East, political instability could produce world oil shortages and significantly higher prices within a few years and

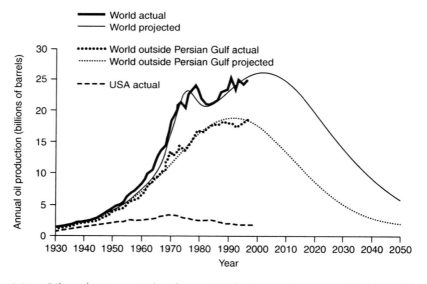

Figure 6.21   *Oil production actual and projected for the world, the world without the Persian Gulf countries and the USA (Reprinted with permission from Kerr, R.A. (1998). The next oil crisis looms large – and perhaps close. Science, **281**, 1128–1131.) Copyright 1998 American Association for the Advancement of Science.*

certainly in the next few decades (Shell, 2001; Anon 2002). Roberts (2003) goes as far as to say that the United States' "physical and economic infrastructure is so highly car dependent that the US is pathologically addicted to oil. . . . The US economy needs oil like a junkie needs heroin and Iraq has 112 billion barrels, the largest supply in the world outside Saudi Arabia". There has been a major effort to reduce fossil fuel consumption in order to reduce greenhouse gas emissions and their climatic impacts, but the geological constraints on future oil availability have not been fully explained to the public. There is also an urgent and compelling need for research and development of alternative energy sources including expansion of renewables and development of hydrogen fuel cells (Rifkin, 2002). Apart from biofuels, most of these are abiotic energy sources and we therefore see here another example of how geodiversity provides us with flexibility. Historians are likely to judge the Age of Oil as having been "a two-century binge of profligate energy use" (Kerr, 1998, p. 1128).

Other non-renewable resources may also have a limited life. Table 3.2 shows the lifetimes of some of the more common minerals, and one view is that we are likely to see world shortages of virgin supplies of some minerals before the end of the present century. As discussed in Chapter 3, many metals have specific uses in modern society and are the subject of global trade. At present the USA imports 100% of the manganese it consumes, 95% of the cobalt, 73% of the chromium and 92% of the platinum (Marshak, 2001). These are included in a list of over 30 "strategic" minerals that the USA stockpiles in case supplies are cut off. As Cutter & Renwick point out, four countries control 70% of the world's bauxite production, six countries produce 90% of all manganese and nearly 95% of platinum is produced by South Africa and Russia. One company, Norilsk Nickel in northern Siberia, produced 62% of the world's palladium, 27% of the world's nickel, and 22% of the world's platinum in 1994, sufficient to influence world prices. Mine closure or trade restrictions could easily lead to market shortages and dramatic price increases (Cutter & Renwick, 1999). Thus, importing of minerals does not imply that national supplies have become exhausted since there are strategic and economic issues to be considered. However, most of the best quality iron ore in the United States has been used and as a result most ore is now imported from other countries (Cutter & Renwick, 1999). Where minerals do run short, lower grade ores are likely to be utilised, though these require greater energy inputs to refine each tonne of metal, produce greater quantities of waste spoil and therefore have greater environmental impacts.

But, there are other arguments to be considered. Cutter & Renwick (1999) emphasise the importance of economic factors and believe that "we will never "run out" of any minerals. We may find that a particular mineral has become too expensive to justify using it, and, in that sense it may become unavailable", but some of the resource will still exist. Lomberg (2001) is sceptical about the threat of mineral extinction for three reasons:

- Firstly, new finds will be made that will extend the life of minerals like oil. This has been the history of oil exploration and thus "we now have more reserves than ever before. . .development in reserves by far outstrips development in demand" (p. 123–124).
- Secondly, shortages create higher prices and incentives to seek new reserves, extract and use resources more efficiently and recycle. For example, traditional methods of oil extraction remove less than half the oil in the reservoir, and other methods could

be used to extract much of what remains. Furthermore, cars have become more fuel efficient and contain half as much metal as they did 30 years ago.

- Thirdly, we find substitutes and better technologies. "We do not demand oil as such but rather the services it can provide. Most often we want heating, energy or fuel, and this we can obtain from other sources" (p. 126). For example, according to Lomberg (2001) there is 242 times more shale oil than conventional petroleum resources and 8 times more than all fossil fuels combined, though he admits that the price will be higher.

Lomberg's (2001, p. 147–148) conclusion is that "we are not likely to experience any significant scarcity of raw materials in the future" and this is also the view of Cutter & Renwick (1999). Nonetheless, we are dealing with non-renewable resources where demand is increasing in many cases and we ought to take reasonable steps to minimise our use of them or use them more wisely. Geoconservation often focuses on scientific, aesthetic or other values but conservation of these utilitarian geomaterials should also be a subject of interest to geoscientists as part of a sustainable future.

6.5.3   Sustainable use of geomaterials

The principles of a sustainable strategy for geological materials must be to reduce the use of virgin resources. This can be achieved by:

- minimisation,
- reuse,
- recycling,
- substitution, downgrading and replication, and
- fiscal instruments.

Together these measures represent a "dematerialisation" of society (ENDS Report, 2001, p. 31). Each of the measures will be outlined in turn.

Minimisation

This means reducing to a minimum (a) the amount of geological material extracted from the ground, (b) the amount of land disturbance and (c) the amount of waste produced. Taking energy fuels as an example of (a), the amount of coal and oil burnt could be reduced by using more fuel-efficient systems, switching to renewable energy sources, employing better insulation and passive solar design, reducing the need to travel, and so on. As well as energy saving, the principle can be applied in other fields. For example, tin plating now uses less than a third of the amount of tin used in the 1950s. Thinner steel and aluminium cans have been introduced giving a huge reduction in metal usage given that 75 billion cans are sold in the United States each year (Jones & Hollier, 1997). Miniaturisation of electronic equipment also means that smaller amounts of metal are used. But despite these trends, demand is still rising as population increases. Bristow (1994, p. 86) believes that "Conservation of mineral resources by the use of technically advanced mining and mineral processing systems is just as important as many other aspects of conservation". In other words, it is important to make the most efficient use of the mineral resource in the ground.

Minimising the amount of land disturbance can be achieved by excavating deeper, not sterilising land by dumping waste on top of further valuable resources, and using modern technology in mineral planning and exploitation. Where there is a choice in quarry siting, the least environmentally damaging site should be selected, other things being equal. In the United Kingdom, the Quarry Products Association has pledged to review its current permissions to extract material within National Parks and only apply for new permissions and extensions where there is an overriding national need or potential benefits for the Park concerned. Scottish Natural Heritage has undertaken a GIS-based approach to evaluating the sensitivity of different components of the natural environment to mineral extraction in the Midland Valley of Scotland where development pressures are greatest (Scottish Natural Heritage, 2000a).

One of the means of minimising the waste produced is to improve the sustainability of the mining sector by developing new technologies that reduce the impact of waste and toxic by-products. For example, tailings are waste materials produced from metal concentration processes such as flotation. The chemical composition and quantities of these waste materials and by-products often have a major environmental impact when they are stored behind tailing dams, particularly when the dams fail. Increasingly, there are efforts to line the disposal areas, solidify them before surface disposal, or use them to backfill quarries. An even more sustainable solution is given in Box 6.6.

Box 6.6 *Solvent Extraction and Electrowinning of Metals*

One method of reducing the amount of tailings and the impact of their disposal is to change the types of ores exploited from sulphide ores to weathered oxide ores that have been partially refined by natural geological processes (Monhemius, 2002). This trend began in the copper industry about 30 years ago and about 25% of the world's copper is now produced from these copper oxide ores by the processes of solvent extraction and electrowinning. These produce marketable copper plates at the mine site and the only significant by-product is a coarse recyclable sediment. There are therefore huge reductions in ore transport, waste materials, toxic by-products and costs.

The process begins with sulphuric acid leaching of the copper oxide to produce copper sulphate and water:

$$CuO + H_2SO_4 = CuSO_4 + H_2O$$

The copper is then extracted from the copper solvent by electrolysis or electrowinning

$$2CuSO_4 + 2H_2O = 2Cu + 2H_2SO_4 + O_2$$

It can be seen therefore that the process is largely self-sustaining with the only by-products being oxygen and spent ore. The sulphuric acid is recycled and only needs occasional topping up. Similar methods are now being applied to zinc silicates and nickel laterites, the Goro mine in New Caledonia, being an example of the latter source.

Reuse

This is about reusing products, as opposed to recycling, which refers to reusing materials, and there is already considerable reuse of geologically derived products. Gemstones and jewellery are usually traded or handed down from generation to generation, largely because of the significant value (economic or sentimental) involved. Diamonds are indeed forever! Building stone, bricks, roof tiles, slates and so on are frequently reused in subsequent buildings, and in some countries this is encouraged by the presence of second-hand building material suppliers. Parts from scrapped vehicles are resold at auto-salvage yards. Soil stripped from site developments is stored and reused either on or off site.

Recycling

This is quite common and applies on a large-scale to glass, aluminium and other metals, as well as demolition rubble, power station ash, road planings, quarry waste and spent ores (Blunden, 1991; Howard Humphries & Partners, 1994; McKirdy, 1996). Glass recycling reduces the amount of virgin sand deposits that need to be excavated and may have other environmental benefits. Similarly for aluminium recycling, which also reduces by 95% the amount of energy involved in transporting and processing the bauxite from the source countries. Over 30% of US aluminium, 40% of its iron and steel and 50% of its lead come from recycled sources (Cutter & Renwick, 1999). However, over 95% of titanium used in the United States is in paint pigments and recycling this would be impossible. Building demolition, road planing prior to resurfacing, and site excavation, produce large quantities of mixed inert material, which can have a wide variety of uses, including fill materials, site restoration and foundation hard-core. Importantly, recycling also reduces disposal problems.

Substitution

This involves using alternative, perhaps less valuable or more common materials that achieve a similar purpose to the traditional material. Similarly, downgrading means not using high-quality materials when a lower quality will suffice. Higher-grade materials should be conserved for applications in which they are essential. Although in many situations material properties are governed by national standards, which will often exclude the use of, for example, mixed demolition materials, screening and sorting is possible. Some materials such as road planings are fairly "clean" and may meet required standards without further processing. Blunden (1991, p. 126) stresses that although most metals have several substitutes this can often only be done "at the price of losing some unique characteristic in a specific application", or at additional cost or with greater energy use.

Jones & Hollier (1997, p. 43) explain that "the substitution of one material for another at its point of end-use has generally come about either because an alternative can be utilized more cheaply or at no extra cost, or because the alternative is a better-performing material. . . .. If a material is more efficient in producing heat or using less metal to achieve the same outcome, it will prove to be more cost-effective". They give the example of oil and natural gas having greater versatility and a higher calorific value than coal. Aluminium is lighter and stronger than many other metals and can be used more sparingly, for example in electrical wiring, can production and cars. Plastic is substituting for metal, silica optical fibres are replacing copper wire and microwave and satellite systems of communication are replacing both types of cabling.

Also included in this category are fossil replicas for use in museum collections, research and education (Nyborg & Santucci, 2000; Williams, 2001; Dady, 2002).

Fiscal instruments

These are economic measures designed to manage resource use. It has been noted that when the price of a mineral rises, so does the economic attraction of recycling and fiscal instruments can be introduced to stimulate this process. They are designed to discourage the extraction of virgin materials and encourage the use of recycled or secondary materials and new recycling processes. Examples from the United Kingdom include the Landfill Tax, introduced in 1996 to discourage landfilling of waste, including inert waste, and the Aggregates Levy introduced in 2002 at £1.60/tonne, which is designed to reduce the use of virgin aggregate. According to the ENDS Report (2001, p. 29). "Slate and china clay businesses are gearing up to sell their wastes to the construction industry, calculating that the aggregates levy will make them competitive for the first time" (see Box 6.7). Part of the income from the levy (£29.3 million per year) has been used to establish an Aggregates Levy Sustainability Fund (ALSF) in England that will fund research and conservation related to aggregate extraction and sustainable construction and demolition practices (Prosser, 2002b).

Box 6.7 Secondary Aggregate Use Examples from the United Kingdom

- For every tonne of kaolin produced, nine tonnes of waste rock, sand and finer residues remain as "waste". Uses for the finer wastes were devised about 30 years ago (Bristow, 1994), and now plans are being formulated to ship 750,000 tonnes per year of waste sand from china clay extraction in Cornwall, England. Destinations would be throughout the United Kingdom, Ireland and western continental Europe for use in the manufacture of concrete or asphalt. About 500 to 600 million tonnes of mineral waste currently covering over 60 km² of Cornwall in waste hills is potentially available for recycling though some of it is being restored to heathland. In addition, about 20 million tonnes of new waste are produced each year, including 10 million tonnes of sand. Recycling of this material not only saves virgin sand deposits but also reduces the impact on the Cornish landscape, The Aggregates Levy has started to make the shipment of this waste economic though it probably needs to be higher than £1.60/tonne to realise the full potential (ENDS Report, 2001, p. 30). More local alternatives, such as beach replenishment and extraction of mica, topaz and heavy metals for use as industrial minerals, have been suggested (Bristow, 1994).
- Slate production results in huge quantities of waste slate. For every tonne of high-quality slate produced, the amount of waste can be over 100 tonnes. Over 700 million tonnes of slate waste are piled into hills in North Wales and a further 6 million tonnes are produced annually. Only a small amount of this is used locally as aggregate or for road construction and site restoration. The problem has been the high transport costs involved in moving the material to the large markets in England and abroad, but the Aggregates Levy begins to change this. Consequently, the major slate producer in the area applied for permission in 2001 to construct a rail terminal at Blaneau Ffestiniog to enable transportation of the waste for use as sub-base, pipe-bedding, capping and general fill in the north-west and midlands of England. The levy would need to be higher to make transportation further afield economic (ENDS Report, 2001, p. 31).

Another strategy is to introduce "producer-responsibility" codes or Pigouvian taxes, named after the 1920s' British economist Arthur Pigou. These make manufacturers responsible for some or all of the costs of recycling or waste disposal. For example, electrical firms will have to take back their end-of-life goods from 2006 under new

EU legislation. Similarly, 12 million cars end up on European refuse dumps each year (Jones & Hollier, 1997) and from 2007 car manufacturers will be required to take back their scrap cars. As the costs of doing so have to be passed on to the customer, there is an incentive for manufacturers to use recyclable parts and materials, reduce packaging, and so on (Barrow, 1999). BMW and Mercedes-Benz cars have led the way in using recycled and recyclable materials in car manufacture. In turn, recycling of cars is being assisted by new technology such as high-powered fragmentisers that can convert a pre-compacted scrap car into accurately sorted bundles of ferrous, non-ferrous and other materials in under a minute (Jones & Hollier, 1997).

6.5.4 Natural and indigenous geomaterials and methods

Incongruous and authentic landforms were discussed in Section 6.4, but geological materials can also be incongruous or authentic depending on the local situation. A major controversy arose in the late 1990s during the rebuilding of the south portico of the British Museum in London, which had been demolished in the nineteenth century. The tender had specified "oolitic limestone, Portland base bed or similar", but it had been understood that Portland stone was being supplied. Following construction, it was realised that a French limestone, Anstrude Roche Claire, had been used, creating a major controversy because of the lack of a perfect match with the rest of the museum. However, the existing stone is weathered and "it has to be remembered that the south portico is of very recent construction and not made to exactly the original design by Smirke" (Anderson, 2000, p. 92).

The use of natural stone declined following the First World War as concrete and concrete products started to gain ascendancy. In Scotland, McMillan *et al.* (1999, p. 3) note that "in the last 30 years this decline has to some extent been reversed with the recognition that, used appropriately, the natural product is not only aesthetically more pleasing but more durable, enabling architects to build on that unique character of Edinburgh". Historic Scotland (1997) demonstrated that it is also less energy intensive and its use is therefore more sustainable than concrete. Building stone has a high residual value on demolition whereas concrete can only be crushed for aggregate (MacFadyen & McMillan, 2002).

The resurgence of interest in the use of natural stone for construction and cladding in Scotland has led to the opening of new quarries and reopening of many long-abandoned ones. Of some 20 stone quarries in production in 1998, 8 were producing dimensioned sandstone and 6 were working flagstone for paving. Trial extraction of slate from Ballachullish has been agreed upon, the first since the last quarry closed in 1955. To encourage these developments, a Scottish Stone Liaison Group and Natural Stone Institute have been formed. The new Scottish Parliament building at Holyrood in Edinburgh was to have been faced with Chinese granite but it has now been agreed that grey Kemnay granite from Aberdeenshire will be a more politically correct choice. Similar developments in Wales have led to the formation of a Welsh Stone Forum (Fforwm Cerrig Cymru) in 2003 with the aim of "promoting understanding of the use of natural stone as a sustainable material in the Welsh environment".

In rural areas, the importance of using appropriate designs and materials for buildings, walls, paths and car parks is being increasingly recognised, through the production of design guides, and so on. The Countryside Commission (now Agency) has published technical guidance on *Design in the countryside* in England (Countryside Commission

1993a, p. 12) that "argues forcefully for the retention of regional diversity, local distinctiveness and harmony between buildings, their settlements and the landscape" and urges all planning authorities "to endorse the principle that new development in the countryside must reflect and respect the diversity and distinctiveness of the local landscape character". Part of this involves the use of local geological materials and many design guides now encourage the use or reuse of local building stone. Scottish Natural Heritage has produced a design guide for *Car Parks in the Countryside,* which promotes the use of local materials, unmetalled parking areas, access tracks and walls. "Where such materials are available locally they can ensure that the hard surfaces are sympathetic to the local soils and geology" (Scottish Natural Heritage, 2000b, p. 53). The guidance believes that it is essential "that the local construction method/style is employed. This demands that traditional walling skills are utilised to ensure the new walls complement rather than detract from earlier workmanship". The Millennium Wall at the UK National Stone Centre in Derbyshire, United Kingdom, comprises 19 different rock types and wall styles from different parts of the country, thus illustrating local diversity of materials and methods.

6.6 Land-use Planning Systems

Most developed countries have some type of land-use planning system established in recognition that unhindered development is likely to be detrimental to the environment. Laws and regulations have therefore been introduced to control the siting, design and impacts of development. In principle, most of these systems could be used, to a greater or lesser extent, to safeguard geodiversity, though their full potential at an international level has yet to be recognised. Protected areas can be notified to local planning authorities so that their status can be taken into account when planning applications, within or adjacent to them, are submitted and decisions made. Similarly, planning policies can be used to give general recognition and protection to types of protected areas. An example of a notified area being saved from a housing development is described in the Netherlands by Gonggrijp (1993b). An area of abandoned meanders and river terraces, close to the town of Roermond and included as a site in the Netherlands' *Nature Policy Plan* (Gonggrijp, 1993a), was threatened by urban expansion. However, the listing in the plan saved the site from development thanks to the intervention of a government minister.

Despite this success, the fact remains that in many countries there is little national influence on planning decisions, which are mostly made locally. Concern is being expressed about the cumulative impact of local decision-making in the US Rockies and the "tyranny of small decisions". As a result, there are calls for better planning policies, legal instruments and development incentives to manage future growth (e.g. Baron *et al.,* 2000). In those countries with decentralised planning systems, the challenge is to motivate local decision makers to take geoscience issues fully into account.

Similarly, Tasmania, Australia introduced a *Resource Management and Planning System* (RMPS) in 1993. A requirement to determine whether a development will affect any sites on the Tasmanian Geoconservation Database is now written into most Development Plans and Environmental Management Plans. However, forestry activities and mineral exploration are not covered and instead have their own codes of practice and planning systems. There are concerns that integrated planning is being hampered at state and local levels, that there are inadequate resources to undertake

Table 6.8 *Protection Zones identified for the Eastern Slopes of the Canadian Rockies (Government of Alberta, 1977)*

Protection	1. Prime Protection
	2. Critical Wildlife
Resource	3. Special Use
Management	4. General Recreation
	5. Multiple Use
	6. Agriculture
Development	7. Industrial
	8. Facility

regional conservation and strategic resource management, and that the effectiveness of local authorities in dealing with local conservation and planning issues is variable. Consequently, improved planning procedures to protect natural diversity have been recommended (DPIWE, 2001b).

Many authorities have introduced planning zones, which attempt to delimit areas of land for certain land uses, leaving other areas for conservation. An early example of this type of approach was Alberta's "Eastern Slopes" (of the Canadian Rockies) study (Government of Alberta, 1977), which introduced three categories and eight zones (see Table 6.8). The Prime Protection Zone, for example, comprised the mountain summits and high plateaux above about 2000 m thus protecting the mountain scenery and sensitive environments of this zone. Developments that would not be permitted in this zone included mineral exploration and development, petroleum and natural gas exploration and development, commercial timber operations, domestic grazing, cultivation, industrial development, residential development and off-highway vehicle use.

Box 6.8 gives another example from Canada of how planning protection was enacted in the Niagara Escarpment area of Ontario in response to threats from aggregate quarrying and urban expansion.

Box 6.8   *Planning in the Niagara Escarpment, Canada*

The Niagara escarpment in Ontario, Canada is a spectacular large-scale landform stretching 750 km from Niagara Falls to Tobermory on Lake Huron. Its cliffs contain the most extensive Silurian stratigraphy in North America (Davidson *et al.*, 2001). It was recognised in the 1970s that the whole of the escarpment was at risk from aggregate production and uncontrolled housing development. As a result, the *Niagara Escarpment Planning and Development Act* (1975) was passed, delineating the whole escarpment as a special planning area and establishing the Niagara Escarpment Commission (NEC) to oversee planning and development control within the plan area. An inventory of the most significant features, including geological sites and landforms was compiled and used to designate nature reserves (e.g. Hockley Valley, Mono Cliffs and Cabot Head), parks and special planning zones along the length of the Escarpment. "As a result, the overall character of the Escarpment, its geological features, and the lands in its vicinity have been afforded lasting protection" (Davidson *et al.*, 2001, p. 230).

In the United Kingdom, the planning system is heavily controlled centrally and is divided into planning policy and development control. Planning policies are contained firstly in government circulars and guidance (e.g. Planning Policy Guidance

Notes (PPGs in England & Wales or NPPGs in Scotland) or Mineral Planning Guidance (MPGs); see www.planning.odpm.gov.uk). For example, UK Government Advice in England & Wales is that statutorily designated sites, including Sites of Special Scientific Interests (SSSIs), "should be protected from damage and destruction with their important scientific features conserved by appropriate management" (PPG9 on *Nature Conservation*, 1994, para 12), and that "development proposals in or likely to affect them must be subject to special scrutiny" (PPG9, 1994, para 29). PPG7 (on *The Countryside...*, 1997) points out that "Government policy is that the best and most versatile agricultural land should be protected" (para 2.17), "the countryside should be safeguarded for its own sake and non-renewable and natural resources should be afforded protection" (para. 2.14). PPG7 also supports attempts to recognise the value of nature conservation in the wider countryside, including the landscape character approach described above. This technique "should help in accommodating necessary change without sacrificing local character. It can help ensure that development respects or enhances the distinctive character of the land and the built environment" (para 2.15). Other PPGs with significant implications for earth science include PPG14 (*Development on Unstable Land*), PPG20 (*Coastal Planning*), PPG23 (*Planning and Pollution Control*) and PPG25 (*Development and Flood Risk*).

Government guidance is then carried through into Regional Planning Guidance or Regional Spatial Strategies or incorporated by local authorities into their Structure Plans, Unitary Development Plans and Local Plans. Most of these plans therefore offer protection to SSSIs and other statutory designated areas, and many now include protection for Regionally Important Geological Sites (RIGS) (see Section 5.10.4). For example, Policy QE6 of the Draft RPG for the West Midlands aims to "conserve and enhance the geodiversity and biodiversity" of the Region. Box 6.9 gives a number of examples of this approach in the United Kingdom.

Box 6.9   *Examples of UK Planning Policies*

Mineral Planning Guidance Note 1 (MPG1) states that:

- Minerals should be conserved, as far as possible, whilst securing an adequate supply to meet the needs of the economy and society.
- Environmental impacts caused by mineral operations and the transport of minerals should be kept to an acceptable minimum through good operational and management practices.
- The production of waste should be minimised and minerals should be used efficiently, taking appropriate account of the use of high-quality materials and opportunities to recycle wastes.
- Restoration of sites should preserve or enhance the overall quality of the environment after working has ceased and, where appropriate, make contributions to improved habitats and biodiversity.
- Areas of designated landscape, nature conservation or heritage value should be protected, as far as possible, from mineral development.
- The unnecessary sterilising of mineral resources by using the land permanently for other purposes should be avoided.

Continued on page 320

Continued from page 319

The *Lake District National Park Local Plan* has a policy (BE1) on the use of geological materials:

"The National Park Authority will, where appropriate, require materials for walls and roofs of development to be in keeping with local vernacular tradition. Stonework should be random rubble or traditionally coursed and jointed with conventional mortar joints and mix. Slate for roofs should be an appropriate colour, texture and size and normally laid in diminishing courses" (Lake District National Park Authority, 1998).

Purbeck Local Plan Policy CA2 states that "Development that would have an adverse effect on the nature conservation interest of an SSSI, either directly or indirectly, will not be permitted unless:

1. the proposed development is required to meet a clearly identified national need, or the proposed development is of strategic importance to the district;
2. there are no acceptable, less damaging, solutions; and
3. the effect would be minor and would not result in a significant reduction to the nature conservation interest of the site...." (Purbeck District Council, 1997).

The *South Norfolk Local Plan* has a policy on Landscape Character (see above) which states that "Development will not be permitted where it would significantly harm the identified assets important to the character of the landscape". In the case of topographic change, Policy BEN5 states that any landscaping scheme should "seek to ensure that any land modelling proposed as associated with uses such as golf courses, landfilling, etc. is sensitive to the local topographical character in terms of height, gradient, scale and shape", and should "reflect the local landscape character and distinctiveness" (South Norfolk Council, 2003).

Dudley Borough Council, Supplementary Planning Guidance (SPG) for Potential Development of a Site (2000) contains a flowchart for assessing the geological significance of development sites (Fig. 6.22)

Section 54A of the *Town and Country Planning Act* (1990), states that local authorities must determine planning applications with regard to adopted development plans unless there are "material considerations" that dictate otherwise. Thus, the UK "plan-led" system, generally gives a high level of protection to both designated and non-statutory geological sites and to the wider countryside. A good example of the effectiveness of the UK system is the recent refusal of planning consent for coastal defences at the Birling Gap in East Sussex (Box 6.10).

Other ways in which the UK planning system can assist in geoconservation include the use of restoration conditions in the granting of consents for mineral extraction and landfill sites, including contours on the final landform and proposals for the handling and storage of soil. However, Bridgland (1994) makes the point that such restoration conditions make it difficult to preserve geologically interesting quarry faces because the restoration programme is agreed to before quarrying even begins. If important faces become exposed during quarrying "it is thus necessary to persuade the various interested parties to modify existing plans – a procedure that might prove expensive...." (Bridgland, 1994, p. 88). He therefore calls for geologists to have a greater influence at the planning stage to maximise the research and teaching potential of exposures created by future quarrying. Specifically, he points out that planning authorities in

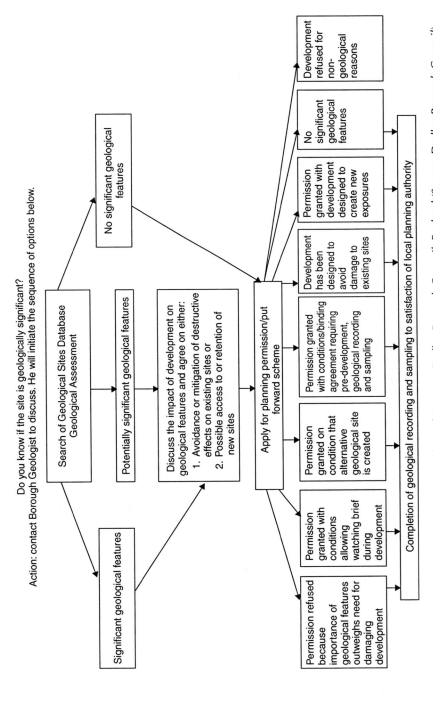

Figure 6.22 *Key to decision-making regarding geological impacts, Dudley Borough Council, England (Source: Dudley Borough Council)*

Box 6.10   *Birling Gap Planning Application and Appeal, England*

One very important example of a proposed development affecting an SSSI occurred at the Birling Gap in East Sussex, England. Local residents submitted a planning application to construct sea defences to prevent their cottages, a hotel and a boathouse from the threat of coastal erosion. They claimed that construction of revetments at the foot of the cliff would slow the rate of erosion and prolong the lifetime of the buildings. The plan was contested by many organisations including English Nature, the National Trust, the Sussex Downs Conservation Board and the Quaternary Research Association and was refused permission by the local authority. The applicants appealed and a public inquiry was held in 2000. Supporting the local authority's case for refusal, English Nature argued that the continued erosion of the cliffs was important to maintain the local beach material supply and thus protect the rest of the cliffline, retain the local cliff exposures of the truncated chalkland dry valley with periglaciated chalk surface and fill of Late Quaternary sediments and conserve the local spectacular coastal scenery. The objectors argued that the revetments would be intrusive and would themselves eventually be undermined by erosion on their flanks.

The Inquiry Inspector agreed with the objectors, believing that the revetments would harm the beauty of the area and have an adverse impact upon the local nature conservation interests. The Inspector confirmed the importance of GCR and SSSI status and made the point that it is not just scientists who put a value on designated sites since they operate within existing legislation authorised by society at large (Prosser, 2001a).

---

the United Kingdom now routinely require developer funding of archaeological rescue digs prior to mineral excavation (through the use of legal agreements), and makes the case for the same type of arrangement to apply to geology. "It seems anachronistic that our geological heritage doesn't enjoy the same treatment. Extraction frequently destroys potentially important sediments just as it destroys archaeological remains, yet there is no systematic mechanism for ensuring that the geological evidence is recorded". (Bridgland, 1994, p. 90).

An important series of publications on *Environmental geology in land-use planning* was produced for the UK government in 1998. One of the recommendations was that the British Geological Survey should be consulted on relevant planning applications, though this has not been accepted by government (Thompson *et al.*, 1998).

6.7   Environmental Impact Assessment

This is an approach, often linked to land-use planning systems, which has the potential to assess the impact of major development proposals on geodiversity, and also the impact of geoprocesses on the development. It is therefore an *ad hoc* rather than a strategic approach to the identification of geodiversity threats and the needs for prevention or mitigation, but in areas where detailed information is limited, it may be a useful tool in geodiversity conservation that deserves to be better known and used. For full details see Wood (1995), Weston (1997), Glasson *et al.* (1999), Harrop & Nixon (1999). More systematic attempts at environmental assessment can be provided by Strategic Environmental Assessment (SEA) procedures and these are also briefly described below.

Environmental Impact Assessment (EIA) originated in the United States under the *National Environmental Policy Act* (1969) and has gradually been adopted by many countries around the world, albeit in different forms. EIA is a systematic and integrative process for assessing the possible impacts of major developments prior to a decision being taken on whether the proposal should be given permission to proceed (Wood, 1995). If carried out effectively, it should prevent environmentally unacceptable projects from being implemented and mitigate the environmental effects of proposals that are approved. The information assembled during the assessment is published as an Environmental Impact Statement (EIS), which accompanies a planning application and allows consultees and the public to understand the impacts of what is being proposed and the measures taken to reduce these impacts. In essence, EIA means that decisions on all major developments can only be taken in the foreknowledge of their likely environmental consequences.

In the United States, it was quickly assimilated into state and local statutes and since then a host of other industrialised countries have adopted similar procedures, including Canada (1973), Australia and New Zealand (1974), France and Germany (1976), the Netherlands (1981) and Japan (1984). The European Community formally established a Directive on EIA in July 1985, and the procedure was adopted by the United Kingdom in 1988.

Most development projects can be broken down into six major phases (Rivas *et al.*, 1995a, b):

- preliminary reconnaissance – site selection involving preliminary EIA;
- site investigation – verification of the technical suitability of the selected site, with detailed EIA studies;
- detailed planning and design – architectural and engineering design of the project, with feedback to EIA and development of mitigation measures;
- construction – during project construction, monitoring of impacts occurs with evaluation and, if necessary, redesign of mitigation measures;
- operation – during the operation of the project, monitoring of impacts, evaluation of mitigation measures and redesign as necessary is continued;
- decommissioning – after the lifetime of the project, there may be provision or scope for site restoration or further mitigation.

Hodson *et al.* (2001, p. 170) recently commented that "Avoiding significant development impacts on soil ultimately protects the whole of the ecosystem from degradation. . . . (but) relatively few types of development have significant impacts on geology". While one can thoroughly agree with the importance of soil protection, it is difficult to support the comment on geology (or geomorphology), given the impacts outlined in Chapter 4. Development can have significant effects on many aspects of the geological and geomorphological environment, and in turn, the geological environment can impact on developments, particularly through the operation of hazardous processes that may seriously affect the development. Thus, before development proceeds, the geological environment needs to be assessed in the broadest sense, including both direct and indirect impacts (Rivas *et al.*, 1995a). It is important that this work takes place in the early phases of the project (site selection, site investigation and design) since retrospective work is usually expensive and rarely provides the best solutions (Erikstad & Stabbetorp, 2001). Earth scientists can also be involved in the later phases (construction, operation and decommissioning) to monitor impacts, assess the effectiveness of mitigation and redesign elements of the scheme as appropriate.

The main steps in the EIA process are set out in Fig. 6.23.

- *Screening.* Screening is the process of deciding whether an EIA is required. Only major projects currently require environmental assessments, and these are specified in national legislation. For example, Schedule 1 of the United Kingdom's 1988 Regulations (Amended, 1997) lists the type of projects where an environmental assessment is mandatory. These include oil refineries, power stations and special waste incinerators. Schedule 2 lists those where one may be necessary depending on the environmental impacts as perceived by the local authority, for instance holiday village near to an SSSI or large poultry farm.
- *Scoping.* Scoping is the process of deciding what are the main environmental issues that need to be investigated, and what issues do not need to be examined in as much detail.
- *Baseline studies.* This is the process whereby the applicant/consultant collects information on the elements of the existing environment relevant to the impacts identified during scoping. The aim is to establish current environmental conditions and consider the character, extent, importance and vulnerability of the various components of the environment. In the geosciences, this may involve desk and field studies of geology, landform, processes and soils.
- *Policies and Plans.* Policies and plans are also examined at this stage to determine the international, national, regional and local policies relevant to the project. This search will also allow identification of protected sites, designated areas, and so on.
- *Impact Predictions and Assessments.* This is where detailed work is carried out to identify and assess the impacts of the project on the existing environment. For impacts of, and on, the geological/geomorphological environment, appropriate specialists should be employed. The stage requires the characteristics of the project to be known (e.g. use of natural resources, soil stripping and storage, land remodelling, waste generation and management, etc) and the impact on receptors to be assessed. Impacts may be direct or indirect; short, medium or long term; reversible or irreversible; permanent or temporary; beneficial or adverse; singular or cumulative. The magnitude or physical extent of the impacts should be quantified where possible.
- *Mitigation.* Mitigation is where measures are introduced to avoid, reduce or recompense some or all of the identified adverse impacts. The most satisfactory method is avoidance, for example by redesign. This emphasises the point that EIA should be an iterative process of refinement of the proposals to reduce the environmental impacts. Reduction may be achieved not only by redesign but also by screen planting or bunding. The latter can itself have adverse impacts (see above). Recompense might involve acceptance of the impacts but attempting to compensate by other measures, for instance, providing habitats elsewhere.
- *Alternatives.* Proposals involving EIAs may be required to explain what alternatives (e.g. sites, designs, etc) have been considered and reasons for rejection.

Among the earth science aspects that will need to be assessed are the following:

- natural hazards that may impact on a project – could be recorded on a hazards/process map;
- impact of the project on soils and geological materials – as Rivas *et al.* (1995a) state, "Any human activity that consumes, sterilises or degrades these resources,

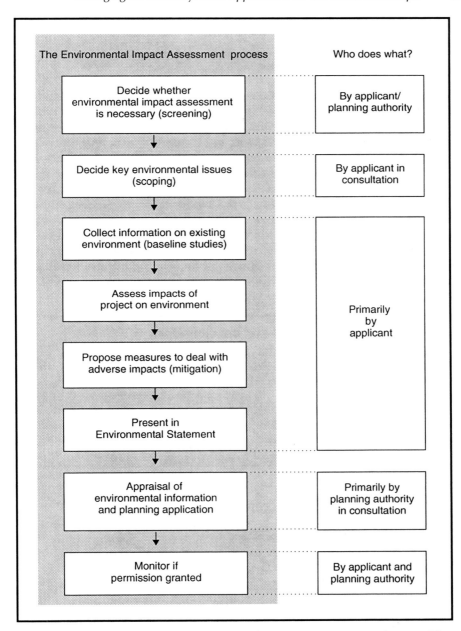

Figure 6.23   *The main steps in the EIA process (After Essex County Council, (2001) The Essex Guide to Environmental Impact Assessment. Essex County Council, Chelmsford, by permission of Essex Country Council)*

would represent a negative impact" and should be reported as such, even though sustainable use of these resources may be regarded as acceptable. Sterilisation refers to development that permanently or temporarily prevents the usage of a resource. For example, building houses over a mineral-rich vein would sterilise it in that mining the vein is prevented, at least in the short to medium term. Degradation normally

refers to pollution of the resource. Geological and soil resources can be recorded on geology/soils maps;
- impact of project on other georesources, including designated sites and areas, landform, landscape aesthetics, processes, hydrology, and so on.

These can be brought together in a Geological Impact Assessment (GIA) that is subsequently integrated into the EIA (Rivas *et al.*, 1995a). Assessments can be carried out in various ways. Rivas *et al.* (1995b) give a case study of a motorway in northern Spain, Bergonzoni *et al.* (1995) describe its use in regional park planning in central Italy and Conacher (2002) questions the appropriateness of EIA to problems of land salinisation and degradation in western Australia. Box 6.11 is an example of a planned military area in the Netherlands and Box 6.12 explains the development of an Oil and Gas Management Plan/EIS for National Park Units in Texas, USA. Impacts have traditionally been assessed using tools such as the Leopold Matrix (Leopold *et al.*, 1971) prepared by the United States Geological Survey, but nowadays more sophisticated GIS systems are used. In Tasmania, Australia, the relevant government department believes that the standard of EIAs has been variable in quality and depth and recommends that EIAs should include on-site assessments of the flora, fauna and geodiversity (DPIWE, 2001b).

Box 6.11   EIA of Military Area at Ede, the Netherlands

The Dutch Ministry of Defence proposed to use part of the Ginkelse Heide area, near Ede in the central Netherlands as a military training area (Asch & van Dijck, 1995). The area comprises four sub-areas shown on a schematic geomorphological map in Fig. 6.24.

- Sijsselt – a forested hill area comprising an ice-push moraine, with podsol soils;
- Ginkelse Heide – an area extending from the toe of the moraine in the west to a glaciofluvial fan in the east where the sand has been blown into low aeolian coversand ridges;

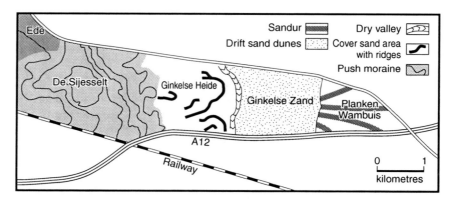

Figure 6.24   *Schematic geomorphological map of the proposed military area at Ede, the Netherlands (After Asch, T.W.J. & van Dijck, S.J.E. (1995) The role of geomorphology in environmental impact assessment in the Netherlands. In Marchetti, M., Panizza, M., Soldati, M. & Barani, D. (eds) Geomorphology and Environmental Impact Assessment. CNR–CS Geodinamica Alpina e Quaternaria, Milan, 63–70, by permission of Centro di Studio)*

Continued on page 327

___ *Continued from page 326* ___

- Ginkelse Zand – an area of aeolian sand dunes;
- Planken Wambuis – a wooded continuation of the glaciofluvial fan.

The plan required the following areas:

- an open area of at least 225 ha for free movement of vehicles and men and construction of tracks;
- a forested area for woodland manoeuvres and bivouacking, with tracks;
- forest edges for digging of shallow defensive holes, again with tracks;
- a 10-m-wide perimeter track connecting all training areas;
- a parking place for loading, unloading and cleaning of vehicles.

The wooded areas of Sijsselt and Planken Wambuis were excluded as open areas because tree felling was prohibited. These areas were assessed as suitable for woodland manoeuvres, and so on. Geomorphologists quickly realised that the other two areas were extremely sensitive to the impacts of soil and sediment disturbance, if intensively used. The following direct and indirect impacts were predicted from intensive use of the two areas:

- destruction of vegetation cover;
- destruction of soil profiles;
- wind erosion of soil and sand and dust hazard;
- destruction of stabilised dunes and ridges;
- soil compaction and disturbance of soil water balance;
- water erosion.

Box 6.12   *Oil & Gas Management Plan/EIS for the Lake Meredith and Alibates Flint Quarries, Texas, USA*

Lake Meredith National Recreation Area and Alibates Flint Quarries National Monument are two units of the USA National Parks System located between two major structural basins in the Texas Panhandle (Fig. 6.25) Important Triassic, Miocene, Pliocene, Pleistocene and Holocene fossils have been discovered in and around these units over the years but no systematic study had ever been made to determine whether oil and gas operations have adversely affected the palaeontological resources of the parks. The baseline data on which to assess the impact of future extensions of oil and gas operations was lacking.

Having identified the need for better baseline information, NPS palaeontologists undertook a comprehensive inventory of the palaeontological resources of the protected areas through literature reviews, museum searches and field surveys. Significant resources identified included Upper Triassic amphibians, reptiles and petrified wood, Miocene – Pliocene root casts, silicified grasses, insect burrows, mammal bone beds and a mastadon tooth, and Pleistocene mammals including a complete skull of the giant bison *Bison latifrons*. A palaeontology resource sensitivity map has been produced (Fig. 6.25) and the important sites have been recorded in a GIS database and are to be monitored periodically.

The final stage has been the development of an oil and gas management plan/EIS identifying circumstances in which a palaeontological survey will be necessary and the procedures for carrying out such a survey. For example, in high-probability fossil areas where ground disturbance is planned, a full palaeontological survey and proposals to minimise fossil disturbance would be required. Guidance is also given on procedures

___ *Continued on page 328* ___

── Continued from page 327 ──────

when unanticipated fossil discoveries occur during approved operations or fossils are damaged within previously identified sites (Santucci *et al.*, 2001a).

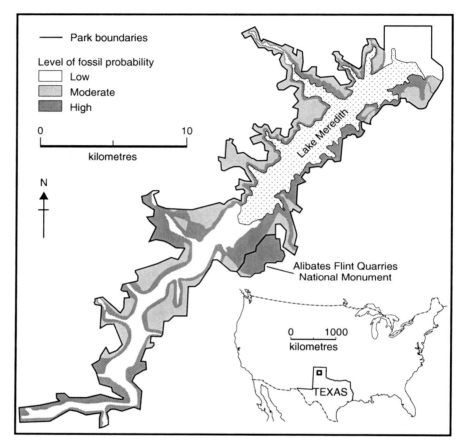

Figure 6.25   *Palaeontology Resource Sensitivity Map of the Lake Meredith and Alibates Areas, Texas, USA (After Santucci, V.L., Hunt, A.P. & Norby, L. (2001a) Oil and gas management planning and the protection of palaeontological resources. Park Science, 21, 36–38)*

A further development is the use of Strategic Environmental Assessment (SEA), which is the environmental impact assessment of public policies, plans and programmes (PPPs). According to Therivel & Thompson (1996), this can be an effective way of helping to meet the twin goals of delivering sustainable development and monitoring nature conservation. "Issues such as significant impacts on our nature conservation resource and inappropriate operations which are likely to exceed the environmental carrying capacity would be assessed at a stage where change can be accommodated without unnecessary cost" (Therivel & Thompson, 1996, p. 5). The European Union adopted the Strategic Environmental Assessment Directive (2001/42/EC) in June 2001 and Member States were required to incorporate it in their national legislation by July

2004 (Holstein, 2002). González *et al.* (1995, p. 180) outline procedures for incorporating geomorphological factors into EIAs of Master Plans.

6.8   Policies, Strategies, Audits and Charters

A very large range of other land policies and strategies can assist in conserving and managing the wider geodiversity resource, and only a few examples can be discussed here. Each country has its own systems of land management that can affect the physical environment and these will include agricultural policy, forest/woodland policy, soil strategies, river or catchment management plans, coastal management plans, and so on. In addition, a number of ideas for geology audits, GAPs and heritage charters are being implemented.

6.8.1   Agricultural policy

Since large areas of land are in agricultural use, agricultural policy and practice can have a huge impact on the physical landscape (see Section 4.8). Over the last 30 years or so, agricultural policy in western Europe, through the Common Agricultural Policy (CAP), has subsidised intensive farming systems, overproduction of food (hence butter mountains and wine lakes) and small farms too unprofitable to survive in a free market. All this has had a detrimental effect on the environment. Since farmers are paid to produce more, there has been an endless drive to use every last acre of land, to produce two crops per year and therefore to add huge doses of chemical fertilizers and pesticides to the land. The result has been soil and water pollution and loss of biodiversity in the countryside as hedges are removed and wild land lost.

Reform of the CAP has proved difficult, not least because of vested farming interests and the huge social and economic impact that subsidy withdrawal causes. Even with subsidies, many farmers have struggled to operate economically and have left agriculture, so farm amalgamation has occurred. At the same time, subsidies for food production are being reduced, while payments for sustainable land management are being increased. These agro-environment schemes encourage more traditional land management practices and environmental protection and although mainly aimed at enhancing biodiversity, they also have the potential to support geodiversity conservation and management through, for example, reducing diffuse pollution, retaining landscape distinctiveness and protecting soils (DEFRA, 2002).An example is the Tir Gofal scheme in Wales (Box 6.13)

Box 6.13   *The Tir Gofal Agro-environment Scheme in Wales*

Tir Gofal is an agro-environment scheme managed by the Countryside Council for Wales in partnership with many other agricultural, forestry and land-management organisations and is co-funded by the EU. It is the latest in a series of initiatives aimed at "helping to maintain the fabric of the countryside" (Countryside Council for Wales, 1999, p. 3). It is a whole farm scheme, offering a 10-year agreement with a five year break clause. It comprises four elements:

1. Land management – mandatory compliance involving management of key habitats. Among the geoconservation measures included are

Continued on page 330

__ *Continued from page 329* __

- retain all existing traditional field boundaries (hedges, walls, banks, slate fences (see Fig. 6.26). . ...;
- safeguard rock features and geological sites;
- protect ponds, streams and rivers with a 1-m buffer strip (increased to 10 m for operations involving farmyard manure or slurry)''. Optional agreements for creation of agreed habitats or features can also be included.

2. Creating new permissive access – voluntary options including providing access for educational purposes;
3. Capital works – payment for capital schemes that protect and manage habitats or support access provision;
4. Training for farmers – including habitat management and skills such as drystone walling.

There is significant scope here for farmers to involve themselves with geoconservation as well as biodiversity (Countryside Council for Wales, 1999).

Figure 6.26   *Slate Fencing, Nant Francon Valley, Wales*

The Environment Agency in England and Wales is encouraging farmers to produce voluntary Whole Farm Plans, which include agricultural chemical usage and measures to reduce diffuse pollution, soil protection and management of river bank erosion. Pilot projects have been successful and the work is now being expanded (e.g. Environment Agency, 2001). In Australia, Carroll *et al.* (2001) describe a "Neighbourhood Catchment" approach to land and stream management whereby discussions take place between government scientists and all landowners within a catchment in order to encourage sustainable land-use practices for the good of all.

6.8.2 Soil conservation

Goudie (2000, p. 160) refers to soil as "one of the thinnest and most vulnerable human resources and one upon which, both deliberately and inadvertently, humans have had a very major impact. Moreover, such an impact can occur with great rapidity in response to land-use change, new technologies or waves of colonization". The implications have now been recognised by many decision-makers and soil management and conservation have therefore become a very active field for policy development (Stocking, 2000). Goudie (2000, p. 199) classifies the main ways in which soil cover may be conserved (Table 6.9), but note that some of these are landscape intrusive.

Brady & Weil (2002) comment on the need for a global perspective on soils. "Changes in soil productivity in one area affect food security and food prices, as well as biodiversity and water quality, in both nearby and distant places. This growing global perspective is paralleled by the growing acceptance of the ecosystem concept as the prime basis for decisions on natural-resource management" (Brady & Weil, 2002, p. 871). After noting that in the United States, Europe and East Asia, soil quality has declined as a result of intensive agricultural practices and from land application of waste materials, they promote the concept of soil health. They define this as "the capacity of a soil to function within (and sometimes outside) its ecosystem boundaries to sustain biological productivity and diversity, maintain environmental quality, and promote plant and animal health" (Brady & Weil, 2002, p. 873). This will only be achieved by better soil management strategies.

A European Soil Charter was developed in the 1990s (Harcourt, 1990), which recognises soil as a finite and valuable resource that requires protection (Box 6.14). This is being further developed by the European Commission as a thematic strategy on soil protection due in 2004 including legislation on soil monitoring and measures to address issues such as erosion, decline of soil organic matter and soil contamination (Commission of the European Communities, 2002).

Table 6.9   *Some traditional practical methods of soil conservation (after Goudie, 2000)*

1. Revegetation:
    (a) deliberate planning
    (b) suppression of fire, grazing, etc. to allow regeneration
2. Measures to stop stream bank erosion
3. Measures to stop gulley enlargement
    (a) planting of trailing plants, etc.
    (b) weirs, gabions, dams, etc.
4. Crop management
    (a) maintaining cover at critical times of year
    (b) rotation;
    (c) cover crops
5. Slope runoff control
    (a) terracing
    (b) deep tillage and application of humus
    (c) transverse hillside ditches to interrupt runoff
    (d) contour ploughing
    (e) preservation of vegetation strips (to limit field width)
6. Prevention of erosion from point sources like roads
    (a) intelligent geomorphic location
    (b) channeling of drainage water to non-susceptible areas
    (c) covering of banks, cuttings, etc. with vegetation
7. Suppression of wind erosion
    (a) soil moisture preservation
    (b) increase in surface roughness through ploughing up clods or by planting windbreaks.

Box 6.14   *European Soil Charter (After Harcourt, 1990)*

(a) Soil is one of humanity's most precious assets. It allows plants, animals and man to live on the Earth's surface.
(b) Soil is a limited resource that is easily destroyed.
(c) Industrial society uses land for agriculture as well as for industrial and other purposes. A regional planning policy must be conceived in terms of the properties of the soil and the needs of today's and tomorrow's society.
(d) Farmers and foresters must apply methods that preserve the quality of the soil.
(e) Soil must be protected against erosion.
(f) Soil must be protected against pollution.
(g) Urban development must be planned so that it causes as little damage as possible to adjoining areas.
(h) In civil engineering projects, the effects on adjacent land must be assessed during planning, so that adequate protective measures can be reckoned in the cost.
(i) An inventory of soil resources is indispensable.
(j) Further research and interdisciplinary collaboration is required to ensure wise use and conservation of the soil.
(k) Soil conservation must be taught at all levels and be kept to an ever-increasing extent in the public eye.
(l) Governments and those in authority must purposefully plan and administer soil resources.

In the United Kingdom, a *Code of Good Agricultural Practice for the Protection of Soils* was published in 1993 (MAFF, 1993) and some legislation to safeguard soils is in place (Puri *et al.*, 2001). The Royal Commission on Environmental Pollution (1996) recommended that a soil strategy be drawn up on the basis of the following principles:

- Soils must be conserved as an essential part of life support systems.
- Soils should be accorded the same priority in environmental protection as air or water.
- Integrated environmental management must include soil sustainability as a key element.
- Where practicable, contaminated sites should be recovered for beneficial use.
- Further contamination of soils from any source should be avoided.

As a result, a draft Soil Strategy for England was produced (see Box 6.15) and Scotland has also produced a *Soil Protection Strategy* (see Gordon & MacDonald, 2003) as a result of several recent studies and reports (e.g. Gordon, 1994b; Taylor, 1995: Taylor *et al.*, 1996; Gauld & Bell, 1997: Adderley *et al.*, 2001). Most Scottish soil classes are, fortuitously, included within existing SSSIs, so that a measure of protection is already in place, but Gauld & Bell (1997) believe that greater attention should be given to soils in SSSI management plans. Wales is also developing a soil conservation strategy. There also needs to be greater attention to sustainable land management in the wider countryside and to how the planning system can safeguard soils. A *Soil Protection Act* (1987, updated 1994) has been introduced in the Netherlands with the goal of ensuring the health and multifunctionality of its soils.

Box 6.15   *A Soil Strategy for England*

A draft of this strategy was published in 2001 (DEFRA, 2001). The strategy is divided into three sections, each of which follows a pressure-state-response approach in which the pressures on the resource are identified, the current state of the resource is described and the proposed policy responses are specified. This aims to achieve the following:

- "to manage the extent of the soil resource in ways which ensure that we can meet our present and future land-use needs". This section of the strategy recognises that soil is being lost through development, erosion, mineral extraction and other factors. The proposals include less development on greenfield sites, advice to land managers on reducing soil erosion and good practice guidance on soil, planning conditions for quarry restoration and research on soil creation from waste materials.
- "to manage the diversity of soils, concentrating particularly on our most valued soils, so that the right balance of soil types is available to meet current and future needs for soil to support our ecosystems as well as landscape, agricultural and cultural functions". The strategy includes the designation of protected ecosystems and landscapes, and planning controls to protect agricultural soils.
- "to maintain and improve the quality of our soils at a level where soil function is not impaired, to ensure that we can meet our current and future social, environmental and economic needs". The strategy includes guidance for farmers on good agricultural practice, maintenance of the Nitrate Vulnerable Zone system, encouragement for organic farming and agro-environment schemes, and controls on the spreading of waste on land.

The final sections of the strategy outline new proposals for soil monitoring, soil indicators and targets, and soil research. The strategy is to be reviewed after five years.

6.8.3   Geology audits and Geodiversity Action Plans

Geology audits (Geoaudits) are comprehensive surveys of the geological resources of an area aimed at developing, supporting and promoting active geological conservation. They summarise current knowledge about the geology of the area and aim to make the data more accessible to local authorities, planners, conservation bodies, statutory organisations, educational establishments and the general public.

Although they inevitably include issues of site conservation, they can and should include wider objectives including the threats to the wider georesource and maintenance of geological databases for the area. An example is the Peterborough Geology Audit, UK (Peterborough Environment City Trust, 1999) (see Box 6.16)

Geology Audits may be extended to become Geodiversity Action Plans (GAPs). Signing the UN Convention on Biodiversity committed the signatory nations to producing and implementing a Biodiversity Action Plan (BAP). Many countries decided that implementation of the national strategy was best carried out at the local level, for example, by local authorities or wildlife groups, with the result that LBAPs are being developed. Carson (1996) argued that BAPs could not ignore the abiotic foundations and the more local the plan, the greater the need to consider such details. For example, the BAP for Devon's rivers and wetlands incorporates targets and actions for fluvial features as an essential prerequisite for the conservation of some of the region's most valuable habitats (Carson, 1996).

Box 6.16   Peterborough Geological Audit, UK

The Peterborough Geology Audit includes the following sections:

- Geological description of the area, solid and drift.
- Shorter descriptions of landscape, soils, working quarries, important historical geological sites, museum collections, urban geology and archaeology.
- Methods of data assembly and site selection.;
- Threats to the georesource, including mineral extraction, land restoration, loss of soils, landfilling and loss of temporary exposures.
- A description of 20 important sites in the Peterborough area including one SSSI, one National Nature Reserves (NNR), six RIGS, and the remainder as locally valuable sites. For each of the RIGS, the current status of the site is given and actions are listed to maintain or improve site conditions. The other sites each have a brief statement of objectives.
- A set of targets under five headings: raising awareness of geological conservation, site surveying and documentation, longer term site protection, site management and enhancement and geological collections in Peterborough. Each target lists a lead group(s), the funding source (mostly "to be arranged") and a target date.
- Details of how the objectives of the audit relate to the other local processes, including the local planning system (see above), Local Agenda 21, UK Biodiversity Action Plan, English Nature's Natural Areas approach (see above).
- Proposals for site monitoring.

Gray (1997a, p. 323) commented that "In the long term, perhaps one day we will see a. ... GAP for the UK to rank alongside its biological counterpart". A national GAP still seems a long way off, but as a result of local initiatives and support from English Nature, a few UK local authorities have already produced Local Geodiversity Action Plans (LGAPs), either separately or as part of their LBAP. A meeting was held in Chester, England in January 2002, at which time, the content and delivery of LGAPs in the United Kingdom was discussed by English Nature, RIGS Groups and others (Burek & Potter, 2002). Subsequently, a report was published and an Internet site established (www.lgaps.org). This indicates that the key factors for developing LGAPs at the local level are as follows:

1. Development of a National GAP to provide an overall framework.
2. Adequate funding.
3. Committed and appropriately trained workers.
4. LGAPs to be the equivalent of LBAPs.
5. Partnership working, including Local Authorities and RIGS groups.
6. People friendly promotion, publicity and policies.

The report also attempts to identify the characteristics of an LGAP as follows:

- Plans should be simple and easy to implement.
- Objectives must be clear and straightforward.
- Targets should be achievable and timed.

- Local audits (see above) should be undertaken first to establish the size and shape of the task.
- Monitoring of Action Plans must be straightforward.

Several LGAPs have already been published or are being prepared, though in some there is a tendency to focus on geological sites rather than a more inclusive approach that values local landscape, active processes and soils. English Nature has funded pilot LGAPs in Cheshire and Warwickshire, and Staffordshire has funded a Geodiversity Officer from an ALSF grant (see Section 6.5.3). Table 6.10 shows an extract from

Table 6.10 *Extract from the Earth Heritage Action Plan section of the Biodiversity Action Plan for Buckinghamshire and Milton Keynes, England*

No.	Action	Lead	Partners	Target/ date	Level of action	Cost (10 years)
1.	Promote sensitive restoration schemes in development	BEHG	LAs	Ongoing	R,L	Staff time
2.	Ensure nature conservation interests are taken into account in management	BEHG	WT, EN	Ongoing	R	Staff time
3.	Introduce management to 10 new sites with educational potential	BEHG	LAs	2005	R,L	£10,000
4.	Review and update RIGS Survey every five years	BEHG	LAs,ERCs	2004	R,L	£10,000
5.	Ensure nature conservation interests of geosites are available to geologists	BNCF	BEHG	Ongoing	R,L	Staff time
6.	Encourage the joint working of ecologists and geologists	BNCF	BEHG	Ongoing	R,L	Staff time
7.	Survey of all geosites to ensure nature conservation included in management	BEHG	LAs	30 sites by 2002	R,L	£15,000
8.	Produce a county-wide strategy for geoconservation and include in other strategies	LAs	BEHG	2005	R	£10,000
9.	Promote the BEHG and seek 500% increase in membership	BEHG	WT	2010	R,L	£10,000
10.	Seek the establishment of an Earth Heritage Development Officer for 3 counties	BEHG	WT	2005	R	£150,000
11.	Promote the use of geoheritage sites for research and education	BEHG	WT	2 leaflets	R,L	£6,000
12.	Promote long-term protection of sites to land owners	BEHG	EN	Ongoing	R	Staff time
13.	Promote the concept of Natural Areas to link conservation and geology	EN	LAs	Ongoing	R	Staff time
14.	Promote responsible use of sites, e.g. fossil collecting	BEHG	EN	Ongoing	R	Staff time
15.	Promote links between neighbouring RIGS groups	BEHG	WT	Ongoing	R	Staff time

BEHG = Buckinghamshire Earth Science Group;
BNCF = Buckinghamshire Nature Conservation Forum;
WT = Wildlife Trust;
EN = English Nature;
LAs = Local Authorities;
ERCs = Environmental Records Centres;
R = Regional actions;
L = Local actions.

the Earth Heritage Action Plan section of the BAP for Buckinghamshire and Milton Keynes, England. In Scotland, SNH has worked with other organisations to produce and ensure that the important links between biodiversity and geodiversity are recognised in LBAPs. One example is the Tayside BAP, which states that "Tayside's complex biodiversity only exists because of its underlying 'geodiversity' (Tayside Biodiversity Partnership, 2002, p. 21).

Once finalised, LGAPs need to be regularly updated and should be cross-referenced and embedded in other regional and local authority plans including LA21 Action Plans, Community Strategies and Local Development Frameworks. There is a real opportunity here for a radical, bottom-up approach to valuing natural physical qualities.

Further recent developments are Company Geodiversity Action Plans (CGAPs). English Nature is encouraging, particularly, mineral companies to draw these up in order to provide a structured means of "identifying resources, setting out good working practices and defining management and promotional methods" (English Nature, 2003, p. 5).

6.8.4   Geoconservation strategies

Along with local audits and action plans described above, national or sub-national strategies for nature conservation, in general, or geoconservation, in particular, are also important. For example, Tasmania's Draft Nature Conservation Strategy recognises that "Geodiversity forms the foundation for all living things" and "protecting these non-living elements, including their ability to develop and change at natural rates, is fundamental to nature conservation" and "is also directly relevant to all aspects of land management" (DPIWE, 2001b).

The United Kingdom's Nature Conservancy Council produced an important national Earth heritage conservation strategy in 1990 (Nature Conservancy Council, 1990) and following the break up of the Council, the devolved agencies have subsequently updated this for their parts of the country. For England, English Nature published updates in 1995 and 2000 and Box 6.17 gives the key points from the latter.

Box 6.17   Key Points in English Nature's Earth Heritage Conservation Strategy "The Past is the Key to the Future" (English Nature, 2000a)

Learning from the past, we will

- ensure that the Earth heritage SSSI series remains scientifically credible and is safeguarded through the continued provision of competent site management advice at national and local levels;
- promote a sustainable approach to the management of Earth heritage resources – wherever possible the resources available now should be available for the future;
- undertake a major site enhancement initiative to improve the condition and status of Earth heritage SSSIs;
- continue to support the Joint Nature Conservation Committee in the publication of the GCR volume series and the maintenance of a Great Britain-wide GCR database.

Continued on page 337

__ *Continued from page 336* __

Enjoying the present, we will

- undertake a major promotion and interpretation initiative celebrating the rich and varied Earth heritage within the SSSI series and network of NNR;
- further develop and exploit the links between Earth heritage, landscape and wildlife and between our Earth heritage and the built environment;
- widely disseminate information about managing Earth heritage sites through the publication of the Earth Heritage magazine and the inclusion of articles in a wide range of literature;
- explore the possibility for increased sponsorship of Earth heritage conservation by the industry;
- develop new training programmes to raise awareness of Earth heritage and Earth heritage conservation amongst English Nature staff and targeted external audiences.

Influencing the future, we will

- continue to support the Earth heritage voluntary movement (including RIGS, the Geoconservation Commission, the Geologists' Association and the RSNC), work more closely with key organisations involved with Earth heritage conservation, and strengthen our involvement at a European level;
- encourage the sharing of knowledge and the promotion of Earth heritage conservation in its widest context;
- widen the application of our Earth heritage expertise to influence changes in policy and practice in key industry sectors such as minerals, waste and energy;
- promote the use of Earth heritage knowledge in understanding future environmental change, and use this knowledge to influence future development of nature conservation policy;
- ensure that Earth heritage conservation issues are incorporated in key conservation policy documents and continue to identify changes in legislation beneficial to Earth heritage conservation.

6.8.5 Natural heritage charters

The most impressive Charter incorporating geodiversity is the *Australian Natural Heritage Charter* (1996; revised 2002). This ground-breaking document is based on a number of fundamental principles, the most important of which is respect for the country's natural heritage. These principles are

- intergenerational equity – the present generation should ensure that the health, diversity and productivity of the environment is maintained or enhanced for the benefit of future generations;
- existence value – living organisms, earth processes and ecosystems may have value beyond the social, economic or cultural values held by humans;
- uncertainty – accepts that our knowledge of natural heritage and the processes affecting it is incomplete, and that the full potential significance or value of natural heritage remains unknown;

- precautionary principle – where there are threats or potential threats of serious or irreversible environmental damage, lack of full scientific certainty should not be used as a reason for postponing measures to prevent environmental degradation.

The Charter applies both inside and outside protected areas. It has a section on definitions (Article 1 of the Charter), which includes definitions of "geodiversity" and "earth processes" and these terms are then threaded through many of the subsequent 43 Articles of the revised edition. Box 6.18 gives some examples of these Articles.

Box 6.18   Examples of Articles of the Australian Natural Heritage Charter (2000) Relevant to Geodiversity Conservation

Article 2 – The basis for conservation is the assessment of the natural significance of a place, usually presented as a statement of significance.

Article 3 – The aim of conservation is to retain, restore or reinstate the natural significance of a place.

Article 5 – Conservation is based on respect for biodiversity and geodiversity. It should involve the least possible physical intervention to ecological processes, evolutionary processes and earth processes.

Article 11 – Elements of geodiversity and biodiversity that contribute to the natural significance of a place should not be removed from the place unless this is the sole means of ensuring their survival, security or preservation and is consistent with the conservation policy.

Article 12 – The destruction of elements of habitat or geodiversity that form part of the natural significance of a place is unacceptable unless it is the sole means of ensuring the security of the wider ecosystem or the long-term conservation of the natural significance.

Article 20 – Reinstatement is appropriate only if there is evidence that the species or habitat elements or features of geodiversity that are to be reintroduced have existed there naturally at a previous time, returning them to the place contributes to retaining the natural significance of that place, and processes that may threaten their existence at that place have been discontinued.

Article 21 – Enhancement is appropriate only if there is evidence that the introduction of additional habitat elements, elements of geodiversity or individuals of an organism that exist at that place are necessary for, or contribute to, the retention of the natural significance of the place.

Article 22 – Where organisms or elements of geodiversity are introduced to a place for the purpose of enhancement, the individuals introduced to the place should not alter the natural species diversity, genetic diversity or geodiversity of the place if that would reduce its natural significance.

Article 23 – Enhancement in existing natural systems should be limited to a minor part of biodiversity or geodiversity of a place and should not change ecosystem processes nor constitute a majority of the habitats or features of geodiversity of the place.

Article 31 – Work on a place should be preceded by research and by review of the available physical, oral, documentary and other evidence about the existing biodiversity and geodiversity......

Article 33 – Evidence of the existing biological diversity, geodiversity and any other significant features of the place... should be recorded before any disturbance of the place.

The inclusion of "geodiversity" in several Articles and the overall approach of the Charter represent a model that other countries could usefully follow. Since the Charter was prepared by the Australian Heritage Commission (the government conservation body), it demonstrated that there is clear government support to the principles of geodiversity conservation.

The Charter is supported by the *Natural Heritage Places Handbook* (Australian Heritage Commission, 1999), which clarifies the 10-step process for natural heritage conservation in Australia, ranging from identifying sites and obtaining information about them through determining the natural significance of the place, developing a conservation policy and plan, to implementing the plan and monitoring the results (see Fig. 6.27). Box 6.19 is an example of a conservation policy for the Red Rock Craters and Lakes, near Alvie in Victoria.

Box 6.19   Conservation Policy for Red Rock Craters and Lakes, Victoria, Australia

Information about the place. Red Rock is a complex of well-preserved maar craters and scoria cones covering 550 hectares. The maars are broad craters that extend below the general ground level. They are surrounded by low rims with steep inner walls and gentle slopes away from the craters. The maars have scalloped edges suggesting that they were formed by multiple eruptions. The deposits associated with them consist of volcanic ash with particles of the size of sand and gravel with abundant large volcanic bombs and blocks. The scoria cones are formed after the maars. They are small, steep-sided volcanoes, that have been progressively superimposed on each other. They consist of scoria fragments and minor layers of ash and fused lava spatter.

Statement of significance. The Red Rock area is a well-preserved example of a complex volcanic landform, of which only a few are known to exist in Australia. It is a significant site for both geological and limnological scientific and educational purposes. Red Rock has an exceptional range of eruptive phenomena. It is a notable multi-orifice volcano with an exceptional range of pyroclastic materials and structures. Red Rock has a complex relationship to the water table and has a wide range of lakes, which have been extensively studied.

Condition and other management information. The Red Rock volcanoes are well preserved, having suffered only negligible erosion. Current land use includes seven major quarries in the area, but most are not operational. Tenure is mainly freehold.

Compatible uses. Most uses are compatible if they do not cause erosion or changes in the natural drainage patterns of the place. Compatible land uses would be tourism, grazing, catchment protection activities and limited quarrying (if this is restricted to places that do not demonstrate unique features essential to understanding the significance of the place). Where there are exposures of the geology, educational use could be made of the area.

Desirable future condition. The place should remain in a condition in which the geomorphology of the landform remains intact and the natural processes of change are not accelerated, directly or indirectly, by human activity. Extractive industry should avoid any places that demonstrate unique features not well represented elsewhere in the place. Tenure is not a critical factor for management, and existing land uses are appropriate to maintain the natural heritage significance identified.

Proposed conservation procedure. A conservation plan can be developed and incorporated in the local planning scheme to guide decisions on management and on future developments and activities proposed for the area.

6.9   Communication, Interpretation and Education

These are very important activities, not only to ensure that the scientific interest of specific sites is understood by the widest possible audience of opinion-formers and

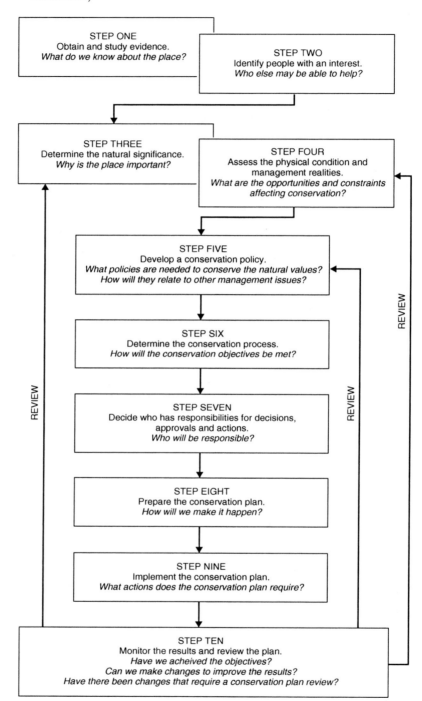

Figure 6.27   *The 10-step process for identifying and managing natural heritage places in Australia (After Australian Heritage Commission, (1999) Natural Heritage Places Handbook. Australian Heritage Commission, Canberra)*

decision-makers but also in creating a more knowledgeable public who understand the importance of geoconservation and will therefore be more likely to support its aims. Daly *et al.* (1994, p. 210) argued that this could be done by

- using terms such as "national heritage" rather than "scientific interest";
- using terms such as "local heritage thereby developing a sense of pride and consequently local approval, support and guardianship";
- by enabling people to profit from conservation, for example, through geotourism (e.g. Geoparks; Section 5.5.2);
- by broadening conservation values to include cultural, historical and educational aspects.

Pemberton (2001a) believes that public awareness can be increased by stressing the links between geodiversity and biodiversity. "This would assist people to value the non-living environment. It would facilitate a greater appreciation of natural diversity and provide a pathway for the general public to better understand the complexities and wonders of our geological history". Unfortunately, as Kiernan (1996, p. 217) points out, "Too often there is no more than a token reference to the landscape, sometimes quite inaccurate, as an anchor for interpretation of plants and animals or for expositions of recreational opportunities. In part this probably reflects the fact that few park management agencies employ any earth science expertise...". And even when the site is clearly geological or geomorphological in nature, there are many examples of missed opportunities for interpretation. For example, there is an excellent guided trail through the impressive lava tubes at São Vicente on the Portuguese island of Madeira (Fig. 6.29), but when visited in 2002, there was no clear explanation from the guide about how the tubes were formed, no literature about the site and no exhibition on the local geology despite space being available.

But despite these disappointments, a wide range of activities is being employed to improve awareness of geology and geomorphology amongst the general public, and it is not possible to give more than a flavour of the work being undertaken, which includes the following:

- Museums; for example, the Hall of Planet Earth in the American Museum of Natural History, New York, the Royal Tyrell Museum, Drumheller, Alberta, Canada and the Natural History Museum, London, UK.
- Visitor centres; for example, the Dynamic Earth visitor centre in Edinburgh, Scotland (Monro & Davison, 2001); the Fuencaliente Volcanoes Visitor Centre on La Palma in the Spanish Canary Islands; the George C. Page Museum at the La Brea Tar Pits, Los Angeles, USA.
- Theme parks; for example the Sand World theme park at Travemünde, Germany; the Crystal Palace Geology Park in London (Doyle, 2001); the west coast fossil park in South Africa (Roberts, 2002); the Gletschergarten in Lucerne, Switzerland. Many disused mines are now also tourist attractions, including the Wieliczka salt mine in Poland (Hallett, 2002) and the Great Orme Copper Mine in North Wales, which is entirely run by amateur geologists and other enthusiasts.
- Site interpretation boards (Hose, 1999; for example see Fig. 5.22); in selecting which sites to interpret, the Countryside Council for Wales has developed a site interpretation strategy that ranked all Geological Conservation Review sites in Wales on the

basis of a large number of criteria, including ownership and access, public safety, scientific themes and interest, visitor numbers, aesthetic value and impact and site fragility. This has enabled a priority list for interpretation to be drawn up and is currently being implemented.

- Books, interpretation leaflets, maps, postcards, and so on. For example, Scottish Natural Heritage publishes a *Landscape Fashioned by Geology* series of booklets that describe the geology of areas of Scotland for the general public. The Geological Survey of Sweden has produced two pilot geotourist maps, comprising a simplified bedrock map and a Quaternary/landform map.
- Displays of fossil and mineral specimens.
- Field guides. For example, the "Landscapes from Stone" project in the north of Ireland (McKeever and Gallagher, 2001) undertaken as self-guided car and cycle routes.
- Geological pedestrian trails, either self-guided (Fig. 6.28) or guide-led (Fig. 6.29), for example, through lit limestone caves or lava tubes. The Portuguese Ministry of

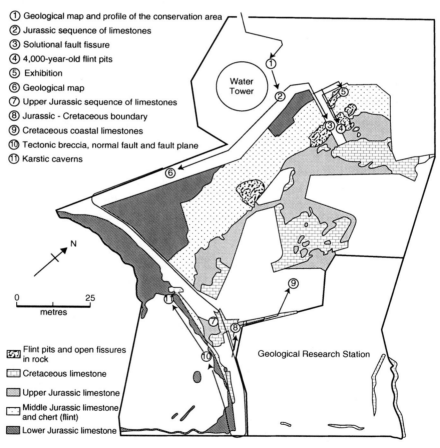

Figure 6.28 *Self-guided trail, Tata geological conservation area, Hungary. (Map compiled by Jozsef Fulop; Reproduced from Fulop, J. (1975) The Mesozoic basement horst blocks of Tata. Geologica Hungarica Series Geologica. Hungarian Geological Institute, Budapest, 16 (see Haas & Hamor, 2001))*

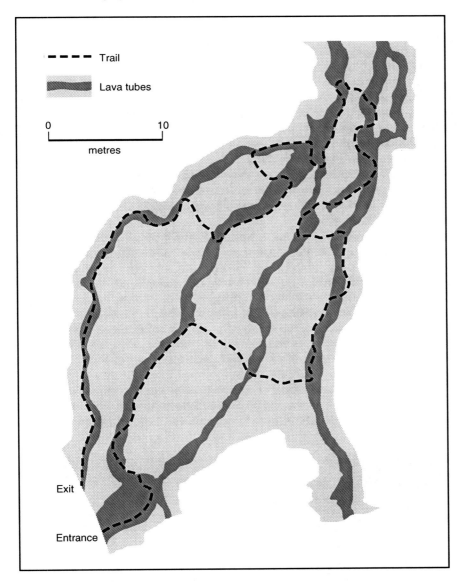

Figure 6.29   *Guided trail through the lava tubes at São Vicente, Madeira, Portugal*

Science and Technology sponsors guided visits for the general public in an initiative called "Geology in the Summer" (Brilha, 2002). Miller (2003) outlines guided walks around Arthur's Seat in Edinburgh, Scotland, to explain its geology.

- Expert-led "hands-on" geological activities such as fossil or mineral collecting. For example, at Fossil Butte National Monument, Wyoming, USA, visitors can accompany a ranger to a fossil quarry in the lake sediments of the Eocene Green River Formation. Seeing fossil fish revealed by splitting rock layers is not only exciting but also allows interpretive opportunities, including the processes of fossilisation, stratigraphical relationships and the methods of palaeontology (V. Santucci, personal communication).

- Geological highway guides or road trails, for example, the American Association of Petroleum Geologists publishes a set of 11 Geological Highway Maps for groups of US states. Figure 6.30 is an example of a geological highway in Wyoming, USA, where signs provide a traverse through the Precambrian and Palaeozoic rocks for interested drivers.
- Geological train trails. A number of these exist in the United Kingdom, including the Llangollen Steam train trail in Wales, "Explore the Severn Valley Railway" and Eskdale railway trail in Cumbria, UK.
- Audio-visual presentations; these are shown at many visitor centres (see above).
- Television programmes, for example, *Walking with Dinosaurs, Walking with Beasts, Earth Story, Wild New World* and a host of natural hazard and disaster series.
- Web sites and CD-ROMs.
- Conferences, activities and events, for example, the biennial *Scottish Geology Week* organised by SNH and the *Earth Alert* conferences of the UK Geologists' Association.
- Children's clubs and educational initiatives, for example, the United Kingdom's *Rockwatch* geology club and Hanson's *Material World* (Hicks, 2003); school-quarry industry links in Canada (Milross & Lipkewich, 2001); Earth Science Teachers' associations in various countries promote the teaching of the geosciences in schools and colleges and are active in commenting on curricula proposals; the Joint Earth Science Education Initiative has a website with many useful ideas and activities (www.jesei.org).
- Links with industrial archaeology, landscape history, the arts and other fields.
- Professional training and continuing professional development, for example, of tour guides (Gillen, 2001) and US National Park officers (Santucci, 2000a).

Figure 6.30 *Example of an information sign on the geological highway over the Bighorn Mountains, Wyoming, USA*

- Student training; for example, a conservation course in a UK Biology Department now includes a geodiversity module (Burek, 2001) and there are several other geodiversity and geoconservation courses that already exist or are being introduced in the United Kingdom and other countries.
- Community awareness programmes. Cameron *et al.* (2001) describe the work of the Bulimba Creek Catchment Co-ordinating Committee in Australia who undertake a range of practical techniques to maintain the natural assets of the area. The methods include water quality monitoring, riparian assessment, community awareness programmes, strategic assessment and the identification of threatening processes. Although much of the work is ecologically based, some of it is related to geomorphological, hydrological and anthropogenic processes, and is a model for community engagement with the local environment.
- State of the environment reports, community strategies, conservation charters, and so on.
- Urban geology; with the majority of the population living in cities, there is a strong case for promoting the geological interest of the fabric of buildings, graveyards and parks, disused quarries, road cuttings, river bank exposures and landforms in urban areas (Prosser & Larwood, 1994; Zinko, 1994; Doyle & Bennett, 1998; MacFadyen & McMillan, 2002). One of the ways of achieving this is through urban geology trails and walks, for example, in London (e.g. Robinson, 1984, 1985), Edinburgh, Chester, Bristol, Exeter, Gloucester, Bath and Malvern in the United Kingdom. There is even a trail associated with a single building, viz Worcester Cathedral.
- Political influence. This is extremely important if geoconservation legislation and policies are to be promoted. A good example is the United Kingdom's "All Party Parliamentary Group for the Earth Sciences" established in 2000 by Prof. Allan Rogers, former MP for Rhondda in Wales. This has regular meetings to learn about topical earth science issues from experts in fields such as radioactive waste disposal, flooding, coastal protection, and so on.

6.10   Conclusions

In this chapter, we have seen some of the ways in which geoconservation is being extended beyond protected areas, spatially into the wider landscape and thematically into natural-resource management. It has been argued by some that the protected area approach described in Chapter 5 works well for densely populated lowland areas and for particular themes such as igneous rocks or palaeontology, whereas the wider approaches outlined in this chapter are more relevant for less populous, upland areas and landscapes. For example, Daly *et al.* (1994) argued that in Britain and Ireland, the preponderance of low relief areas, poor rock exposure and high level of threat led to a strongly protected area approach, whereas in Norway with its high relief, good rock exposure and generally low level of threat, there has been an emphasis on protecting landscapes and landform systems. However, they point out that in Norway, Quaternary deposits and geomorphological features are often locally threatened, particularly in the more populous southern coastal zone of the country, and in Scotland, there are analogies with the situation in most of Norway. They therefore conclude that the differences in current approach appear more to do with pragmatism and tradition rather than fundamental differences in the types of features or landscapes. It is easy to agree with this conclusion and to emphasise that geoconservation needs both site-based and wider approaches if it is to maximise its effectiveness.

Doyle & Bennett (1998, p. 56) believe that "However, well framed the statutory powers of site protection may be, however much effort is expended on site safeguard.... it is clear that these will be ineffective without a public will to conserve". They argue that what is needed is a "twin-strand approach" in which one strand aims to establish a protected-area network within the context of a natural areas framework, while the other strand raises awareness of geology and its conservation. As Chapters 5 and 6 have shown, both these strands are highly complex, but it is important that geoscientists become less insular and utilise the many opportunities that arise in public life to promote the interests of geoconservation.

It should also be clear from this chapter that the widening scope of geoconservation is taking it increasingly into traditional fields of land resource management and applied geology and geomorphology, and to more integrated approaches to environmental management. Chapter 7 will return to this issue after first discussing some comparisons between geodiversity and biodiversity.

7

Comparing and Integrating Geodiversity and Biodiversity

7.1 Introduction

This chapter outlines some important issues relating to geodiversity and biodiversity. First, it discusses some criticisms that have been levelled at the use of the term "geodiversity" but generally rejects them. Secondly, it discusses how geodiversity might be measured and monitored, this being an important current focus for biodiversity. Thirdly, it assesses whether geological extinction is a threat in the same way as in biology. Fourthly, having established geodiversity as an important and independent field of nature conservation, it goes on to explore the paradox that it needs to be better integrated with biodiversity. But it then goes further to argue for this integrated nature conservation to be part of a truly holistic approach to land management planning, more generally. Finally, it examines the possibility of conflicts between geodiversity and biodiversity conservation and discusses ways of resolving these conflicts.

7.2 Criticisms of "Geodiversity"

We began this book with a discussion of how the term "geodiversity" had emerged in response to the spectacular success of "biodiversity" in regalvanising efforts in wildlife conservation. Although this can be seen as "jumping on the biodiversity bandwagon", it is also the case that geoconservationists have come to realise that "diversity" is a basic guiding principle underlying all nature conservation, not just bioconservation. As noted in Section 5.19, the concept of representative sites has been long recognised in geoconservation but the obvious point that what is needed is a representative sample of geodiversity had not dawned until the focus on biodiversity made it obvious.

But as we noted in Chapter 1, there have been criticisms of the use of the term. These criticisms may be summarised as follows:

1. We "may be attempting to draw too strong a parallel between sites, landscape features and processes in biology and geology", particularly, given "the amounts of time commonly involved" (Joyce, 1997, p. 39).
2. There is "a contrast between an active ecosystem involving living organisms which may come to an end completely e.g. by the death of the organisms, and the non-living processes of the landscape or earth's subsurface, where processes may stop, restart, and change their nature with no parallel to the death or extinction of living organisms" (Joyce, 1997, p. 39).

Geodiversity: valuing and conserving abiotic nature   Murray Gray
© 2004 John Wiley & Sons, Ltd   ISBNs: 0-470-84895-2 (HB); 0-470-84896-0 (PB)

3. The heritage significance of a geological site, landform or region may, in some cases, "lie not in its diversity but in its uniformity. Examples might be a gibber plain, an active dune field, a thick and uniform sedimentary sequence, a mass extinction fossil site" (Joyce, 1997, p. 39). Similarly, Stock (1997, p. 42) believes that we should "avoid the notion that 'geo-diversity' has some objective quality to earth scientists as 'biodiversity' may have to biologists and ecologists. Diversity of landscapes (say) is not necessarily an over-riding priority in heritage".

4. There appears to be "no scientific advantage in trying to force geodiversity into being of the same conceptual type as biodiversity, which is based on biological and ecological theory involving species, genera, families, etc. Indeed, there could be considerable intellectual disadvantages in such usage" (Stock, 1997, p. 42).

In response to these criticisms, it is accepted that the nature of biodiversity and geodiversity are not identical. Kiernan, (1996, p. 12) for example, notes that "there are important differences between the way landforms come into being and are related to one another and the evolutionary relationships that exist between biological organisms". But it is not accepted that this precludes the use of the term "geodiversity". The important issue is whether it fulfils a particular role or leads to new insights, and it is the contention of this book that exploring and valuing the planet in terms of its geodiversity achieves precisely these objectives.

It is also important to recognise a glass that is half full rather than half empty; there are important similarities between biodiversity and geodiversity. For example, the concept of soil "orders" is established in soil classification as in biological classification (Kiernan, 1997a, p. 4). The terms "species" and "varieties" have long been used for minerals and are an essential part of palaeontology. We can also recognise landform assemblages as the equivalent of plant or animal assemblages. As Kiernan (1996, p. 12) points out, "There are different types and assemblages or systems of landforms just as there are species and communities of plants and animals, and, like plants and animals, there are some landforms that are common and others that are rare, some that are robust and some that are fragile. . . . The case for conserving geodiversity is just as compelling as the case for conserving biodiversity, whether undertaken for intrinsic or for utilitarian reasons, whether judged on scientific, moral or economic terms".

Joyce's comment on timescales is unclear since it should be noted that processes in biology and geology can both occur on very short or very long timescales. The point about uniformity being impressive is understood but is matched in biology. The spectacle of a dune field stretching as far as the eye can see or in repeated geological patterns (Fig. 7.1) are matched in biology by swarms of monarch butterflies, flocks of flamingos on African lakes or herds of caribou in arctic North America. As Sharples (2002a) has pointed out, what is important is the overall geodiversity of the planet and not of particular systems.

In Chapter 3, we noted that geodiversity underpins biodiversity and is its life-support system, therefore creating an inextricable link between the two. Geodiversity also gives us a strong integrating framework for the earth sciences, enabling us to view rocks, minerals, sediments, fossils, landforms, soils and processes as comprising a geosphere with a range of roles and values that are in need of protection and conservation. There is also the political point made by Semeniuk (1997, p. 54) that "it is necessary to make strong connections between biodiversity and geodiversity if politicians, planners and decision makers are to be convinced that geodiversity has a place in general conservation".

Figure 7.1 *Repetition of one aspect of geodiversity can be impressive. Tertiary basalt columns cut across by marine erosion, Gaint's Causeway, Northern Ireland*

Table 7.1   *The main measures of biodiversity*

α-diversity	Species numbers in each habitat
ß-diversity	Proportion of habitats in which a given species is present
γ-diversity	Total number of species in a region
Habitat diversity	Number of habitats in an area
Age diversity	Frequency distribution of ages of a species in an area or habitat
Genetic diversity	A measure of clonal variation in a species

As described in Section 3.5, societies have often made major progressive leaps by exploiting the abiotic materials they discovered in the natural environment (cf Stone Age, Iron Age, Bronze Age, Computer Age). Just as biologists often argue the case for conserving plants and animals because of their future potential uses to society, a similar argument can be applied to the abiotic world, and this provides a further reason for conserving geodiversity.

7.3   Measuring Geodiversity

A huge amount of work has been carried out worldwide in measuring biodiversity, drawing up lists of threatened species and implementing habitat action plans and species-recovery programmes. In comparison to this, the measurement of geodiversity has barely begun. Biologists refer to at least six measures of biodiversity (Table 7.1). This section discusses some ideas, initiatives and issues related to the measurement of geodiversity.

7.3.1   Spatial diversity

Vincent (in Prosser, 2002c) believes that we should measure geodiversity on an index based on grid squares, thus answering questions about which area of land is the most geodiverse. But it is unclear how this could be done for all elements of geodiversity given the differences between biodiversity and geodiversity. For some aspects, it would be relatively easy. In palaeontology, for example, we could simply catalogue the fossil species discovered and establish rarity and vulnerability at international, national or more local levels. As we saw in Section 4.11, it is likely that we would discover many threatened and vulnerable fossil species. Similarly, international stratotypes have been identified and their loss would be fairly obvious. In the case of minerals, the task would be less straightforward because of the many mineral species and secondary species, but nonetheless, it could be done in much the same way. In the case of rocks, landforms, processes and soils, we would have to rely on internationally agreed classification systems. These exist for rocks and soils, but not yet for landforms and processes, though there are many commonly used landform and process names. Hardest of all would be landscapes, since these are ill defined and continuous, but there are similarities here with habitat definition and measurement. However, in both this situation and elsewhere, we start confronting issues of scale.

An important distinction between geodiversity and biodiversity also becomes evident here. In the geosciences, we place great value on interpretations of, for example, rocks and landforms and these may change through time as ideas and theories develop. Measurement of geodiversity must therefore involve more than simply counting the number of different rocks and landforms and must also include the interpretations that we place upon them as well as the interrelationships between them.

7.3.2   Geoindicators

Geoindicators are measures of change in geological or geomorphological resources or systems, and are the equivalent of bioindicators. They might include changes in the area of wetland, the quality of surface or groundwater, glacier mass balance or rates of coastal erosion. For example, the US National Park Inventory and Monitoring Program has established a system using "vital signs" as measurable indicators of changes to resource condition, particularly for palaeontological resources (Koch *et al.*, 2002). Repeat photography is being used by USGS research scientists at the Glacier National Park to assess glacier shrinkage in response to climate change.

The Geoindicators initiative (GEOIN) is an international project directed by the International Union of Geological Sciences (IUGS) to monitor and assess rapid geological change (100 years or less). Box 7.1 shows the 27 selected Geoindicators with some being simple, single parameters (e.g. shoreline position) and others being a complex of measures (e.g. frozen ground activity). There is also a standard checklist aimed at achieving standardisation of reporting (Berger & Iams, 1996). Several examples of the implementation of the scheme are included in a special issue of *Environmental Geology* (Berger & Satkunas, 2002).

Separate attempts are also being made in Tasmania to monitor some sites by taking regularly repeated measurements of critical parameters at a site to determine whether the integrity of a feature or process is being adversely affected by disturbance, or alternatively, is being maintained or restored by appropriate management practices (C. Sharples, personal communication).

The concept of physical process change is, of course, very familiar to gemorphologists, but the novelty is that it is now being used to provide information for ecosystem monitoring and environmental management, both within and beyond protected areas (Berger, 1996). This is a further indication that ecologists are recognising the importance of geoprocesses to the health of ecological systems.

Box 7.1   *Geoindicators Selected by the IUGS (and Some Environmental Changes They Reflect)*

Cryosphere

- Frozen ground activity – surface and groundwater hydrology, permafrost melting, thermokarst, glacier fluctuations – precipitation, insolation, river runoff

Arid and semi-arid zones

- Desert surface crust and fissures – aridity
- Dust storm magnitude, duration and frequency – dust transport, aridity, land use
- Dune formation and reactivation – wind speed and direction, moisture, sediment availability
- Wind erosion – land use, vegetation cover

Marine and coastal zones

- Coral chemistry and growth patterns – surface-water temperature, salinity
- Relative sea level – coastal subsidence and uplift, fluid withdrawal, sedimentation and compaction
- Shoreline position – coastal erosion, land use, sea levels, sediment transport and deposition

Lakes

- Lake levels and salinity – land use, streamflow, groundwater circulation

Rivers and streams

- Streamflow – precipitation, basin discharge, land use
- Stream channel morphology – sediment load, flow rates, land use, surface displacement
- Stream sediment storage and load – sediment transport, flow rates, land use, basin discharge

Wetlands

- Wetland extent, structure and hydrology – land use, biological productivity, streamflow

Surface and groundwater

- Surface water quality – land use, water–rock–soil interactions, flow rates
- Groundwater quality – land use, pollution, rock and soil weathering, acid precipitation, flow rates
- Groundwater chemistry in the unsaturated zone – weathering, land use
- Groundwater level – abstraction, recharge
- Karst activity – groundwater chemistry and flow, vegetation cover, fluvial processes

Continued on page 352

Continued from page 351

Soils

- Soil quality – land use, chemical, biological and physical soil processes
- Soil and sediment erosion – surface runoff, wind, land use

Natural hazards

- Landslides and avalanches – slope stability, mass movement, land use
- Seismicity – natural and human-induced release of earth stresses
- Volcanic unrest – near-surface movement of magma, heat flow, magmatic degassing

Other

- Sediment sequence and composition – land use, erosion and deposition
- Surface displacement – land uplift, faulting, fluid extraction
- Surface-temperature regime – heat flow, vegetation cover, land use

7.3.3   Geological extinction?

In Section 3.5, we saw that there is economic value in the diversity of geological materials. In Section 6.5, we noted that the material resources of the planet are finite and the principles of sustainable development dictate that we should use these resources wisely for the sake of future generations who might also want to use them. This is very important since most geological resources are non-renewable, or renewable only on very long timescales. According to Milton (2002, p. 123), "Biodiversity conservation is... about sustaining the widest possible variety of living things – protecting rare and vulnerable organisms and their habitats, trying to prevent extinction". The question arises therefore, whether there is a geological equivalent of biological extinction. This will occur where a particular type of rock, sediment, mineral, fossil, landform, process, landscape or soil has been so heavily eroded, quarried, utilised, impacted or degraded that it is in danger of disappearing completely. One example of this is the highest quality, pure white marble from the famous Carrara marble quarries in North Italy that has been utilised by sculptors through the centuries, for example, Michaelangelo's *David* in Florence. This resource has recently become exhausted and lower-quality marble is now being used (*The Times*, 26 March 2001).

In Section 6.5, we discussed the over-exploitation of mineral resources and in Section 4.11, we saw how over-collecting, including the use of hammers and power tools, has devastated many important fossil and mineral sites. In some cases, this has led to loss of fossils. Swart (1994, p. 321) believes that over-collecting at the world famous Ediacara Fossil Reserve in South Australia has lead to the virtual destruction of not just the site but also the loss of many of the fossil species. This immediately raises three particular issues:

1. Most of the fossil species from the site have been curated in museums or exist in private collections. Therefore, we must distinguish between field extinction of fossils, minerals and rocks and extinction more generally.

2. It is difficult to prove field extinction since other examples of the same fossils, minerals or rocks may exist at depth at a site, or in geologically unexplored parts of the world, or even beneath the Greenland or Antarctic ice sheets, for example.
3. Again we run into issues of scale, since extinction locally and nationally may also be regarded as important, as well as global extinction.

In comparing biological and geological extinction, Pemberton (2001a) points out that "in a lot of instances, rare or threatened species can be propagated or bred in captivity. On the contrary, many geo features have formed under conditions, climatic or geological, that are now inactive. They are essentially relict or 'fossil' features which, once disturbed, will never recover or will be removed forever". In similar vein, Kiernan (1991) comments that since many landforms are relict features, "They are irreproducible, at least over a human timescale. Conservation of genetic material in laboratories or botanical gardens is not an option in conservation of these landforms, only recognition and appropriate management of the stocks we have now". The fact is that most georesources are non-renewable and it is simply not possible to recreate many aspects of geodiversity once they are lost. For example, if the international reference section of the Ordovician–Silurian boundary at Dobb's Linn in Scotland is lost, it is lost forever. This means that the case for conserving geodiversity is often stronger than that for biodiversity.

7.3.4   Increasing geodiversity

It is estimated that there are still millions of wildlife species awaiting discovery, description and classification. To some extent this is also true for geodiversity. As was noted in Section 1.4, as well as losing geodiversity, it is also possible to see it increase because of field research, mapping and discoveries or by restoration (Section 6.3).

New mineral discoveries are still being made. For example, at the Broken Hill, silver–lead–zinc ore deposit in western New South Wales, Australia, geoscientists from Museum Victoria have reported over 70 secondary mineral species, three of which are new (mawbyite, segnitite and kintoreite). At the Lake Boga Granite in northern Victoria, amongst the 50 phosphate species are at least two new species (ulchrite and bleasdaleite) and at the nearby Wycheproof Granite, two new zirconium minerals have been discovered (wycheproofite and selwynite). These cases illustrate the scope for the recognised geodiversity of minerals to increase significantly in future. Since most of these new minerals are likely to be rare, it follows that they could be easily lost without geoconservation.

Similarly in palaeontology, discoveries of new species are very common. For example, Xu *et al.* (2003) have recently described a new species of basal dromaesaurid dinosaur (*Microraptor gui*) that had four wings and could probably glide, representing an intermediate stage towards active, flapping-flight. There is no doubt that there are probably thousands of new fossil species still undiscovered. As noted above, other elements of geodiversity have less clear classification systems and new names, for example, for rocks, soils and landforms are likely to be introduced in future.

7.4   Integrating Geodiversity and Biodiversity

Having put forward the case for valuing and conserving geodiversity in this book, it is essential to end it with a plea for greater integration of nature conservation objectives, recognising the roles of both geodiversity and biodiversity. This is not to say that there is no risk in linking geodiversity too closely to biodiversity since the former may come to be seen as simply the foundation of biodiversity rather than being valuable in its own right. This should be avoided. As Sharples (2002a) states, "geodiversity has important conservation values of its own, independent of any role in sustaining living things".

Several authors have, however, reached the conclusion that integrated nature conservation is important and some have begun using terms such as biophysical or geoecological management, clumsy though these are. Pemberton (2001a) has made the point that "Treating the entire natural environment as an intricately related system is quite logical from a land management and conservation perspective and quite obvious from the way the natural environment works.... To be serious and consistent about nature conservation there needs to be a return to the approach where the entire natural environment is considered not just the aboveground or living part of the environment". As Carson (1996, p. 8) points out, the Victorian "naturalists" understood the value of integration but "Somewhere the two were separated.... Working towards biodiversity offers an opportunity to remarry these separate, yet complimentary, facets of conservation". One example of a strategy aimed at achieving just such integration in nature conservation management is the draft *Nature Conservation Strategy* for Tasmania, Australia. "By considering all the elements of nature conservation in a holistic way this Strategy is unique in Australia" (DPIWE, 2001b, p. i).

At the global scale, recognition of the interdependency of geological and biological systems leads us inevitably back to the Gaia philosophy, pioneered by Lovelock (1979, 1995) and outlined in Chapter 3. As Manning (2001, p. 20–21) puts it, "...we owe to James Lovelock the general recognition that life has not been merely a passenger on the planet; it has affected certain crucial processes throughout most of Earth's history". This integrated approach to geology and biology has recently found favour with scientists throughout the world through the development of Earth Systems Science, which investigates the linkages through geological time between lithosphere, biosphere, atmosphere and hydrosphere. A special meeting on this theme, jointly organised by the Geological Society of London and the Geological Society of America, was held in Edinburgh, Scotland in 2001 (Kump, 2001). This new thinking has been stimulated by new observational tools (e.g. remote sensing), new observational platforms (e.g. satellites), new analytical methods (e.g. chemical and biochemical analyses) and new theories and models, aided by increased computational power (Lamb & Singleton, 1998; Boulton, 2001).

At the more local level, Hopkins (1994) commented that the ecological research community has paid little attention to the geologically influenced patterns of plant distributions, partly because experimental investigation and interpretation has proved so difficult. However, he suggests that "if biodiversity is to be developed scientifically, it is now timely for ecologists, geologists and geomorphologists to collaborate more closely on the interactions between species habitats and geological phenomena" (p. 3). In other words, an important first step in achieving integration of biodiversity and geodiversity is to establish more clearly, the scientific links between the two.

Detailed research of this type is beginning to be done. Moles & Moles (2002), for example, have carried out detailed geochemical work on the soils of part of The Burren National Park, Ireland, and then related the variations to biodiversity (Moles & Moles, 2003). At the landscape scale (km²), distribution patterns of vegetation and species are linked to bedrock geology and the distribution of glacial sediments.

At the landform element scale (ha), weathering and mass movement processes have created a pattern of alkaline soils in exposed limestone till near drumlin crests, and more acid soils in colluvium on lower slopes. This is illustrated in Fig. 7.2. At a patch scale ($10s\,m^2$), geological influences on vegetation include the occurrence of patches of clay-rich soil derived from glacially transported shale boulders. Finally, at the site scale ($1-10\,m$), boulder-strewn surfaces inhibit access by grazing animals (hares, goats, cattle) thus resulting in more successful seed production than in exposed grassland. Their conclusion is that effective conservation of biodiversity in The Burren requires management action to preserve soil and sediment heterogeneity in order to maintain biodiversity.

A second example of research linking geodiversity with biodiversity concerns the dynamics of fluvial floodplains and their restoration (Fig. 7.3). Richards *et al.* (2002) have noted that declining floodplain biodiversity reflects the influence of river management methods and changes of land use that have separated rivers from their riparian zones and floodplains. They argue that the existence, maintenance and restoration of total floodplain ecosystem diversity (i.e. the combination of habitat, species and age diversities) reflect the continued functioning of the channel dynamics at the reach scale. Smaller-scale processes are also likely to be restored if the freedom of the channel to migrate and adjust its pattern is reinstated. However, current simulation models rely on the maintenance of a stable channel as a model boundary condition. They therefore argue that there is a need for interdisciplinary simulation modelling of the interaction between channel and ecosystem dynamics. For example, successful manipulation of the flood pulse for ecological restoration requires more information on

- the optimal timing of relationships between seed production and flooding;
- critical soil-water potentials for the germination of seeds;
- critical rates of soil water decline to optimise seedling performance; the response of community dynamics to hydrological processes.

Restoration of geomorphological dynamics will require more reliable data on

- the relationship between rates of channel change, floodplain turnover and habitat diversity for different channel patterns;
- the balance of seed-based regeneration and vegetative reproduction in relation to erosion, and sedimentation and their effects on genetic diversity;
- the effects of different riparian vegetation on bank erosion rates;
- the rates and patterns of sedimentation within riparian vegetation during overbank flow.

Richards *et al.* (2002, p. 575–576) conclude that there is "a need for closer collaboration between aquatic and terrestrial biologists and fluvial geomorphologists, to inform the choice of restoration aims, policy and practice, and to ensure that the research and data needs are met for restoration of the appropriate dynamics at the appropriate scale....". As well as plant diversity, Sadler *et al.* (2004) have demonstrated the importance of the diversity of exposed riverine sediments and the flood pulse for beetle diversity in England and Wales.

Sharples (2002b) also foresees the need for more research to identify in detail the role of geodiversity in sustaining ecological systems and the extent to which changes in abiotic variables affect biodiversity:

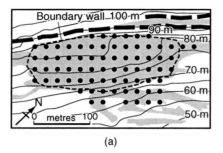

(a)

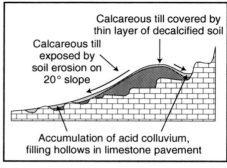

Calcareous till covered by
thin layer of decalcified soil

Calcareous till
exposed by
soil erosion on
20° slope

Accumulation of acid colluvium,
filling hollows in limestone pavement

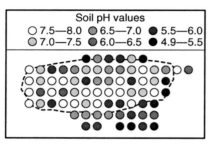

Soil pH values

○ 7.5—8.0   ◑ 6.5—7.0   ● 5.5—6.0
◎ 7.0—7.5   ◕ 6.0—6.5   ● 4.9—5.5

(b)

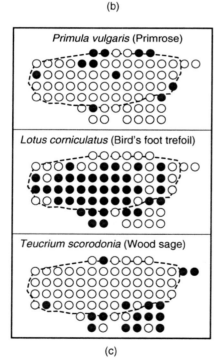

Primula vulgaris (Primrose)

Lotus corniculatus (Bird's foot trefoil)

Teucrium scorodonia (Wood sage)

(c)

Figure 7.2 *Relationship between geodiversity and biodiversity in part of The Burren National Park, Ireland. (Reproduced with permission from Norman Moles)*

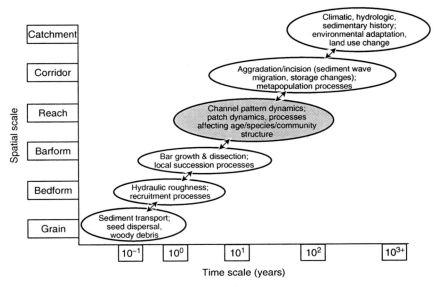

Figure 7.3 *Relationships between geodiversity and biodiversity in fluvial environments (After Richards et al., 2002 Geomorphic dynamics of floodplains: ecological implications and a potential modelling strategy. Freshwater Biology, Blackwell Science Ltd)*

"... if we accept the value of geodiversity in ecological processes then we need to determine the role which particular elements of geodiversity play in ecological processes. Such a determination is a matter of scientific research and monitoring, which allows us to determine whether the disturbance, degradation or destruction of a 'geo-phenomenon' will result in an unacceptable degree of change or degradation to the broader natural environment and ecological processes of which it is part; if so then it is a thing of significant geoconservation value, which should be managed accordingly to avoid such detrimental effects" (Sharples, 2002b, p. 6).

Gordon *et al.* (2001) stress the need for an integrated "geo-ecological" approach to the conservation management of fragile upland environments that are sensitive to both natural and human impacts, and the same issues apply to mountains in the Czech Republic and Sweden (Gordon *et al.*, 2002).

7.5 Integrated Land Management

It is also necessary to go beyond the integration of geodiversity and biodiversity to see nature conservation as an important part of land management and planning in the widest sense. Usher (2001) uses three spheres, "geosphere", "biosphere" and "anthrosphere" to illustrate the integrated nature of the issues that confront nature conservation (Fig. 7.4) emphasising the need for integration of natural and cultural landscapes and management objectives. Valuing and conserving geodiversity must be at least an equal partner in this approach: in fact, it is difficult to envisage a workable land management approach that does not include geoconservation. As Gordon *et al.* (1994, p. 188) argue, "Conservation solutions that are sympathetic not only to the intrinsic scientific interest of the geomorphological resource, but also to the dynamics and inherent sensitivity of geomorphological systems, are essential elements in any holistic strategy for sustainable environmental management".

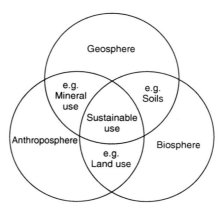

Figure 7.4   *The overlapping nature of the geosphere, biosphere and anthrosphere (Reproduced by permission of Scottish Natural Heritage)*

Figure 7.5   *The Integrated Coastal Zone Management Project for beach and dune systems in Donegal, Ireland*

Examples of holistic approaches both in theory and in practice are beginning to emerge. The Pan-European Biological and Landscape Diversity Strategy (Council of Europe, 1996) emphasises the value of landscapes as a mosaic of cultural, ecological and geological features. Integrated Catchment Management and Integrated Coastal Zone Management (Sorensen, 1993, 2000; Westmacott, 2002) are now widely applied. An example of the latter is the management project being carried out on the beach and dune systems of Donegal, Ireland, by Donegal County Council and the University of Ulster (Fig. 7.5). Similarly, English Nature has proposed a new integrated approach to coastal management (Covey & Laffoley, 2002) and Scottish Natural Heritage's *Natural Heritage Futures* Project (Scottish Natural Heritage, 2002) has also aimed for integration. The latter sees sustainable use of natural resources by including such measures as giving priority to reuse of brownfield sites, promoting public transport, cycling and walking, achieving greater energy efficiency and recycling materials. However, it is important that these new approaches involve a fundamental rethinking of what a fully integrated approach should be rather than a bolting-on of a few token geoconservation words and phrases. Boxes 7.2, 7.3 and 7.4 give examples of fully integrated land-use planning projects from Canada, Australia and Scotland.

Box 7.2   Integrated Land Management Strategy, Oak Ridges Moraine, Ontario, Canada

The Oak Ridges Moraine is a large interlobate moraine deposited at the margin of the Laurentide ice mass in southern Ontario during the Late Wisconsinian Stage of the Pleistocene. The moraine is 2 to 20 km wide, 50 to 250 m high and extends for over 160 km, parallel to the Lake Ontario shoreline. It performs several functions:

- It has a "diverse and complex landscape character, which offers an aesthetic contrast to a regional landscape of lacustrine plains dominated by agricultural and urban development" (Frazer & Johnson, 1997, p. 18).
- It forms the headwaters and aquifer recharge area for the region's surface-water and groundwater supplies.
- It contains very large sand and gravel deposits close to the developing conurbations of southern Ontario.
- It has significant biodiversity.

The moraine has been increasingly subject to development pressures in the Greater Toronto area where the population is predicted to rise from 4.25 million in 1997 to 6.5 million by 2022. Because of these development pressures on a valuable geological, ecological and landscape resource, the Provincial Government undertook a long-term planning study of the moraine as a landscape unit and involved local residents by forming a Citizens Advisory Committee to work alongside a Technical Working Committee. The aim was to try to gain consensus on the future sustainable planning of the area and to produce an integrated land-use plan. Geology and geomorphology were prominent in the formulation of the plan. More than 12 Earth Science Areas of Natural and Scientific Interest (ANSIs) have been selected to protect significant features, but the plan provides conservation strategies for the entire moraine landscape. "Any land use changes requires a Landform Conservation Plan which must convince the approval body that the essential landform character will be maintained" (Frazer & Johnson, 1997, p. 19). The Landform Conservation System is accompanied by a Natural Heritage System to sustain biodiversity, and a Water Resources System to sustain surface and groundwater resources. Local planning policies have been amended to require municipalities to have regard for natural heritage values in judging planning applications. This is an example of the development of integrated land-use strategies, landscape character assessment, planning policies and partnership working (Davidson *et al.*, 2001).

Box 7.3   Integrated Catchment Management at Mole Creek Karst, Tasmania, Australia

The Mole Creek karst area lies in northern Tasmania (Fig. 7.6) and many of the rivers flow underground. The area has been subject to a number of problems characteristic of

Continued on page 360

Continued from page 359

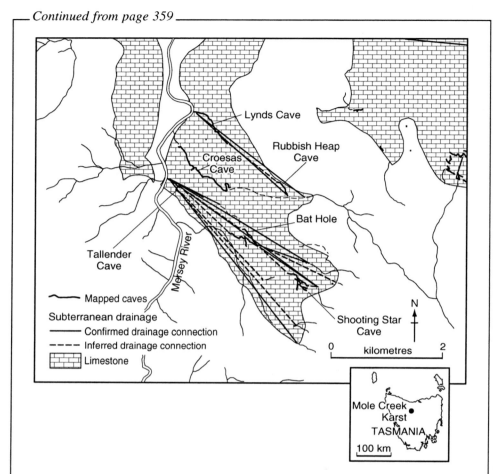

Figure 7.6 _Map of the Mole Creek Karst Area, Tasmania, Australia. (Map Compiled by and reproduced with permission from Rolan Eberhard)_

karst in agricultural areas (see Fig. 4.8), including vegetation removal, soil erosion and cave sedimentation, stock access to streams and reductions in water quality, tipping of chemical drums and stock carcasses in caves and sinkholes, and stream damming and diversion (Eberhard & Houshold, 2001). In addition, a number of potentially damaging operations have been proposed in recent years, including logging and intensive pig rearing. Landownership is mixed with private ownership being predominant in the lower more fertile lands and public ownership in the higher areas. However, the situation is more complex with public areas also occurring on lower areas, including the Mole Creek Karst National Park, and these complexities of landownership make land management more difficult where surface operations can damage geomorphological interests, both underground and downstream.

As a result of these problems, Tasmania's Department of Primary Industries, Water and Environment (DPIWE) in co-operation with Australia's National Heritage Trust, is developing an integrated catchment management strategy for the Mole Creek karst. The elements of the strategy include the following:

Continued on page 361

_ Continued from page 360 _

- *Inventory and zoning.* This involves systematic documentation and GIS storage of geomorphology and hydrology by mapping of surface landforms and caves and water tracing with dyes. This is followed by sensitivity zoning, a map-based tool for modelling the management implications of spatial variations in geomorphology, hydrology and ecology across the area.
- *Cross tenure protocol.* Agreement is being sought from private owners and public agencies for a protocol to promote sustainable land management approaches and practices in the karst area in the knowledge of its special value and sensitivity. A project development committee has been established with representatives from state organisations and agencies, local authorities, farmers and other landowners.
- *Education.* Community awareness is being promoted through publications, media publicity and formation of a "karstcare" group to start restoring cave degradation resulting from past recreational activities inside caves.
- *Ground works.* This includes fencing off karst features and streams, revegetation of cleared land and construction of off-stream water storage to avoid stock access to streams (Eberhard & Houshold (2001).

Box 7.4   Cairngorms River Project, Scotland

The Cairngorms Rivers Project was established by a partnership of organisations in the Central Highlands of Scotland in the late 1990s in order to help develop an integrated and sustainable management strategy, which balances the economic, social and environmental pressures on the Cairngorms rivers. These pressures include

- economic pressures relating to fishing, distilling, agriculture, forestry and tourism;
- social pressures relating to the value of rivers in supporting working communities and providing recreation and amenity opportunities for local people and visitors; and
- environmental pressures such as land-use change within the catchments and the potential for climate change.

In Phase 1 of the Project cover, 130 existing management plans and policies drawn up by the large number of organisations involved in managing the Cairngorms rivers were collated. "Each has its own remit, objectives, priorities and timescales" and there was concern about the proliferation of rivers-based initiatives in the area. The need for greater integration and collaboration was therefore compelling (Walker, 2000). A hierarchy of plans was proposed with European, UK and Scottish legislation and policy providing the fundamental framework for a Cairngorms Partnership Rivers Strategic Management Plan. This would then underpin individual river, area or sectoral plans, which would in turn provide the basis for individual estate, enterprise, farm and forest plans (Fig. 7.7).

An important feature identified in Phase 1 was the need for integrated river basin management rather than the traditional sectoral, localised and fragmented approach, often on a reach-by-reach basis. This has been given stimulus by the EU Water Framework Directive, which will require a whole catchment approach, the setting of environmental objectives for all surface-water and groundwater resources, and the management and protection of water quantity, water quality and physical habitat. The creation of the Cairngorms National Park is a further major management initiative that had to be taken into account.

_ Continued on page 362 _

Continued from page 361

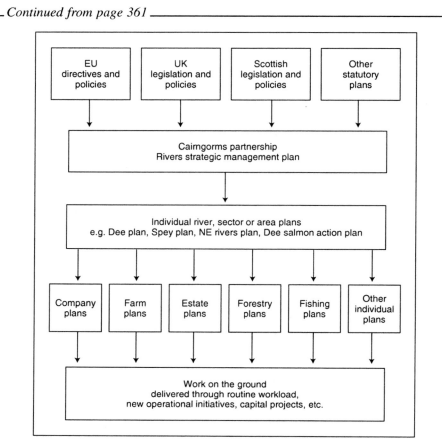

Figure 7.7   *Hierarchy of Legislation and Plans to show the place of the Cairngorms Partnership Rivers Strategic Management Plan (Reproduced with permission from Walker, S. (2000) Cairngorms Rivers Project: Phase 1 Report. University of Aberdeen, Aberdeen)*

Phase 1 first identified the key issues facing the Cairngorms rivers. These included water quality, water quantity and abstraction, groundwater/surface-water interactions, impact of land-use change especially afforestation, sediment erosion and transport, flood risk, landscape value of rivers, river engineering and climate change. Opportunities for enhanced management were then recognised including

- mechanisms to agree to improvements in abstraction agreements;
- need for a strategic cross-sectoral forum for river management interests;
- appropriate tools to assess the value of management action before/after implementation;
- need for a water focus in Cairngorms National Park proposals;
- need for a co-ordinated programme of river maintenance;
- need to encourage a more consistent approach to "in-river" construction works.

Phase 2 is addressing the formulation of the strategic management plan.

7.6 Potential Geodiversity/Biodiversity Conflicts

Having established the case for greater integration of geodiversity and biodiversity, it would be wrong to suggest that this will always be easy and straightforward. There are likely to be situations in which the interests of the two will probably conflict. Conflict is a common occurrence in nature conservation, for example, where there have been animal introductions that disturb existing populations (Milton, 2002). Recently, hedgehogs, introduced into the Scottish Outer Hebrides in the 1950s, have been culled because of their alleged cause of declining numbers of ground-nesting wading birds' eggs.

Some biologists have taken the view that since biodiversity is about "sustaining the widest possible variety of living things" (Milton, 2002, p. 123), the aim of Biodiversity Action Plans should be to maximise biodiversity, perhaps by restructuring landscapes to provide a greater range of habitats. For example, it has been suggested that in river restoration, the main aim should be to maximise the types of habitat in order to maximise species richness. However, these measures can conflict with the objective of retaining and restoring authentic landforms and respecting local landscape character and is certainly contrary to Articles 21 to 23 of The Australian Natural Heritage Charter (see Box 6.18). In other words, it needs to be appreciated by biologists and others that the aim of river restoration is not just about enhancing habitats and increasing species richness but is more fundamentally about restoring the river because of the existence and aesthetic values of a natural river. There are many examples of pool and riffle sequences being introduced into sluggish rivers where they quickly become silted up. There therefore needs to be a greater understanding of geomorphological processes amongst landscape planners and an integrated approach to landscape management.

There is also a tendency for some biologists to see natural geomorphological processes as threatening to biodiversity, and therefore to argue for preservation of the *status quo*. A recent UNEP-WCMC Report (2002) on mountain areas saw South America's Mountains as "particularly vulnerable to 'destructive earthquakes' with approximately 88% of the mountain land area deemed at risk". Similarly, much has been written on the impact of climate change on wildlife, but climate and sea level have been changing continuously, albeit rather more slowly, since the world formed and plant and animal life has had to adapt. Indeed, it could be argued that continuous planetary change and the operation of earth surface processes have been important drivers in creating heterogeneity and new habitats and in delivering modern biodiversity. Habitats may change, but change is part of nature.

A further example of the conflicts that may arise is illustrated by a situation that has arisen at an area known as Blakeney Freshes in Norfolk, England (Box 7.5).

In order to resolve these conflicts, it will be important to assess the separate and relative values associated with the biodiversity and geodiversity concerned. In doing so, there will inevitably be situations where it becomes more important to conserve the element of geoheritage. There may also be situations where the operation of a geological process threatens an element of geoheritage (e.g. coastal erosion destroying a restricted fossil site), and again a careful assessment on a case-by-case basis will have to be used to resolve the issue. Conflicts between geoheritage and archaeological/historical sites also occur, one example being Holt Castle in Wales where, works to shore up the collapsing walls would be likely to obliterate the Holt Castle Quarry RIGS site with its exposure of the Holt–Coddington fault (C. Burek, personal communication).

Box 7.5   *Management Options Proposed by the Environment Agency, North Norfolk, UK*

The North Norfolk coast of the United Kingdom is posing a number of problems for coastal management. The area is of significant ecological importance with SSSI, Nature Reserve, Ramsar site, SPA and cSAC designations. In the area near Cley, the coastline is dominated by a barrier beach (Blakeney Ridge) of shingle. Landward of this are freshwater marshes (Cley Marshes and Blakeney Freshes) protected from river flooding by embankments. At the landward edge of the marshes are the villages of Salthouse and Cley linked by the coast road (Fig. 7.8).

Two problems occur in this area. First, the Blakeney Ridge is raised and steepened each year by bulldozing the shingle in order to make it high enough to prevent sea flooding of the villages, road and marshes. This is costly and produces an artificially steep bank. Even so, inundation occurs every few years. Secondly, the Blakeney Ridge is moving landward at *c*.1 m/year so that it is encroaching on the tidal River Glaven near Blakeney Chapel. At this point, the river is being squeezed between the Ridge and the embankment to the south (Fig. 7.8). It has already suffered blockages during storm events and could be permanently blocked within the next 20 years. This would threaten the freshwater ecology of the Blakeney Freshes. The conflicts in this area are therefore between allowing the coastline to evolve naturally or continuing to try to protect freshwater habitats and private property in the villages.

Solutions proposed to the first problem in the 1990s were as follows:

1. Construct a clay bank parallel to the Blakeney Ridge that would be a more stable structure capable of protecting the villages and most of the freshwater marsh whilst allowing the Ridge to evolve naturally.
2. Construct two separate embankments enclosing Salthouse and Cley and allow the coast and marshes to the north to evolve naturally as saltwater environments.
3. Raise the existing A149 road on an embankment serving both as a defence for the villages and communication during flood events.

After consultation, the Environment Agency opted for the first option (Fig. 7.8) but this has been reassessed because of its expense and landscape impact and has now been abandoned in favour of more local flood alleviation schemes.

A number of options were also proposed by the Environment Agency to solve the second problem (see Fig. 7.8):

1. Do nothing	Allow channel to continue along existing alignment without artificial intervention.
2. Minimum dredge	Dredge existing channel when required and re-profile shingle.
3. Defend channel	Install sheet piling to limit southward shingle movement.
4. Train channel	Strengthen embankments to encourage new outfall through Ridge.
5. New channel	New 1-km channel bypassing pinch point, and new embankment on the south side of new channel.
6. New channel	New 2-km channel through middle of Blakeney Freshes on line of old creek. New embankments on either (a) both sides or (b) south side only.
7. Breach of spit	Create engineered breach of the Ridge to create a new outfall.

Continued on page 365

_ Continued from page 364 _

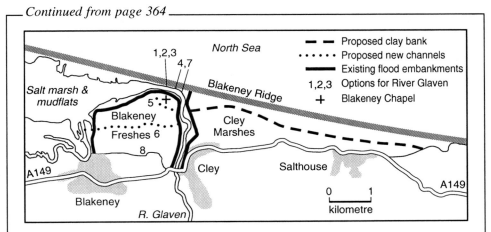

Figure 7.8 *Management options for the Blakeney Freshes and Cley Marshes, Norfolk, England (Reproduced by permission of the Environment Agency)*

8. Managed realignment   Create new channel as in Options 5 or 6 but with no embankments, thus allowing saltwater flooding of Blakeney Freshes.

Options 1 to 3 are likely to be ineffective and/or expensive in the long term. Options 4 and 7 are geomorphologically unsound since they would disrupt existing coastal systems, including longshore drift and would be expensive to maintain. Options 5 and 6 sustain at least part of the freshwater marsh habitat but from a geomorphological point of view, Option 8 is preferable, thus allowing the coastal system to operate in the most natural way. However, this would turn the ecologically important freshwater marsh into a saltmarsh and is opposed by ecologists. Nonetheless, it is beginning to be realised that the long-term protection of the freshwater marshes is likely to be unsustainable as sea level rises and that the long-term goal (>100 years) should be to restore the entire North Norfolk coastline to natural processes. A Coastal Habitat Management Plan (ChaMP) is being produced as required in cases of protected sites where conservation of existing interests *in situ* is not possible due to natural changes to dynamic shorelines.

7.7 Conclusions

Despite objections from some geologists, the term and concept of "geodiversity" appear to have validity and to reflect much common practice. We have barely begun to measure geodiversity but it is important to do so, so that we may know our heritage and whether it is threatened. Geoconservation should not only be established as an independent field of nature conservation but also needs to be integrated with biodiversity conservation and both need to become part of land management strategies more generally, resolving any conflicts along the way by careful assessment of values.

8

Towards a Vision
for Geodiversity Conservation

8.1   Valuing and Conserving Geodiversity

Several conclusions arise from this book:

- The Earth has not only a huge biodiversity but also an enormous geodiversity and this diversity is of value both intrinsically and in utilitarian ways that are barely appreciated by human societies.
- Not all of the geodiversity of the planet needs to be conserved: only those elements that are seen as being significant for one or more of their values. This proportion of total geodiversity is known as the geoheritage.
- There are significant threats to this geoheritage from a variety of human activities either directly or through influencing the rates of operation of earth surface processes. Different aspects of geodiversity vary in their sensitivity and vulnerability to different types of threat.
- Protection of geoheritage is a complex task involving decisions on sites to be included, boundary identification, site documentation, monitoring and enforcement. Most countries now agree that the principal aim of geoconservation should be to maintain a representative sample of a country's or area's geodiversity. While there are significant efforts at protection and management at the international level and in several nations, sub-nations and regions, some countries have barely begun the processes of geoconservation.
- But even this is unlikely to be sufficient to protect geodiversity and a number of approaches are being used to assess and conserve the wider geodiversity resource and to make decision makers and the public aware of this geoheritage, for as Hillel (1991, p. 9) has put it, "we cannot protect what we do not understand".
- Geoconservation needs to be established as an independent subject in its own right and also to be integrated with wildlife conservation in a holistic approach to nature conservation, and beyond that, to comprehensive and sustainable land management and natural resource strategies.

As part of this wider approach, Crofts (2001) discusses the following six principles for sustaining the Earth's resources:

1. Accept that natural change is inevitable.
2. Work with natural functions and processes.

Geodiversity: valuing and conserving abiotic nature   Murray Gray
© 2004 John Wiley & Sons, Ltd   ISBNs: 0-470-84895-2 (HB); 0-470-84896-0 (PB)

3. Manage natural systems within the limits of their capacity.
4. Manage natural systems in a spatially integrated manner.
5. Use non-renewable Earth resources wisely and sparingly at a rate that does not restrict future options.
6. Use renewable resources within their regeneration capacity.

Although it is possible to add, subtract or modify these principles, they present the basis for a future strategy for sustaining geodiversity and ought to be embedded in each country's nature conservation strategy.

The cover of this book shows the Devil's Tower in Wyoming, USA, which illustrates many of the principles in this book. It demonstrates a diversity of mineralogy, jointing, talus features and wildlife habitats. It is of value in a great many ways – cultural, aesthetic, functional and scientific. It has threats from the impact of climbing but it has been protected as a National Monument and with a Climbing Code of Practice. Ultimately, however, it will disappear through the operation of natural mass movement processes by which the columns topple and accumulate as talus.

8.2   A Vision for Geodiversity Conservation

So what should our vision be for geodiversity and geoconservation for, say, 2025, bearing in mind that the world had barely heard of the terms "biodiversity" or "sustainable development" 20 years ago? First, we need to establish our motivation for valuing and conserving abiotic nature. Gordon & Leys (2001b, p. 7) help us with this by pointing out that the justification for geoconservation rests on two main principles:

- the duty to future generations to preserve our heritage so that it may become theirs, and
- the direct benefits for mankind and the natural world from the sustainable use of natural resources.

"Earth science conservation, narrow in outlook and appeal and with a focus on sites, may be viewed as being at a threshold. There is now a wider vision that recognises the integration of Earth and biological systems...and with that recognition, there are opportunities to secure an essential role in sustainable management of the natural heritage and natural resources, and to promote wider public understanding and appreciation" (Gordon & Leys, 2001b, p. 6). Everything we have, we inherited from the past, and everything we use or lose deprives future generations of it. That, in a nutshell, is the case for valuing and conserving abiotic nature.

We might then set an agenda from the international to the local level to implement the elements of best practice outlined in this book. The following is written in the present tense as though in 2025, following the style used by Scottish Natural Heritage (2002) for its Vision statements in the Natural Heritage Futures programme.

- International geoconservation has been re-examined by UNESCO in collaboration with the IUGS, IUCN and other organisations and major efforts have been made to recognise, value, designate and conserve a representative selection of the world's geodiversity. Hundreds of geological and geomorphological sites are now included on the World Heritage Sites list. However, increased development pressures and conflicts over resources have increased the number of World Heritage Sites under threat. A global Geosites network has also been established as an internationally agreed network of several thousand sites recognised for their scientific importance. After a slow start, the Geoparks initiative expanded rapidly and many countries have had to limit the numbers because their territories were becoming almost continuous geoparks. A global Convention on Geodiversity has been signed and a European Directive on Geodiversity has been introduced.

- At the national level, many countries have adopted Natural Heritage Charters similar to the Australian one (Box 6.18) so that geodiversity is valued on a par with biodiversity. The terms "geodiversity", "geoconservation", "geoheritage" and "geoindicators" have been included in many national dictionaries. Most countries have decided to identify a representative selection of the nation's geodiversity and designate these as protected areas. The model being followed is the Geological Conservation Review (GCR) network in Britain (Section 5.10.3), but it recognises more than scientific values. It is proving more difficult to organise "the community, political, legislative, technical and financial support necessary to establish protection mechanisms and make them work" (Kiernan, 1991). Countries are aware that they should
  - produce full documentation on each site, as illustrated by the GCR Site Documentation Programme in Scotland (see Section 5.10.3);
  - specify which activities are permissible within these areas and which are not, in the same way as Alberta and Ontario in Canada have done (Figs. 5.18 and 5.19);
  - legislate in relation to geological theft or damage and enforce the law, with appropriate fines;
  - regularly monitor site condition (see Table 5.7 for an example from Scotland);
  - enhance sites by keeping them clear of vegetation and talus (e.g. the Face Lift scheme in England);
  - where appropriate, produce suitable facilities and interpretive materials as outlined in Section 6.9 (visitor centres, displays, books, maps, leaflets, interpretive panels, geological trails etc.).

Both static and dynamic sites are being recognised, and some countries have established protection for specific types of feature, along the lines of the heritage/scenic rivers in the United States and Canada, caves in the United States and Austria or limestone pavements in the United Kingdom (see Chapter 5).

Many countries have also adopted a "georegionalisation approach", such as those outlined in Section 6.2 for the United Kingdom, Alberta, Tasmania and Italy. This has not only provided a consistent framework for the site conservation network but has also allowed recognition of national landscape diversity. The related descriptions include the geological and geomorphological attributes and values of each spatial unit, and these are being protected from inappropriate developments such as landform remodelling (Section 6.4). Landscape restoration (Section 6.3), the use of natural and generally indigenous geomaterials (Section 6.5.4) and the continued operation of natural dynamic geosystems (Section 6.3) are being encouraged. All these initiatives are being supported by national legislation and policies to conserve the georesource, codes of conduct, land-use planning policies (Section 6.6), environmental impact assessments (which adequately cover geodiversity issues; see Section 6.7) and land management strategies (Section 6.8). The latter have increasingly adopted a holistic approach though not all of them have yet fully integrated geological, biological and cultural objectives (Section 7.6). Geodiversity is included within the "State of the Environment" reports of many countries and a number of them have introduced Geodiversity Action Plans. Governments are working in partnership with mineral and quarrying companies to minimise the use of virgin materials, the environmental impact of mining and the production of waste materials, and the use of fiscal instruments has assisted the sustainable use of natural resources (Section 6.5). Oil production has peaked and prices are rising sharply. There is public outrage that they had not been told that oil was finite and criticisms of governments for not doing more to develop alternative energy sources (Section 6.5.2).

Significant academic research has been undertaken to establish the relationships between geodiversity and biodiversity at all scales from global Earth Systems Science to patch scale dynamics. Several university course modules and MSc degrees in Geodiversity have been established, along with the *Journal of Geodiversity and Geoconservation*. Several books on particular areas or topics, such as *Geodiversity of Tasmania* and *Coastlines of the World* have been published. Sir David Attenborough has presented a television series on *Life on Earth II: the diversity of the fossil record* and several "coffee table" books have been produced in the style of Vernon's (2000) *Beneath our feet: the rocks of planet Earth.*

- At the local level, many regional, county and local governments have established geology audits and geodiversity action plans (Section 6.8.3) for their areas. There is increased use of local designations, such as the US State Parks and Scientific and Natural Areas and the UK Regionally Important Geological Sites (Section 5.10.4), together with land-use plans to protect important sites and landscapes (see Section 6.6). They are doing more than previously to ensure that quarries, river and coastal restoration schemes are appropriate and authentic, and in particular, that quarry restoration conditions retain the geological interest of the quarry (Section 6.3.1). Field managers in various organisations are increasingly receiving training in geoconservation so that they are able to implement policies and codes of practice aimed at conserving geodiversity.

- At the community and individual level, there is increasing appreciation of local geology and landscape and there have been several community campaigns to retain these and prevent inappropriate changes in their physical character. Many more geological societies and clubs have been formed worldwide to encourage local interest and thousands of people now go on geodiversity courses and geotourism holidays and undertake geoconservation work in their spare time. Earth science teaching in schools has stimulated a greater interest in the geosciences amongst the young and more are studying geoscience subjects and are being employed in geoconservation work.

As imaginary as the uniform world with which the book started perhaps, but not impossible with the right strategy, the sufficient resources and the political motivation. It is up to geoscientists to recognise that the concept of geodiversity can

- give us a strong basis for valuing the abiotic world;
- provide the main criterion for geoconservation;
- act as an integrating mechanism for the geosciences;

- promote the importance of the geosciences through their role in sustaining many elements of both modern society and biodiversity;
- promote the role of the geosciences in integrated land resource management.

The Irish Nobel Laureate poet, Seamus Heaney, has talked about "giving glory to things because they are", and for the final words of this book, we quote from the words of the Scottish poet Hugh MacDiarmid. Contemplating the beach shingle above the shoreline on one of the Shetlands Islands, he wrote these important words:

"I look at these stones and know little about them,
But I know their gates are open too,
Always open, far longer open, than any bird's can be,
That every one of them has had its gates wide open far longer
Than all birds put together, let alone humanity....
What happens to us
Is irrelevant to the world's geology
But what happens to the world's geology
Is not irrelevant to us.
We must reconcile ourselves to the stones,
Not the stones to us....
Let men find the faith that builds mountains
Before they seek the faith that moves them".

Hugh MacDiarmid, (1956) *On a raised beach*

References

Abernethy, B. & Wansborough, T.M. (2001) Where does river rehabilitation fit in catchment management? In Rutherford, I., Sheldon, F., Brierley, G. & Kenyon, C. (eds) *Third Australian Stream Management Conference*. Co-operative Research Centre for Catchment Hydrology, Monash, 1–6.

Abrahams, A.D. & Parsons, A.J. (eds) (1994) *Geomorphology of Desert Environments*. Chapman & Hall, London.

Achuff, P.L. (1994) *Natural Regions, Subregions and Natural History Themes of Alberta: A Classification for Protected Areas Management*. Government of Alberta, Edmonton.

Adams, W.M. (1996) *Future Nature: A Vision for Conservation*. Earthscan, London.

Adams, W.M. (1998) Landforms, authenticity and conservation value. *Area*, **30**, 168–169.

Adderley, W.P., Davidson, D.A., Grieve, A.C., Hopkins, D.W. & Salt, C.A. (2001) *Issues Associated with the Development of a Soil Protection Strategy for Scotland*. University of Stirling Report, 81.

Alcalá, L. (1999) Spanish steps towards geoconservation. *Earth Heritage*, **11**, 14,15.

Alcalá, L. & Morales, J. (1994) Towards a definition of the Spanish palaeontological heritage. In O'Halloran, D., Green, C., Harley, M., Stanley, M. & Knill, J. (eds) *Geological and Landscape Conservation*. Geological Society, London, 57–61.

Alexander, D. (1993) *Natural Disasters*. UCL Press, London.

Alexandrowicz, Z. (1993) Earth science conservation in Poland. *Earth Science Conservation*, **33**, 7–9.

Alexandrowicz, Z. (ed) (1999) *Representative Geosites of Central Europe*. Polish Geological Institute, Special Papers 2, Warsaw.

Alexandrowicz, Z. (2002) Report of the Polish ProGEO committee activities. *ProGEO News*, **4**, 8–10.

Allan, J.A. (1995) *Stream Ecology*. Chapman & Hall, London.

Allen, P.A. (1997) *Earth Surface Processes*. Blackwell Science, Oxford.

Alvarez, L.W., Alvarez, W., Asaro, F. & Michel, H.V. (1980) Extraterrestrial cause of the Cretaceous – Tertiary extinction. *Science*, **208**, 1095–1108.

Amadio, V., Amadei, M., Bagnaia, R., Di Bucci, D., Laureti, L., Lisis, A., Lugeri, F.R. & Lugeri, N. (2002) The role of geomorphology in landscape ecology: the landscape unit map of Italy, scale 1:250000 (Carta della Natura Project). In Allison, R.J. (ed) *Applied Geomorphology: Theory and Practice*. Wiley, Chichester, 265–282.

Anderson, R. (2000) *The Great Court and the British Museum*. British Museum Press, London.

Anon (2000) Pebble problems. *Earth Heritage*, **13**, 6.

Anon (2002) Oil security and climate agendas find a connection. *ENDS Report*, **333**, 11.

Arendt, A.A., Echelmeyer, K.A., Harrison, W.D., Lingle, C.S. & Valentine, V.B. (2002) Rapid wastage of Alaska glaciers and their contribution to rising sea levels. *Science*, **297**, 382–386.

Arisci, A., De Waele, J., Di Gregorio, F., Ferrucci, I., Follesa, R. & Piras, G. (2002) The human factor in natural and cultural landscapes: the geomining park of Sardinia. *Abstract for Natural and Cultural Landscapes Conference*. Royal Irish Academy, Dublin, 76.

Asch, T.W.J. & van Dijck, S.J.E. (1995) The role of geomorphology in environmental impact assessment in the Netherlands. In Marchetti, M., Panizza, M., Soldati, M. & Barani, D. (eds) *Geomorphology and Environmental Impact Assessment*. CNR-CS Geodinamica Alpina e Quaternaria, Milan, 63–70.

Ascherson, N. (2002) *Stone Voices: The Search for Scotland*. Granta Books, London.

ASH Consulting Group, (1994) *Coastal Erosion and Tourism in Scotland: A Review of Protection Measures to Combat Coastal Erosion Related to Tourist Activities and Facilities*, Scottish Natural Heritage Review, 12.

Ashurst, J. & Dimes, F.G. (eds) (1990) *Conservation of Building and Decorative Stone*. Butterworth-Heinemann, Oxford.

Ashman, M. & Puri, G. (2002) *Essential Soil Science*. Blackwell, Oxford.

Asplund, D. (1996) Energy use of peat. In Lappalainen, E. (ed) *Global Peat Resources*. International Peat Society, Finland, 319–325.

Attenborough, D. (1990) Foreward. *Nature Conservancy Council Earth Science Conservation in Britain: A Strategy*. Nature Conservancy Council, Peterborough, 5–6.

Attfield, R. (1999) *The Ethics of the Global Environment*. Edinburgh University Press, Edinburgh.

Aust, H. & Sustrac, G. (1992) Impact of development on the geological environment. In Lumsden, G.I. (ed) *Geology and the Environment in Western Europe*. Oxford University Press, Oxford, 202–280.

Australian Heritage Commission, (1996) *Australian Natural Heritage Charter*. 1st ed. Australian Heritage Commission, Canberra.

Australian Heritage Commission, (1999) *Natural Heritage Places Handbook*. Australian Heritage Commission, Canberra.

Australian Heritage Commission, (2002) *Australian Natural Heritage Charter*. 2nd ed. Australian Heritage Commission, Canberra.

Bak, P. (1997) *How Nature Works: The Science of Self-organized Criticality*. Oxford University Press, Oxford.

Balogh, S.J. (1997) Mercury and suspended sediment loadings in the lower Minnesota river. *Environmental Science & Technology*, **31**, 198–202.

Barenttino, D., Vallejo, M. & Gallego, E. (eds) (1999) *Towards the Balanced Management and Conservation of the Geological Heritage in the New Millenium*. Sociedad Geológica de Espana, Madrid.

Barnes, J. (1995) *Basic Geological Mapping*. 3rd ed. Wiley, Chichester.

Baron, J.S., Theobald, D.M. & Fagre, D.B. (2000) Management of land use conflicts in the United States Rocky Mountains. *Mountain Research and Development*, **20**, 24–27.

Barrow, C.J. (1999) *Environmental Management: Principles and Practice*. Routledge, London.

Barry, J. (1999) *Environment and Social Theory*. Routledge, London.

Bartley, R. & Rutherfurd, I. (2001) Statistics, snags and sand: measuring the geomorphic recovery of streams disturbed by sediment slugs. In Rutherfurd, I., Sheldon, F., Brierley, G. & Kenyon, C. (eds) *Third Australian Stream Management Conference.* Co-operative Research Centre for Catchment Hydrology, Monash, 15–22.

Barton, M.E. (1998) Geotechnical problems with the maintenance of geological exposures in clay cliffs subject to reduced erosion rates. In Hooke, J. (ed) *Coastal Defence and Earth Science Conservation.* Geological Society, London, 32–45.

Battarbee, R.W., Flower, R.J., Stevenson, A.C. & Rippey, B. (1985) Lake acidification in Galloway: a palaeoecological test of competing hypotheses. *Nature*, **314**, 350–352.

Battarbee, R.W. (1992) Holocene lake sediments, surface water acidification and air pollution. In Gray, J.M. (ed) *Applications of Quaternary Research.* Wiley, Chichester, 101–110.

Batty, M.H. & Pring, A. (1997) *Mineralogy for Students.* 3rd ed. Longman, Harlow.

Bayfield, N. (2001) Mountain resources and conservation. In Warren, A. & French, J.R. (eds) *Habitat Conservation: Managing the Physical Environment.* Wiley, Chichester, 7–38.

Beavis, S. & Lewis, B. (2001) The impact of farm dams on the management of water resources within the context of total catchment development. In Rutherford, I., Sheldon, F., Brierley, G. & Kenyon, C. (eds) *Third Australian Stream Management Conference.* Co-operative Research Centre for Catchment Hydrology, Monash, 23–28.

Beckerman, W. & Pasek, J. (2001) *Justice, Posterity, and the Environment.* Oxford University Press, Oxford.

Bell, F. (1999) *Geological Hazards.* Spon Press, London.

Bell, F.G., Hälbich, T.F.J. & Bullock, S.E.T. (2002) The effects of acid mine drainage from an old mine in the Witbank Coalfield, South Africa. *Quarterly Journal of Engineering Geology and Hydrogeology*, **35**, 265–278.

Belshaw, C. (2001) *Environmental Philosophy: Reason, Nature and Human Concern.* Acumen, Chesham.

Belsky, A.J. (1995) Spatial and temporal landscape patterns in arid and semi arid savannas. In Hansson, L., Hahrig, L. & Merriance, G. (eds) *Mosaic Landscapes and Ecological Processes.* Chapman & Hall, London, 31–56.

Benn, D.I. & Evans, D.J.A. (1998) *Glaciers and Glaciation.* Arnold, London.

Bennett, M.R. (1994) The management of Quaternary SSSI in England, with specific reference to disused pits and quarries. In Stevens, C., Gordon, J.E., Green, C. P. & Macklin, M.G. (eds) *Conserving our Landscape*, English Nature, Peterborough, 36–40.

Bennett, M.R. & Doyle, P. (1997) *Environmental Geology.* Wiley, Chichester.

Bennett, M.R. & Glasser, N.F. (1996) *Glacial Geology: Ice Sheets and Landforms.* Wiley, Chichester.

Bennett, N. (2003) Brighton rocked by cliff dilemma. *Earth Heritage*, **19**, 7.

Bennison, G.M. & Moseley, K. (2003) *Introduction to Geological Structures and Maps.* 7th ed. Hodder Arnold, London.

Bentley, M. (1996) *Moss of Achnacree and Achnaba Landforms.* Earth Science Documentation Series, Scottish Natural Heritage, Edinburgh.

Benton, M. & Harper, D. (1997) *Basic Palaeontology.* Pearson Education, Harlow.

Berger, A.R. (1996) The geoindicator concept and its application: an introduction. In Berger, A.R. & Iams, W.J. (eds) *Geoindicators: Assessing Rapid Environmental Change in Earth Systems.* Balkema, Rotterdam, 1–14.

Berger, A.R. & Iams, W.J. (eds) (1996) *Geoindicators: Assessing Rapid Environmental Change in Earth Systems*. Balkema, Rotterdam.

Berger, A.R. & Satkunas, J. (2002) Introduction to special issue of environmental geology journal on geoindicators. *Environmental Geology*, **42**, 709–710.

Bergonzoni, M., Vezzani, A., Lugaresaresti, J.I., Soldati, M. & Barani, D. (1995) Environmental impact assessment studies in the regional park of Sassi di Roccamalatina (Northern Apennines, Italy). In Marchetti, M., Panizza, M., Soldati, M. & Barani, D. (eds) *Geomorphology and Environmental Impact Assessment*. CNR-CS Geodinamica Alpina e Quaternaria, Milan, 139–156.

Best, M.G. & Christiansen, E.H. (2001) *Igneous Petrology*. Blackwell Science, Oxford.

Bird, E. (2000) *Coastal Geomorphology*. Wiley, Chichester.

Bland, W. & Rolls, D. (1998) *Weathering: An Introduction to Scientific Principles*. Arnold, London.

Bloom, A.L. (1998) *Geomorphology: A Systematic Analysis of Late Cenozoic Landforms*. 3rd ed. Prentice Hall, New Jersey.

Blunden, J. (1985) *Mineral Resources and their Management*. Longman, London.

Blunden, J. (1991) The environmental impact of mining and mineral processing. In Blunden, J. & Reddish, A. (eds) *Energy, Resources and Environment*. Hodder & Stoughton, London, 79–131.

Boardman, J. (1988) Public policy and soil erosion in Britain. In Hooke, J.M. (ed) *Geomorphology in Environmental Planning*. Wiley, Chichester, 33–50.

Boulton, G.S. (2001) The Earth system and the challenge of global change. In Gordon, J.E. & Leys, K.F. (eds) *Earth Science and the Natural Heritage*. Stationery Office, Edinburgh, 26–54.

Bourassa, S.C. (1992) *The Aesthetics of Landscape*. Bellhaven Press, London.

Bowersox, G.W. & Chamberlin, B.E. (1995) *Gemstones of Afghanistan*. Geoscience Press, Tucson.

Bradbury, J., Cullen, P., Dixon, G. & Pemberton, M. (1995) Monitoring and management of streambank erosion and natural revegetation on the lower Gordon river, Tasmanian world heritage area, Australia. *Environmental Management*, **19**, 259–272.

Bradshaw, M. & Weaver, R. (1993) *Physical Geography*. Mosby, St Louis.

Brady, N.C. & Weil, R.R. (2002) *The Nature and Properties of Soil*. 13th ed. Prentice Hall, New Jersey.

Brampton, A.H. (1998) Cliff conservation and protection: methods and practices to resolve conflicts. In Hooke, J. (ed) *Coastal Defence and Earth Science Conservation*. Geological Society, London, 46–57.

Brancucci, G., Burlando, M. & Marin, V. (2002) The role of geological heritage in the Natura 2000 Network (Habitats Directive, 92/43/EEC): A local study in northern Italy (Liguria Region). *Abstract for Natural & Cultural Landscapes Conference*. Royal Irish Academy, Dublin, 16.

Breeze, D. & Munro, G. (1997) *The Stone of Destiny: Symbol of Nationhood*. Historic Scotland, Edinburgh.

Bridge, J. (2002) *Rivers and Floodplains*. Blackwell, Oxford.

Bridges, E.M. (1994) Soil conservation: an international issue. In O'Halloran, D., Green, C., Harley, M., Stanley, M. & Knill, J. (eds) *Geological and Landscape Conservation*. Geological Society, London, 11–15.

Bridgland, D. (1994) The conservation of Quaternary geology in relation to the sand and gravel extraction industry. In O'Halloran, D., Green, C., Harley, M., Stanley, M.

& Knill, J. (eds) *Geological and Landscape conservation.* Geological Society, London, 87–91.

Brierley, G.J. & Fryirs, K. (2001) Creating a catchment-framed biophysical vision for river rehabilitation programs. In Rutherford, I., Sheldon, F., Brierley, G. & Kenyon, C. (eds) *Third Australian Stream Management Conference.* Co-operative Research Centre for Catchment Hydrology, Monash, 59–66.

Brilha, J. (2002) Geoconservation and protected areas. *Environmental Conservation,* **29,** 273–276.

Bristow, C.M. (1994) Environmental aspects of mineral resource conservation in southwest England. In O'Halloran, D., Green, C., Harley, M., Stanley, M. & Knill, J. (eds) *Geological and Landscape Conservation.* Geological Society, London, 79–86.

British Geological Survey, (2001) *Building Stone Resources of the United Kingdom.* British Geological Survey, Keyworth.

Brizga, S. & Finlayson, B. (2000) *River Management: The Australasian Experience.* Wiley, Chichester.

Bromley, R.G. (1996) *Trace Fossils: Biology, Taphonomy and Applications.* 2nd ed. Chapman & Hall, London.

Bronowski, J. (1973) *The Ascent of Man.* BBC Publications, London.

Brookes, A. (1985) River channelisation. *Progress in Physical Geography,* **9,** 44–73.

Brookes, A. (1986) Response of aquatic vegetation to sedimentation downstream from channelisation works in England and Wales. *Biological Conservation,* **38,** 351–367.

Brookes, A. (1988) *Channelised Rivers: Perspectives for Environmental Management.* Wiley, Chichester.

Brookes, A. (1996) River restoration experience in northern Europe. In Brookes, A. & Shields, F.D. (eds) (1996a) *River Channel Restoration.* Wiley, Chichester, 233–267.

Brookes, A., Baker, J. & Redmond, C. (1996) Floodplain restoration and riparian zone management. In Brookes, A. & Shields, F.D. (eds) (1996a) *River Channel Restoration.* Wiley, Chichester, 201–229.

Brookes, A. & Sear, D.A. (1996) Geomorphological principles for restoring channels. In Brookes, A. & Shields, F.D. (eds) (1996a) *River Channel Restoration.* Wiley, Chichester, 75–101.

Brookes, A. & Shields, F.D. (eds) (1996a) *River Channel Restoration: Guiding Principles for Sustainable Projects.* Wiley, Chichester.

Brookes, A. & Shields, F.D. (1996b) Perspectives on river channel restoration. In Brookes, A. & Shields, F.D. (eds) (1996a) *River Channel Restoration.* Wiley, Chichester, 1–19.

Brookes, A. & Shields, F.D. (1996c) Towards an approach to sustainable river restoration. In Brookes, A. & Shields, F.D. (eds) (1996a) *River Channel Restoration.* Wiley, Chichester, 385–402.

Bruce-Burgess, L. (2001) Appraisal: the 'acid test' of river restoration. In Rutherfurd, I., Sheldon, F., Brierley, G. & Kenyon, C. (eds) *Third Australian Stream Management Conference.* Co-operative Research Centre for Catchment Hydrology, Monash, 81–86.

Brunsden, D. & Thornes, J.B. (1979) Landscape sensitivity and change. *Transactions of the Institute of British Geographers,* **NS4,** 463–484.

Brussard, P.F., Murphy, D.D. & Noss, R.F. (1992) Strategy and tactics for conserving biodiversity in the United States. *Conservation Biology,* **6,** 157–159.

Buck, S.J. (1998) *The Global Commons: An Introduction.* Earthscan, London.

Buckeridge, J.S. (1994) Geological conservation in New Zealand: options in a rapidly eroding environment. In O'Halloran, D., Green, C., Harley, M., Stanley, M. & Knill, J. (eds) *Geological and Landscape Conservation* Geological Society, London, 263–269.

Bull, L. & Kirkby, M. (eds) (2002) *Dryland Rivers*. Wiley, Chichester.

Bullard, J. & Livingstone, D. (2002) Interactions between aeolian and fluvial systems in dryland environments. *Area*, **34**, 8–16.

Burd, F. (1995) *Managed Retreat: A Practical Guide*. English Nature, Peterborough.

Burek, C. (2001) Non-geologists now dig geodiversity. *Earth Heritage*, **16**, 21.

Burek, C. & Conway, J. (2000) Limestone pavements: managing a fragile heritage. In Oliver, P. (ed) *Proceedings of the second UK RIGS Conference*. Hereford & Worcester RIGS Group, Worcester, 37–55.

Burek, C. & Legg, C. (1999) Grike-roclimates. *Earth Heritage*, **12**, 10.

Burek, C. & Potter, J. (2002) Minding the LGAPs. *Geoscientist*, **12**(9), 16–17.

Burnett, M.R., August, P.V. Brown, J.H. & Killingbeck, K.T. (1998) The influence of geomorphological heterogeneity on biodiversity. 1. A patch scale perspective. *Conservation Biology*, **12**, 363–370.

Bush, D.M., Pilkey, O.H. & Neal, W.J. (1996) *Living by the Rules of the Sea*. Duke University Press, Durham, London.

Butcher, N.E. (1994) The work of the Lothian and Borders RIGS Group in Scotland. In O'Halloran, D., Green, C., Harley, M., Stanley, M. & Knill, J. (eds) *Geological and Landscape Conservation*. Geological Society, London, 343–345.

Caitcheon, G., Prosser, I., Wallbrink, P., Douglas, G., Olley, J., Hughes, A., Hancock, G. & Scott, A. (2001) Sediment delivery from Moreton Bay's main tributaries: a multifaceted approach to identifying sediment sources. In Rutherfurd, I., Sheldon, F., Brierley, G. & Kenyon, C. (eds) *Third Australian Stream Management Conference*. Co-operative Research Centre for Catchment Hydrology, Monash, 103–108.

Cairney, T. (1995) *The Re-use of Contaminated Land: A Handbook of Risk Assessment*. Wiley, Chichester.

Cairns, J. (1991) The status of the theoretical and applied science of restoration ecology. *The Environmental Professional*, **13**, 186–194.

Campbell, C.A. & Laherrère, J.H. (1998) The end of cheap oil. *Scientific American*, **278**(3), 60–65.

Campbell, M. & McCall, J. (2002) RIGS' first decade. *Geoscientist*, **12**(8), 12–13.

Campbell, S. & Bowen, D.Q. (1989) *Quaternary of Wales*. Nature Conservancy Council, Peterborough.

Campbell, S. & Wood, M. (2002) Scientific vandalism? *Earth Heritage*, **17**, 3.

Cameron, W., Conover, K. & Micic, S. (2001) Tools and techniques used by the Bulimba creek catchment coordinating committee to improve the Bulimba creek. In Rutherford, I., Sheldon, F., Brierley, G. & Kenyon, C. (eds) *Third Australian Stream Management Conference*. Co-operative Research Centre for Catchment Hydrology, Monash, 109–114.

Canup, R.M. & Asphaug, E. (2001) Origin of the moon in a giant impact near the end of the Earth's formation. *Nature*, **412**, 708–712.

Carroll, C., Rohde, K., Millar, G., Dougall, C., Stevens, S., Ritchie, R. & Lewis, S. (2001) Neighbourhood catchments: a new approach for achieving ownership and change in catchment and stream management. In Rutherford, I., Sheldon, F.,

Brierley, G. & Kenyon, C. (eds) *Third Australian Stream Management Conference.* Co-operative Research Centre for Catchment Hydrology, Monash, 121–126.

Carson, G. (1996) Biodiversity's underlying influence. *Earth Heritage*, **6**, 8–9.

Carson, R. (1962) *Silent Spring.* Houghton Mifflin, Boston.

Carvalho, L. & Anderson, N.J. (2001) Lakes. In Warren, A. & French, J.R. *Habitat Conservation: Managing the Physical Environment.* Wiley, Chichester, 123–146.

Catto, N. (2002) Anthropogenic pressures on coastal dunes, southwestern Newfoundland. *Canadian Geographer*, **46**, 17–32.

Christopherson, R.W. (2000) *Geosystems.* 4th ed. Prentice Hall, New Jersey.

Clarkson, E.N.K. (1993) *Invertebrate Palaeontology and Evolution.* 3rd ed. Chapman & Hall, London.

Clarkson, E.N.K. (2001) The palaeontological resource of Great Britain – our fossil heritage. In Bassett, M.G., King, A.H., Larwood, J.G. & Parkinson, N.A. (eds) *A Future for Fossils.* National Museums & Galleries of Wales, Cardiff, 10–17.

Cleal, C., Thomas, B., Bevins, R. & Wimbledon, B. (2001) Deciding on a new world order. *Earth Heritage*, **16**, 10–13.

Cleal, C., Thomas, B., Bevins, R. & Wimbledon, B. (2003) The global geosites project in Great Britain. *Geoscientist*, **13**(2), 16–17.

Clifford, N.J. (2001) Conservation and the river channel environment. In Warren, A. & French, J.R. (eds) *Habitat Conservation: Managing the Physical Environment.* Wiley, Chichester, 67–104.

Clifton-Taylor, A. (1987) *The Pattern of English Building.* 4th ed. Faber & Faber, London.

Cobb, J.B. (1988) A Christian view of biodiversity. In Wilson, E.O. (ed) *Biodiversity.* National Academy Press, Washington DC, 481–485.

Collins, C. (ed) (1995) *The Care and Conservation of Palaeontological Materials.* Butterworth-Heinemann, Oxford.

Commission of the European Communities, (2002) *Towards a Thematic Strategy for Soil Protection.* Commission of the European Communities, Brussels.

Conacher, A.J. (2002) Geomorphology and environmental impact assessment in relation to dryland agriculture. In Allison, R.J. (ed) *Applied Geomorphology: Theory and Practice.* Wiley, Chichester, 317–333.

Constanza, R., d'Arge, R., deGroot, R., Farber, S., Gresso, M., Hannon, B., Limburg, K., Naeem, S., O'Neill, R.V., Paruelo, J., Raskin, R.G., Sutton, P. & van den Belt, M. (1997) The value of the world's ecosystem services and natural capital. *Nature*, **387**, 253–260.

Cooke, R.U., Brunsden, D., Doornkamp, J.C. & Jones, D.K.C. (1982) *Urban Geomorphology in Drylands.* Clarendon Press, Oxford.

Cooke, R.U. & Doornkamp, J.C. (1990) *Geomorphology in Environmental Management.* Oxford University Press, Oxford.

Cooke, R.U., Warren, A. & Goudie, A.S. (1993) *Desert Geomorphology.* UCL Press, London.

Coppold, M. & Powell, W. (2000) *A Geoscience Guide to the Burgess Shale.* Yoho – Burgess Shale Foundation, Field, British Colombia.

Council of Europe, (1996) *Pan-European Biological and Landscape Diversity Strategy.* Nature & Environment, Council of Europe Publishing, Strasbourg, 74.

Countryside Commission, (1993a) *Landscape Assessment Guidance (CCP423).* Countryside Commission, Cheltenham.

Countryside Commission, (1993b) *Golf Courses in the Countryside (CP438)*. Countryside Commission, Cheltenham.

Countryside Commission, (1993c) *Design in the Countryside (CP418)*. Countryside Commission, Cheltenham.

Countryside Commission, (1996) *Countryside Design Summaries (CP502)*. Countryside Commission, Cheltenham.

Countryside Commission, (1998) *Planning for Countryside Quality (CCP529)*. Countryside Commission, Cheltenham.

Countryside Council for Wales, (1999) *Tir Gofal: Agri-environment Scheme for Wales*. Countryside Council for Wales, Bangor.

Covey, R. & Laffoley, D.d'A. (2002) *Maitime State of Nature Report for England: Getting onto an Even Keel*. English Nature, Peterborough.

Cowie, J.W. & Wimbledon, W.A.P. (1994) The world heritage list and its relevance to geology. In O'Halloran, D., Green, C., Harley, M., Stanley, M. & Knill, J. (eds) *Geological and Landscape Conservation*. Geological Society, London, 71–73.

Crawford, H., Baines, J., Black, J., Johns, J., Postgate, N., Robson, E. & Treadwell, L. (2003) Bombing could devastate rich remains of ancient cities. *The Independent*, 5 March 2003, 17.

Creaser, P. (1994a) The protection of Australia's fossil heritage through the protection of movable cultural heritage act 1986. In O'Halloran, D., Green, C., Harley, M., Stanley, M. & Knill, J. (eds) *Geological and Landscape Conservation*. Geological Society, London, 69–70.

Creaser, P. (1994b) Australia's world heritage fossil nomination. In O'Halloran, D., Green, C., Harley, M., Stanley, M. & Knill, J. (eds) *Geological and Landscape Conservation*. Geological Society, London, 75–77.

Creaser, P. (1994c) An international earth sciences conservation convention framework for the future. *Australian Geologist*, **90**, 45–46.

Cribb, S. & Cribb, J. (1998) *Whisky on the Rocks*. British Geological Survey, Keyworth.

Crofts, R. (2001) Sustainable use of the earth's resources. In Gordon, J.E. & Leys, K.F. (eds) *Earth Science and the Natural Heritage*. Stationery Office, Edinburgh, 286–295.

Crowther, P.R. (2000) RIGS and Northern Ireland. In Oliver, P. (ed) *Proceedings of the Second UK RIGS Conference*. Hereford & Worcester RIGS Group, Worcester, 141–151.

Cserny, T. (1994) Geological conservation in Hungary. In O'Halloran, D., Green, C., Harley, M., Stanley, M. & Knill, J. (eds) *Geological and Landscape Conservation*. Geological Society, London, 249–253.

Cullingford, J.B. & Nadin, V. (1994) *Town & Country Planning in Britain*. 11th ed. Routledge, London, New York.

Cuomo, C.J. (1998) *Feminism and Ecological Communities*. Routledge, London, New York.

Cutter, S.L. & Renwick, W.H. (1999) *Exploitation, Conservation, Preservation: A Geographic Perspective on Natural Resource Use*. 3rd ed. Wiley, Chichester.

Czudek, T. & Demek, J. (1970) Thermokarst in Siberia and its influence on the development of lowland relief. *Quaternary Research*, **1**, 103–120.

Dady, P. (2002) Time, tide and fossil replicas. *Earth Heritage*, **17**, 22–23.

Daily, G.C. (ed)(1997) *Nature's Services*. Island Press, Washington DC.

Daly, D. (1994) Conservation of peatlands: the role of hydrogeology and the sustainable development principle. In O'Halloran, D., Green, C., Harley, M., Stanley, M. & Knill, J. (eds) *Geological and Landscape Conservation.* Geological Society, London, 17–21.

Daly, D., Erikstad, L. & Stevens, C. (1994) Fundamentals in earth science conservation. *Mémoires de la Société de Geologique de France*, **165**, 209–212.

D'Andrea, M., Lisi, A. & Lugeri, N. (2002) The Italian geosites inventory data-base. *Abstract for Natural And Cultural Landscapes Conference.* Royal Irish Academy, Dublin, 28.

Dasmann, R.F. (1984) *Environmental Conservation.* 5th ed. Wiley, Chichester.

Davidson, J.P., Reed, W.E. & Davis, P.M. (1997) *Exploring Earth.* Prentice Hall, New Jersey.

Davidson, R.J. (2001) Towards an ecological monitoring program for Ontario's provincial parks and conservation reserves. In Porter, J. & Nelson, J.G. (eds) *Ecological Integrity and Protected Areas.* Parks Research Forum of Ontario, Waterloo.

Davidson, R.J., Kor, P.S.G., Fraser, J.Z. & Cordiner, G.S. (2001) Geological conservation and integrated management in Ontario, Canada. In Gordon, J.E. & Leys, K.F. (eds) *Earth Science and the Natural Heritage.* Stationery Office, Edinburgh, 228–233.

Davies, J. & Pearce, N. (1993) Motorways can seriously improve your exposure. *Earth Science Conservation*, **32**, 22–23.

Davies, J.L. (1980) *Geographical Variation in Coastal Development.* 2nd ed. Longman, Harlow.

Davis, G.H. & Reynolds, S.J. (1996) *Structural Geology of Rocks and Regions.* 2nd ed. Wiley, Chichester.

Davis, J.R. & Fitzgerald, D. (2002) *Beaches and Coasts.* Blackwell, Oxford.

Decker, R.W. & Decker, B.B. (1998) *Volcanoes.* Freeman, New York.

Deer, W.A., Howie, R.A. & Zussman, J. (2001) *Feldspars.* 2nd ed. Geological Society, London.

Deffeyes, K.S. (2001) *Hubbert's Peak: The Impending World Oil Shortage.* Princeton University Press, Princeton.

DEFRA, (2001) *Draft Soil Strategy for England and Wales.* Department of the Environment, Food & Rural Affairs, London.

DEFRA, (2002) *The Strategy for Sustainable Farming and Food: Facing the Future.* Department of the Environment, Food & Rural Affairs, London.

Demek, J. (1994) Global warming and permafrost in Eurasia: a catastrophic scenario. *Geomorphology*, **10**, 317–329.

Detwyler, T.R. (1971) *Man's Impact on the Environment.* McGraw-Hill, New York.

Dixon, G. (1995) *Aspects of Geoconservation in Tasmania: A Preliminary Review of Significant Earth Features.* Report to the Australian Heritage Commission, Occasional Paper 32. Parks & Wildlife Service, Tasmania.

Dixon, G. (1996a) *Geoconservation: An International Review and Strategy for Tasmania.* Occasional Paper 35, Parks & Wildlife Service, Tasmania.

Dixon, G. (1996b) *A Reconnaissance Inventory of Sites of Geoconservation Significance on Tasmanian Islands.* Report to Parks & Wildlife Service, Tasmania and Australian Heritage Commission.

Dixon, G. & Duhig, N. (1996) *Compilation and Assessment of some Places of Geoconservation Significance.* Regional Forest Agreement, Commonwealth of Australia and State of Tasmania.

Dixon, G., Houshold, I., Pemberton, M. & Sharples, C. (1997) Wizards of Oz: geo-conservation in Tasmania. *Earth Heritage*, **8**, 14–15.

Dolman, P. (2000) Biodiversity and ethics. In O'Riordan, T. (ed) *Environmental Science for Environmental Management*. 2nd ed. Pearson Education, Harlow, 119–148.

Domas, J. (1994) Damage caused by opencast and underground coal mining. In O'Halloran, D., Green, C., Harley, M., Stanley, M. & Knill, J. (eds) *Geological and Landscape Conservation*. Geological Society, London, 93–97.

Donnelly, L. (2002) Finding the silent witness. *Geoscientist*, **12**(5), 16–18.

Donner, J.J. (1995) *The Quaternary History of Scandinavia*. Cambridge University Press, Cambridge.

Donovan, S. (2002) A karst of thousands: Jamaica's limestone scenery. *Geology Today*, **18**, 143–151.

Doughty, P. (2002) Grottos, granite and gold. *Geodiversity Update*, **4**, 1–3.

Douglas, I., Greer, T., Sinun, W., Anderton, S., Bidin, K., Spilsbury, M., Suhaimi, J. & Bin Sulaiman, A. (1995) Geomorphology and rainforest logging practices. In McGregor, D.F.M. & Thompson, D.A. (eds) (1995) *Geomorphology and Land Management in a Changing Environment*. Wiley, Chichester, 309–320.

Doyle, P. (1989) Threat to Bartonian stratotype. *Earth Science Conservation*, **26**, 27.

Doyle, P. (1996) *Understanding Fossils*. Wiley, Chichester.

Doyle, P. (2001) Restoring a Victorian vision of 'deep time'. *Earth Heritage*, **16**, 16–18.

Doyle, P. & Bennett, M.R. (1998) Earth heritage conservation: past, present and future agendas. In Bennett, M.R. & Doyle, P. (eds) *Issues in Environmental Geology: A British Perspective*. Geological Society, London, 41–67.

Doyle, P. & Bennett, M.R. (2002) *Fields of Battle: Terrain in Military History*. Kluwer Academic Publishers, Dordrecht.

DPIWE, (2001a) *Draft Reserve Management Code of Practice*. Department of Primary Industries, Water and Environment, Hobart, Tasmania.

DPIWE, (2001b) *Draft Nature Conservation Strategy*. Department of Primary Industries, Water and Environment, Hobart, Tasmania.

Drew, D. (2001) *Burren Karst*. Geographical Association, Sheffield.

Duff, K. (1994) Natural Areas: an holistic approach to conservation based on geology. In O'Halloran, D., Green, C., Harley, M., Stanley, M. & Knill, J. (eds) *Geological and Landscape Conservation*. Geological Society, London, 121–126.

Dutton, Y. (1992) Natural resources in Islam. In Khalid, F. & O'Brien, J. (eds) *Islam and Ecology*. Cassell, London, 51–67.

Eberhard, R. (1994) *Inventory and Management of the Junee River Karst System, Tasmania*. Report to Forestry, Tasmania, Hobart.

Eberhard, R. (1996) *Inventory and Management of Karst in the Florentine Valley, Tasmania*. Report to Forestry, Tasmania, Hobart.

Eberhard, R. (ed) (1997) *Pattern & Process: Towards a Regional Approach to National Estate Assessment of Geodiversity*. Australian Heritage Commission, Canberra.

Eberhard, R. & Houshold, I. (2001) River management in karst terrains: issues to be considered with an example from Mole Creek, Tasmania. In Rutherford, I., Sheldon, F., Brierley, G. & Kenyon, C. (eds) *Third Australian Stream Management Conference*. Co-operative Research Centre for Catchment Hydrology, Monash, 197–204.

Eden, S., Tunstall, S.M. & Tapsell, S.M. (1999) Environmental restoration: environmental management or environmental threat? *Area*, **31**, 151–159.

Eder, W. (1999) Geoparks of the future. *Earth Heritage*, **12**, 21.

Edmonds, R. (2001) Fossil collecting on the west Dorset coast: a new voluntary code of conduct. In Bassett, M.G., King, A.H., Larwood, J.G. & Parkinson, N.A. (eds) *A Future for Fossils*. National Museums & Galleries of Wales, Cardiff, 46–51.

Edmonds, R. (2003) Don't miss AONB opportunity. *Earth Heritage*, **19**, 3.

Ehrlich, P.R. (1988) The loss of diversity: causes and consequences. In Wilson, E.O. (ed) *Biodiversity*. National Academy Press, Washington DC, 21–27.

Ellis, J. (ed) (1993) *Keeping Archives*. 2nd ed. Australian Society of Archivists, O'Connor, Australian Capital Territory.

Ellis, N.V., Bowen, D.Q., Campbell, S., Knill, J., McKirdy, A.P., Prosser, C.D., Vincent, M.A. & Wilson, R.C.L. (1996) *An Introduction to the Geological Conservation Review*. Joint Nature Conservation Committee, Peterborough.

ENDS Report, (2001) Secondary aggregates: gearing up for new market opportunities. *ENDS Report*, **323**, 29–31.

Engel, C. (2003) *Wild Health*. Phoenix Books, London.

English Nature, (1993) *Natural Areas: English Nature's Approach to Setting Nature Conservation Objectives: A Consultation Paper*. English Nature, Peterborough.

English Nature, (1998) *Natural Areas: Nature Conservation in Context (CD-Rom)*. English Nature, Peterborough.

English Nature, (1999) *A Series of Eight Regional Natural Areas Publications*. English Nature, Peterborough.

English Nature, (2000a) *The Past is the Key to the Future*. English Nature, Peterborough.

English Nature, (2000b) *National Nature Reserves, The Future: A Policy Statement*. English Nature, Peterborough.

English Nature, (2002) *Revealing the Value of Nature*. English Nature, Peterborough.

English Nature, (2003) *Geodiversity and the Minerals Industry: Conserving our Geological Heritage*. Entec UK Ltd, Newcastle-upon-Tyne.

Environment Agency, (2001) *Best Farming Practices: Profiting from a Good Environment*. Environment Agency, Peterborough.

Erikstad, L. (1994a) The legal framework of earth science conservation in Norway. *Mémoires de la Société Géologique de France*, **165**, 21–25.

Erikstad, L. (1994b) The building of an international airport in an area of outstanding geological diversity and quality. In O'Halloran, D., Green, C., Harley, M., Stanley, M. & Knill, J. (eds) *Geological and Landscape Conservation*. Geological Society, London, 47–51.

Erikstad, L. (1994c) Quaternary geology conservation in Norway, inventory program, criteria and results. *Mémoires de la Société Géologique de France*, **165**, 213–215.

Erikstad, L. (2000) The linchpin safeguarding Norwegian geological interests. *Earth Heritage*, **13**, 12.

Erikstad, L. & Stabbetorp, O. (2001) Natural areas mapping: a tool for environmental impact assessment. In Gordon, J.E. & Leys, K.F. *Earth Science and the Natural Heritage*. Stationery Office, Edinburgh, 224–227.

Essex County Council, (2001) *The Essex Guide to Environmental Impact Assessment*. Essex County Council, Chelmsford.

Evans, A.M. (1997) *An Introduction to Economic Geology and its Environmental Impact*. Blackwell Science, Oxford.

Evans, D.A. (2003) *Glacial Landsystems*. Arnold, London.

Eyles, N. (ed) (1983) *Glacial Geology; An Introduction for Engineers and Earth Scientists*. Pergamon, Oxford.

Feick, J. & Draper, D. (2001) Valid threat or "tempest in a teapot"? An historical account of tourism development and the Canadian Rocky mountain parks world heritage site designation. *Tourism Recreation Research*, **26**, 35–46.

Fennell, D.A. (1999) *Ecotourism: An Introduction*. Routledge, London.

Fiffer, S. (2000) *Tyrannosaurus Sue*. W.H. Freeman & Co, New York.

FISRWG, (1998) *Stream Corridor Restoration: Principles, Processes and Practices*. Federal Interagency Stream Restoration Working Group, Washington DC.

Ford, D.C. & Williams, P.W. (1989) *Karst Geomorphology and Hydrology*. Unwin Hyman, London.

Fordyce, F. & Johnson, C. (2002) The rock diet. *Planet Earth*, Autumn 2002, Natural Environment Research Council, Swindon, 18–20.

Forster, M.W.C. (1999) *An Overview of Fossil Collecting with Particular Reference to Scotland*. Research, Survey and Monitoring, Report 115, Scottish Natural Heritage, Edinburgh.

Forster, M. (2001) Fossils under the hammer: recent US natural history auctions. In Bassett, M.G., King, A.H., Larwood, J.G. & Parkinson, N.A. (eds) *A Future for Fossils*. National Museums & Galleries of Wales, Cardiff, 98–105.

Fortey, R. (1997) *Life: an Unauthorised Biography*. Flamingo, London.

Fortey, R. (2002) *Fossils: The Key to the Past*. 3rd ed. Natural History Museum, London.

Foster, J. (ed) (1997) *Valuing Nature?* Routledge, London.

Fox, W. (1990) *Towards a Transpersonal Ecology: Developing New Foundations for Environmentalism*. Shambala Publications, Boston.

Francis, P. (1993) *Volcanoes: A Planetary Perspective*. Oxford University Press, Oxford.

Frazer, J.Z. & Johnson, F.M. (1997) Earth science heritage protection for the Oak Ridges Moraine, Ontario, Canada. *Earth Heritage*, **7**, 18–19.

Fredericia, J. (1990) Saturated hydraulic conductivity of clayey tills and the role of fractures. *Nordic Hydrology*, **21**, 119–132.

French, H.M. (1996) *The Periglacial Environment*. 2nd ed. Longman, Harlow.

French, J.R. & Reed, D.J. (2001) Physical contexts for saltmarsh conservation. In Warren, A. & French, J.R. (eds) *Habitat Conservation: Managing the Physical Environment*. Wiley, Chichester, 179–228.

French, J.R. & Spencer, T. (2001) Sea-level rise. In Warren, A. & French, J.R. (eds) *Habitat Conservation: Managing the Physical Environment*. Wiley, Chichester, 305–347.

Fry, N. (1992) *The Field Description of Metamorphic Rocks*. Wiley, Chichester.

Fukuyama, F. (2001) We remain at the end of history. *The Independent*, 11 October 2001, 5.

Fulop, J. (1975) The Mesozoic basement horst blocks of Tata. *Geologica Hungarica Series Geologica*. Hungarian Geological Institute, Budapest, 16.

Funnell, B.M. (1994) A geological heritage coast: North Norfolk (Hunstanton to Happisburgh). In Stevens, C., Gordon, J.E., Green, C.P. & Macklin, M.G. (eds) *Conserving our Landscape*. English Nature, Peterborough, 59–62.

Furley, P. (1996) Brazil cerrado. In Anderson, U.G. & Brooks, S.M. (eds) *Advances in Hillslope Processes*. Wiley, Chichester, 327–346.

Gagen, P., Gunn, J. & Bailey, D. (1993) Landform replication experiments on quarried limestone rock slopes in the English Peak District. *Zeitschrift fur Geomorphologie, Supplement-Band*, **87**, 163–170.

Galloway, W.E. (1975) Process framework for describing the morphologic and stratigraphic evolution of the deltaic depositional systems. In Broussard, M.L. (ed) *Deltas: Models for Exploration*. Houston Geological Society, Houston, 87–98.

Garcia-Cortes, A., Rábano, I., Locutura, J., Bellido, F., Fernández-Gianotti, J., Martin-Serrano, A., Quesada, C., Barnolas, A. & Durán, J. J. *et al.* (2001) First Spanish contribution to the Geosites Project: list of the geological frameworks established by consensus. *Episodes*, **24**, 79–92.

Garcia-Guinea, J. & Calafarra, J.M. (2001) Mineral collectors and the geological heritage: protection of a huge geode in Spain. *European Geologist*, **11**, 4–7.

Garrels, R.M. & Mackenzie, F.T. (1971) *Evolution of Sedimentary Rocks*. W.W. Norton, New York.

Gaston, K.J. (1996) *Biodiversity: A Biology of Numbers and Difference*. Blackwell Science, Oxford.

Gauld, J.H. & Bell, J.S. (1997) *Soils and Nature Conservation in Scotland*. Scottish Natural Heritage Review, No. 62.

Gerrard, J. (2000) *Fundamentals of Soils*. Routledge, London.

Giardino, J.R., Schroder, J.F. & Vitek, J.D. (eds) (1987) *Rock Glaciers*. Allen & Unwin, London.

Gibbons, S. & McDonald, H.G. (2001) Fossil sites as national natural landmarks: recognition as an approach to protection of an important resource. In Santucci, V.L. & McClelland, L. (eds) *Proceedings of the 6th Fossil Resource Conference*. Technical Report NPS/NRGRD/GRDTR-01/01, 130–136.

Gilbert, O. (1996) *Rooted in Stone: The Natural Flora of Urban Walls*. English Nature, Peterborough.

Gillen, C. (2001) Tour guides set out on geological time trail. *Earth Heritage*, **16**, 20–21.

Gippel, C.J. Seymour, S., Brizga, S. & Craigie, N.M. (2001) The application of fluvial geomorphology to stream management in the Melbourne water area. In Rutherford, I., Sheldon, F., Brierley, G. & Kenyon, C. (eds) *Third Australian Stream Management Conference*. Co-operative Research Centre for Catchment Hydrology, Monash, 239–244.

Gisotti, G. & Burlando, M. (1998) The Italian job. *Earth Heritage*, **9**, 11–13.

Glasser, N.F. (2001) Conservation and management of the earth heritage resource in Great Britain. *Journal of Environmental Planning and Management*, **44**, 889–906.

Glasson, J., Therivel, R. & Chadwick, A. (1999) *Introduction to Environmental Impact Assessment*. 2nd ed. UCL Press, London.

Glypteris, S. (1991) *Countryside Recreation*. Longman, Harlow.

Goldie, H.S. (1994) Protection of limestone pavement in the British Isles. In O'Halloran, D., Green, C., Harley, M., Stanley, M. & Knill, J. (eds) *Geological and Landscape Conservation*. Geological Society, London, 215–220.

Gomez, N. (1991) La protection des sites a oeufs de dinosaures de la Sainte-Victoire (Aix en Provence, France) (abstract). *Terra Nova*, **3**, 13.

Gonggrijp, G.P. (1993a) Nature policy plan: new developments in the Netherlands. In Erikstad, L. (ed) *Earth Science Conservation in Europe. Proceedings from the Third*

Meeting of the European Working Group of Earth Science Conservation, 5–16. Norsk Institutt for Naturforskning, Oslo.

Gonggrijp, G.P. (1993b) River meanders or houses: a case study. In Erikstad, L. (ed) *Earth science Conservation in Europe. Proceedings from the Third Meeting of the European Working Group of Earth Science Conservation,* 35–38. Norsk Institutt for Naturforskning, Oslo.

González, A., Diaz de Terán, Francés, E. & Cendrero, A. (1995) The incorporation of geomorphological factors into environmental impact assessment for master plans: a methodological proposal. In McGregor, D.F.M. & Thompson, D.A. (eds) (1995) *Geomorphology and Land Management in a Changing Environment.* Wiley, Chichester, 179–193.

Goodwin, R. (1992) *Green Political Theory.* Polity, Cambridge.

Gordon, J.E. (1994a) Conservation of geomorphology and Quaternary sites in Great Britain: an overview of site assessment. In Stevens, C., Gordon, J.E., Green, C.P. & Macklin, M.G. (eds) *Conserving our Landscape.* English Nature, Peterborough, 11–21.

Gordon, J.E. (ed) (1994b) *Scotland's Soils: Research Issues in Developing a Soil Sustainability Strategy.* Scottish Natural Heritage Review, No. 13.

Gordon, J.E. (1996) *Restoration and Management of Mineral Extraction Sites in Quaternary Landforms and Deposits.* Information & Advisory Note, 41. Scottish Natural Heritage, Edinburgh.

Gordon, J.E., Brazier, V. & Lees, R.G. (1994) Geomorphological systems: developing fundamental principles for sustainable landscape management. In O'Halloran, D., Green, C., Harley, M., Stanley, M. & Knill, J. (eds) *Geological and Landscape Conservation.* Geological Society, London, 185–189.

Gordon, J.E., Brazier, V., Thompson, D.B.A. & Horsfield, D. (2001) Geo-ecology and the conservation management of sensitive upland landscapes in Scotland. *Catena,* **42,** 323–332.

Gordon, J.E., Dvorák, I.G., Jonasson, C., Josefsson, M., Kociánová, M. & Thompson, D.B.A. (2002) Geo-ecology and management of sensitive montane landscapes. *Geografiska Annaler,* **84A,** 193–203.

Gordon, J.E. & Leys, K.F. (eds) (2001a) *Earth Science and the Natural Heritage: Interactions and Integrated Management.* Stationery Office, Edinburgh.

Gordon, J.E. & Leys, K.F. (2001b) Earth science and the natural heritage: developing a more holistic approach. In Gordon, J.E. & Leys, K.F. (eds) *Earth Science and the Natural Heritage.* Stationery Office, Edinburgh, 5–18.

Gordon, J.E. & MacDonald, J. (2003) Towards a soil protection strategy for Scotland: an environmental view. In McTaggart, I. & Gairns, L. (eds) *Diffuse Pollution and Agriculture.* Scottish Agricultural College, Edinburgh, 115–120.

Gordon, J.E. & MacFadyen, C.C.J. (2001) Earth heritage conservation in Scotland: state, pressures and issues. In Gordon, J.E. & Leys, K.F. (eds) *Earth Science and the Natural Heritage.* Stationery Office, Edinburgh, 130–144.

Gordon, J.E. & McKirdy, A.P. (1997) Quaternary landforms and deposits as part of Scotland's natural heritage. In Gordon J.E. (ed) *Reflections on the Ice Age in Scotland.* Scottish Natural Heritage, Edinburgh, 179–188.

Gore, J.A. (1996) Foreword. In Brookes, A. & Shields, F.D. (eds) *River Channel Restoration.* Wiley, Chichester, xiii–xv.

Goudie, A.S. (2000) *The Human Impact on the Natural Environment*. 5th ed. Blackwell, Oxford.

Goudie, A.S., Livingstone, I. & Stokes, S. (1999) *Aeolian Environments, Sediments and Landforms*. Wiley, Chichester.

Goudie, A. & Viles, H. (1997) *The Earth Transformed: An Introduction to Human Impacts on the Environment*. Blackwell, Oxford.

Government of Alberta, (1977) *A Policy for Resource Management of the Eastern Slopes*. Government of Alberta, Edmonton.

Government of Alberta, (1994a) *A Framework for Alberta's Special Places: Natural Regions*, Report 1, Government of Alberta, Edmonton.

Government of Alberta, (1994b) *Alberta Protected Areas: System Analysis (1994)*. Government of Alberta, Edmonton.

Government of Alberta, (1996) *Natural History Overview and Theme Evaluation of the Canadian Shield Natural Region*. Government of Alberta, Edmonton.

Government of Quebec, (2002) *Geological Heritage of Quebec: Geosites in Quebec*. Government of Quebec, Montreal.

Graf, W.L. (1988) *Fluvial Processes in Dryland Rivers*. Springer-Verlag, Berlin.

Graf, W. (1992) Science, public policy, and western American rivers. *Transactions of the Institute of British Geographers*, **NS17**, 5–19.

Graf, W. (1996) Geomorphology and policy for restoration of impounded American rivers: what is "natural"? In Rhoades, B.L. & Thorn, C.E. (eds) *The Scientific Nature of Geomorphology: Proceedings of the 27th Binghampton Symposium in Geomorphology*. Wiley, Chichester.

Grant, A. (1995) Human impacts on terrestrial ecosystems. In O'Riordan, T. (ed) *Environmental Science for Environmental Management*. Longman, Harlow, 66–79.

Gray, J.M. (1975) The Loch Lomond Readvance and contemporaneous sea-levels in Loch Etive and neighbouring areas of western Scotland. *Proceedings of the Geologists' Association*, **86**, 227–238.

Gray, J.M. (1993) Quaternary geology and waste disposal in south Norfolk, England. *Quaternary Science Reviews*, **12**, 899–912.

Gray, J.M. (1996) Environmental policy and municipal waste management in the UK. *Transactions of the Institute of British Geographers*, **NS22**, 69–90.

Gray, J.M. (1997a) Planning and landform: geomorphological authenticity or incongruity in the countryside. *Area*, **29**, 312–324.

Gray, J.M. (1997b) The origin of the Blakeney esker, Norfolk. *Proceedings of the Geologists' Association*, **108**, 177–182.

Gray, J.M. (1998a) Landforms, authenticity and conservation value: a reply. *Area*, **30**, 273–274.

Gray, J.M. (1998b) Hills of waste: a policy conflict in environmental geology. In Bennett, M.R. & Doyle, P. (eds) *Issues in Environmental Geology: A British Perspective*. Geological Society, London, 173–195.

Gray, J.M. (2000) Beyond RIGS: geomorphological conservation in the wider countryside. In Oliver, P. (ed) *Proceedings of the Second UK RIGS Conference*. Hereford & Worcester RIGS Group, Worcester, 9–22.

Gray, J.M. (2001) Geomorphological conservation and public policy in England: a geomorphological critique of English Nature's 'Natural Areas' approach. *Earth Surface Processes & Landforms*, **26**, 1009–1023.

Gray, J.M. (2002a) Landraising of waste in England, 1990–2000: a survey of the geomorphological issues raised by planning applications. *Applied Geography*, **22**, 209–234.

Gray, J.M. (2002b) "Land form" rather than "landforms": geomorphological conservation outside protected areas. *Abstract for Natural and Cultural Landscapes Conference*. Royal Irish Academy, Dublin, 44.

Gray, J.M. & Jarman, D. (2004) Creating "glacial" landforms from waste materials: two UK case studies. *Scottish Geographical Journal*, in press.

Greensmith, J.T. (1989) *Petrology of the Sedimentary Rocks*. Unwin Hyman, London.

Gregory, K.J. (2000) *The Changing Nature of Physical Geography*. Arnold, London.

Griffiths, J.S. (ed) (2001) *Land Surface Evaluation for Engineering Purposes*. Geological Society of London, Engineering Geology Special Publication, No. 18.

Grigorescu, D. (1994) The geo-ecological education and geological site conservation in Romania. In O'Halloran, D., Green, C., Harley, M., Stanley, M. & Knill, J. (eds) *Geological and Landscape Conservation*. Geological Society, London, 467–472.

Groombridge, B. (1992) *Global Biodiversity: Status of the Earth's Living Resources*. Chapman & Hall, London.

de Groot, R.S. (1992) *Functions of Nature*. Wolters-Noordhoff, Groningen.

Gross, R. (1998) *Dinosaur Country: Unearthing Alberta Badlands*. Badland Books, Wardlow, Alberta.

Grube, A. (1992) Earth science conservation in Germany. *Earth Science Conservation*, **31**, 16–19.

Grube, A. (1994) The national park system in Germany. In O'Halloran, D., Green, C., Harley, M., Stanley, M. & Knill, J. (eds) *Geological and Landscape Conservation*. Geological Society, London, 175–180.

Guerin, B. (1992) *Review of Scoria and Tuff Quarrying in Victoria*. Geological Survey of Victoria Report 96.

Gulliford, A. (2000) *Sacred Objects and Sacred Places: Preserving Tribal Traditions*. University Press of Colorado, Boulder.

Gunn, J. (1993) The geomorphological impacts of limestone quarrying. *Catena Supplement*, **25**, 187–197.

Gunn, J. (1995) Environmental change and land management in the cuilcagh karst, northern Ireland. In McGregor, D.F.M. & Thompson, D.A. (eds) *Geomorphology and Land Management in a Changing Environment*. Wiley, Chichester, 195–209.

Gunn, J., Hunting, C. & Cornelius, S. (1994) Conserving the geomorphological features of the Cuilcagh karst, Ireland. In Stevens, C., Gordon, J.E., Green, C.P. & Macklin, M.G. (eds) *Conserving our Landscape*. English Nature, Peterborough, 191–194.

Guthrie, M. (2003) Geodiversity – proving its worth. *Earth Heritage*, **19**, 16.

Haas, J. & Hamor, G. (2001) Geological garden in the neighbourhood of Budapest, Hungary. *Episodes*, **24**, 257–261.

Haigh, M.J. (1978) *Evolution of Slopes on Artificial Landforms – Blaenavon, UK*. Research Paper 183, Department of Geography, University of Chicago.

Haigh, M.J. (2002) Land reclamation and deep ecology: in search of a more meaningful physical geography. *Area*, **34**, 242–252.

Hall, A. (1996) *Igneous Petrology*. 2nd ed. Pearson Education, Harlow.

Hallett, D. (2002) The Wieliczka salt mine. *Geology Today*, **18**, 182–185.

Haltinder, J.P., Kondolf, G.M. & Williams, P.B. (1996) Restoration approaches in California. In Brookes, A. & Shields, F.D. (eds) (1996a) *River Channel Restoration*. Wiley, Chichester, 291–329.

Hamblin, W.K. & Christiansen, E.H. (2001) *Earth's Dynamic Systems*. 9th ed. Prentice Hall, New Jersey.

Hansom, J., Crick, M. & John, S. (2000) *The Potential Application of Shoreline Management Planning to Scotland*. Scottish Natural Heritage Review, No. 121.

Hansom, J. & Gordon, J.E. (1998) *Antarctic Environments and Resources: A Geographical Perspective*. Longman, Harlow.

Harcourt, H. (1990) Quality as well as quantity. *Naturopa*, **65**, 4–6.

Hardie, R. & Lucas, R. (2001) Geomorphic categorisation of streams in the Hawkesbury- Nepean catchment: an application and review of the river styles framework. In Rutherfurd, I., Sheldon, F., Brierley, G. & Kenyon, C. (eds) *Third Australian Stream Management Conference*. Co-operative Research Centre for Catchment Hydrology, Monash, 271–276.

Hardwick, P. & Gunn, J. (1994) The conservation of cave systems in mixed lithology catchments: a case study of the Castleton Karst, England. In Stevens, C., Gordon, J.E., Green, C.P. & Macklin, M.G. (eds) *Conserving our Landscape*. English Nature, Peterborough, 198–202.

Hardy, T. (1891) *Tess of the D'Urbervilles*. Reprinted by Pengiun, Harmondsworth.

Hardyment, C. (2000) *Literary Trails: Writers in their Landscapes*. National Trust, London.

Harley, M. (1994) The RIGS (Regionally Important Geological/geomorphological Sites) challenge- involving local volunteers in conserving England's geological heritage. In O'Halloran, D., Green, C., Harley, M., Stanley, M. & Knill, J. (eds) *Geological and Landscape Conservation*. Geological Society, London, 313–317.

Harlow, P.G.L. (1994) Legislation and attitudes to geological conservation in Queensland, Australia: past, present and future. In O'Halloran, D., Green, C., Harley, M., Stanley, M. & Knill, J. (eds) *Geological and Landscape Conservation*. Geological Society, London, 303–308.

Harrison, S.J. & Kirkpatrick, A.H. (2001) Climatic change and its potential implications for environments in Scotland. In Gordon, J.E. & Leys, K.F. (eds) *Earth Science and the Natural Heritage: Interactions and Integrated Management*. Stationery Office, Edinburgh, 296–305.

Harrop, D.O. & Nixon, J.A. (1999) *Environmental assessment in practice*. Routledge, London.

Hart, D.D. & Poff, N.L. (2002) How dams vary and why it matters for the emerging science of dam removal. *Bioscience*, **52**, 659–668.

Haslett, S.K. (2000) *Coastal Systems*. Routledge, London.

Hatfield, C.B. (1997) Oil back on the global agenda. *Nature*, **387**, 121.

Hawksworth, D.L. (1995) *Biodiversity: Measurement and Estimation*. Chapman & Hall, London.

Haynes, V.M., Grieve, I.C., Price-Thomas, P. & Salt, K. (1998) *The Geomorphological Sensitivity of the Cairngorm High Plateaux*. Research, Survey and Monitoring Report, No. 66, Scottish Natural Heritage, Edinburgh.

Haynes, V.M., Grieve, I.C., Gordon, J.E., Price-Thomas, P. & Salt, K. (2001) Assessing geomorphological sensitivity of the Cairngorm high plateaux for conservation purposes. In Gordon, J.E. & Leys, K.F. (eds) *Earth Science and the Natural Heritage.* Stationery Office, Edinburgh, 120–124.

Hayward, B.W. (1989) Earth science conservation in New Zealand. *Earth Science Conservation,* **26,** 4–6.

Heritage Council, (2002) *Integrating Policies for Ireland's Landscape.* The Heritage Council, Kilkenny.

Hester, R.E. & Harrison, R.M. (2001) *Assessment and Reclamation of Contaminated Land.* Thomas Telford, Tonbridge.

Heywood, V.H. & Watson, R.T. (1995) *Global Biodiversity Assessment.* Cambridge University Press, Cambridge.

Hicks, J. (2003) Packing educational punch. *Earth Heritage,* **19,** 18–19.

Hillel, D. (1991) *Out of the Earth: Civilization and the Life of the Soil.* University of California Press, Berkeley.

Hilson, G. (2002) The environmental impact of small-scale gold mining in Ghana: identifying problems and possible solutions. *Geographical Journal,* **168,** 57–72.

Historic Scotland, (1997) *A Future for Stone in Scotland.* Historic Scotland, Edinburgh.

Hodgson, J.M. (1978) *Soil Sampling and Description.* Clarendon Press, Oxford.

Hodson, M.J., Stapleton, C. & Emberton, R. (2001) Soils, geology and geomorphology. In Morris, P. & Therivel, R. (eds) *Methods of Environmental Impact Assessment.* 2nd ed. Spon Press, London, 170–196.

Holman, I.P., Loveland, P.J., Nicholls, R.J., Shackley, S., Berry, P.M., Rounsevell, M.D.A., Audsley, E., Harrison, P.A. & Wood, R. (2002) *REGIS – Regional Climate Change Impact & Response Studies in East Anglia and North West England.* Department of Environment, Food & Rural Affairs, London.

Holstein, T. (2002) Moving to a SEA green paper. *Town & Country Planning,* **71,** 218–220.

Hooke, J. (1994) Conservation: the nature and value of active river sites. In Stevens, C., Gordon, J.E., Green, C.P. & Macklin, M.G. (eds) *Conserving our Landscape.* English Nature, Peterborough, 110–116.

Hooke, J. (1998) Issues and strategies in relation to geological and geomorphological conservation and defence of the coast. In Hooke, J. (ed) *Coastal Defence and Earth Science Conservation.* Geological Society, London, 1–9.

Hooke, J.M. (1999) Decades of change: contributions of geomorphology to fluvial and coastal engineering and management. *Geomorphology,* **31,** 373–389.

Hooke, R.L. (1994) On the efficacy of humans as geomorphic agents. *GSA Today,* **4,** 217–225.

Hopkins, J. (1994) Geology, geomorphology and biodiversity. *Earth Heritage,* **2,** 3–6.

Horner, J.R. & Dobb, E. (1997). *Dinosaur Lives: Unearthing an Evolutionary Saga.* Harcourt Brace & Co, San Diego.

Hose, T. (1998) Selling coastal geology to visitors. In Hooke, J. (ed) *Coastal Defence and Earth science Conservation.* Geological Society, London, 178–195.

Hose, T. (1999) Selling geology to the public. *Earth Heritage,* **11,** 10–12.

Hoskins, W.G. (1955) *The Making of the English Landscape.* Hodder & Stoughton, London.

Houshold, I., Sharples, C., Dixon, G. & Duhig, N. (1997) Georegionalisation – a more systematic approach for the identification of places of geoconservation significance. In Eberhard, R. (ed) *Pattern & Process: Towards a Regional Approach to National Estate Assessment of Geodiversity.* Australian Heritage Commission, Canberra, 65–89.

Howard, A.D. (1967) Drainage analysis in geologic interpretation: a summation. *American Association of Petroleum Geologists Bulletin*, **51**, 2246–2259.

Howard, H. & Partners (1994) *Demolition and Construction Wastes in the UK*. Department of the Environment, London.

Hubbert, M.K. (1956) Nuclear energy and the fossil fuels. *Drilling and Production Practice*. American Petroleum Institute, Washington DC, 7–25.

Huggett, R.J. (1995) *Geoecology: An Evolutionary Approach*. Routledge, London.

Huggett, R.J. (2003) *Fundamentals of Geomorphology*. Routledge, London.

Hughes, F.M.R. & Rood, S.B. (2001) Floodplains. In Warren, A. & French, J.R. (eds) *Habitat Conservation: Managing the Physical Environment*. Wiley, Chichester, 105–121.

Humphreys, D. (2002) From economic to sustainable development: establishing a new framework for mineral extraction, Paper presented to the *"Minerals 2002" Conference of the Institute of Quarrying*, Macclesfield, England, October 2002.

Hunt, E.L.R. (1988) *Managing Growth's Impact on the Mid-south's Historic and Cultural Resources*. National Trust for Historic Preservation, Washington DC.

Huse, S. (1987) The Norwegian river protection scheme: a remarkable achievement of environmental conservation. *Ambio*, **16**, 5.

Hutton, J. (1795) *Theory of the Earth*. William Creech, Edinburgh.

Hurlstone, P. & Long, M. (2000) Striving for a balance. *Earth Heritage*, **14**, 24–25.

Huxley, J. (1947) *Report of the Committee on Nature Conservation in England and Wales*. HMSO, London.

IPCC, (2001) *Climate Change 2001*. Intergovernmental Panel on Climate Change, Geneva.

Iversen, T.M., Kronvang, B., Madsen, B.L., Markmann, P. & Nielsen, M.B. (1993) Re-establishment of Danish streams: restoration and maintenance measures. *Aquatic Conservation: Marine & Freshwater Ecosystems*, **3**, 1–20.

Iversen, T.M., Madsen, B.L. & Bøgestrand, J. (2000) River conservation in the European Community, including Scandinavia. In Boon, P.J., Davies, B.R. & Petts, G.E. (eds) *Global Perspectives on River Conservation: Science, Policy and Practice*. Wiley, Chichester, 70–103.

Jackli, H. (1979) Opening address. In Schlüchter, C. (ed) *Moraines & Varves*. Balkema, Rotterdam, 5–7.

Jarman, D. (1994) Geomorphological authenticity: the planning contribution. In Stevens, C., Gordon, J.E., Green, C.P. & Macklin, M.G. (eds) *Conserving our Landscape*, English Nature, Peterborough, 41–45.

Jarzembowski, E.A. (1989) Writhlington geological nature reserve. *Proceedings of the Geologists' Association*, **100**, 219–234.

Jeffreys, K. (2000) Face lift achieves enhancements. *Earth Heritage*, **14**, 20–21.

Jeffries, (1997) *Biodiversity and Conservation*. Routledge, London.

Jennings, J.N. (1985) *Karst Geomorphology*. Blackwell, Oxford.

Jenny, H. (1941) *Factors of Soil Formation*. McGraw-Hill, New York.

Jerie, K., Houshold, I. & Peters, D. (2001) Stream diversity and conservation in Tasmania: yet another new approach. In Rutherford, I., Sheldon, F., Brierley, G. & Kenyon, C. (eds) *Third Australian Stream Management Conference*. Co-operative Research Centre for Catchment Hydrology, Monash, 329–335.

Jermy, A.C., Long, D., Sand, M.J.S., Stork, N.E. & Winser, S. (eds) (1996) *Biodiversity Assessment: A Guide to Good Practice*. HMSO, London.

Jiang, P. (1994) Conservation of national geological sites in China. In O'Halloran, D., Green, C., Harley, M., Stanley, M. & Knill, J. (eds) *Geological and Landscape Conservation*. Geological Society, London, 243–245.

Johansson, C.E. (ed) (2000) *Geodiversitet i Nordisk Naturvård*. Nordisk Ministerråad, Copenhagen.

Johnson, B.R. (1997) Monitoring peatland rehabilitation. In Parkyn, L., Stoneman, R.E. & Ingram, H.A.P. (eds) *Conserving Peatlands*. CAB International, Oxford, 323–331.

Jones, G. (1996) Planning lead needed when golf boom is back on track. *Planning for the Natural and Built Environment*, **1175**, 26–27.

Jones, G. & Hollier, G. (1997) *Resources, Society and Environmental Management*. Paul Chapman Publishing, London.

Joyce, E.B. (1994) Identifying geological features of international significance: the pacific way. In O'Halloran, D., Green, C., Harley, M., Stanley, M. & Knill, J. (eds) *Geological and Landscape Conservation*. Geological Society, London, 507–513.

Joyce, E.B. (1995) *Assessing the Significance of Geological Heritage*. A methodology study for the Australian Heritage Commission.

Joyce, E.B. (1997) Assessing geological heritage. In Eberhard, R. (ed) *Pattern & Process: Towards a Regional Approach to National Estate Assessment of Geodiversity*. Australian Heritage Commission, Canberra, 35–40.

Joyce, E.B. (1999) Different thinking. *Earth Heritage*, **12**, 11–13.

Jungerius, P.D., Matundura, J. & van de Anger, J.A.M. (2002) Road construction and gulley erosion in west Pokot, Kenya. *Earth Surface Processes & Landforms*, **27**, 1237–1247.

Karis, L. (2002) The Russian perspective. *ProGEO News*, **1**, 8.

Karr, J.R., Allan, J.D. & Benke, A.C. (2000) River conservation in the United States and Canada. In Boon, P.J., Davies, B.R. & Petts, G.E. (eds) *Global Perspectives on River Conservation*. Wiley, Chichester, 3–39.

Katz, E. (1992) The ethical significance of human intervention in nature. *Restoration and Management Notes*, **9**, 90–96.

Kelk, B. (1992). Natural resources in the geological environment. In Lumsden, G.I. (ed) *Geology and the Environment in Western Europe*. Clarendon Press, Oxford, 34–138.

Kemp, T.S. (1999) *Fossils and Evolution*. Oxford University Press, Oxford.

Kenny, J.A. & Hayward, B.W. (eds) (1993) *Inventory of Important Geological Sites and Landforms in the Northland Region*. Geological Society of New Zealand, Miscellaneous Publication, 67.

Kerr, J.W. (1990) *Frank slide*. Baker Publishing Ltd, Calgary.

Kerr, R.A. (1998). The next oil crisis looms large – and perhaps close. *Science*, **281**, 1128–1131.

Khalid, F. & O'Brien, J. (eds) (1992) *Islam and Ecology*. Cassell, London.

Kiernan, K. (1991) Landform conservation and protection, *Fifth Regional Seminar on National Parks and Wildlife Management, Tasmania 1991*, resource document, Tasmanian Parks, Wildlife & Heritage Department, Hobart, 112–129.

Kiernan, K. (1994) *The Geoconservation Significance of Lake Pedder and its Contribution to Geodiversity*. Unpublished Report to the Lake Pedder Study Group.

Kiernan, K. (1995) *An Atlas of Tasmanian Karst*, 2 vols, Report 10 (2 vols), Tasmanian Forest Research Council, Hobart.

Kiernan, K. (1996) *The Conservation of Glacial Landforms*. Forest Practices Unit, Hobart.

Kiernan, K. (1997a) *The Conservation of Landforms of Coastal Origin*. Forest Practices Board, Hobart.

Kiernan, K. (1997b) Landform classification for geoconservation. In Eberhard, R. (ed) *Pattern & Process: Towards a Regional Approach to National Estate Assessment of Geodiversity*. Australian Heritage Commission, Canberra, 21–34.

King, A.H. & Larwood, J.G. (2001) Conserving our most 'fragile' fossil sites in England: the use of OLD25. In Bassett, M.G., King, A.H., Larwood, J.G. & Parkinson, N.A. (eds) *A Future for Fossils*. National Museums & Galleries of Wales, Cardiff, 24–31.

King, R.B. (1987) Review of geomorphic description and classification in land resource surveys. In Gardiner, V. (ed) *International Geomorphology 1986 Part II*. Wiley, Chichester, 384–403.

Kirkbride, M.P., Duck, R.W., Dunlop, A., Drummond, J., Mason, M., Rowan, J.S. & Taylor, D. (2001) *Development of a Geomorphological Database and Geographical Information System for the North West Seaboard: Pilot Study*. Scottish Natural Heritage Commissioned Report BAT/98/99/137, Scottish Natural Heritage, Edinburgh.

Klein, C. (2002) *Mineral Science*. 22nd ed, Wiley, Chichester.

Knighton, A.D. (1998) *Fluvial Forms and Processes*. Arnold, London.

Knill, J. (1994) An international earth heritage convention – convenience or confusion. *Earth Heritage*, **1**, 7–8.

Knutz, A. (1984) The production of road construction material from till. *Striae*, **20**, 99–100.

Koch, A.L., Santucci, V.L. & McDonald, H.G. (2002) *Developing Palaeontological Resource Monitoring Strategies for the National Park Service*. Paper presented to the Geological Society of America Annual Meeting, Denver.

Kondolf, G.M. & Downs, P.W. (1996) Catchment approach to planning channel restoration. In Brookes, A. & Shields, F.D. (eds) (1996a) *River Channel Restoration*. Wiley, Chichester, 129–148.

Korhonen, R. & Lüttig, G.W. (1996) Peat in balneology and health care. In Lappalainen, E. (ed) *Global Peat Resources*. International Peat Society, Finland, 339–345.

Kump, L. (2001) Gaia theory gets a boost at Edinburgh. *Gaia Circular*, Winter 2001/Spring 2002, 19–20.

Kuzmiak, D.T. (1991) A history of the American environmental movement. *Geographical Journal*, **157**, 265–278.

Lake, P.S. (2001) Restoring streams: re-building and re-connecting. In Rutherford, I., Sheldon, F., Brierley, G. & Kenyon, C. (eds) *Third Australian Stream Management Conference*. Co-operative Research Centre for Catchment Hydrology, Monash, 369–372.

Lake District National Park Authority (1998) *Lake District National Park Local Plan*. Lake District National Park Authority, Kendal.

Lamb, S. & Singleton, D. (1998) *Earth Story: The Shaping of our World*. BBC Books, London.

Lancaster, N. (1995) *Geomorphology of Desert Dunes*. Routledge, London.

Landscape Institute, (1995) *Guidelines for Landscape and Visual Impact Assessment*. E. & F.N. Spon, London.

Langford, N. (1870) In Haines, A. (ed) *The Discovery of Yellowstone Park, Journal of the Washburn Expedition to the Yellowstone and Firehole Rivers in the Year 1870.* Republished in 1972 by University of Nebraska Press, Lincoln.

Lappalainen, E. (1996) General review on world peatland and peat resources. In Lappalainen, E. (ed) *Global Peat Resources.* International Peat Society, Finland, 53–56.

Larwood, J. (2003) The camera never lies. *Earth Heritage*, **19**, 10–11.

Larwood, J.G. & King, A.H. (2001) Conserving our most 'fragile' fossil sites in England: the use of OLD25. In Bassett, M.G., King, A.H., Larwood, J.G. & Parkinson, N.A. (eds) *A Future for Fossils.* National Museums & Galleries of Wales, Cardiff, 24–31.

Larwood, J.G. & Prosser, C.D. (1996) The nature of the urban geological resource: an overview. In Bennett, M.R., Doyle, P., Larwood, J.G. & Prosser, C.D. (eds) *Geology on your Doorstep.* Geological Society, London, 19–30.

Larwood, J. & Prosser, C. (1998) Geotourism, conservation and society. *Geologica Balcanica*, **28**, 97–100.

Lees, G. (1997) *Coastal Erosion and Defence: II Coastal Erosion and Coastal Cells.* Information & Advisory Note 72. Scottish Natural Heritage, Edinburgh.

Leopold, A. (1949) *A Sand Country Almanac.* Oxford University Press, New York.

Leopold, S.A. (1959) *Wildlife in Mexico.* University of California Press, Berkeley.

Leopold, L.B., Clark, F.E., Hanshaw, B.B. & Blasley, J.R. (1971) *A Procedure for Evaluating Environmental Impact.* US Geological Survey Circular 645, Department of the Interior, Washington DC.

Levy, A. & Scott-Clark, C. (2001) *Stone of Heaven.* Weidenfeld & Nicholson, London.

Leys, K.F. (2001) Voluntary earth heritage conservation in Scotland: developing the RIGS movement. In Gordon, J.E. & Leys, K.F. (eds) *Earth Science and the Natural Heritage.* Stationery Office, Edinburgh, 279–282.

Limestone Pavement Action Group, (1999) *Managing our Fragile Heritage.* Limestone Pavement Action Group, Cumbria.

Livingstone, I. & Warren, A. (1996) *Aeolian Geomorphology.* Longman, Harlow.

Lloyd, J.W. (1986) Hydrogeology and beer. *Proceedings of the Geologists' Association*, **97**, 213–219.

Lomberg, B. (2001) *The Skeptical Environmentalist: Measuring the Real State of the World.* Cambridge University Press, Cambridge.

Lovejoy, T.E. (1997) Biodiversity: what is it? In Reaka-Kudla, L., Wilson, D.E. & Wilson E.O. (eds) *Biodiversity II: Understanding and Protecting our Biological Resources.* Joseph Henry Press, Washington DC, 7–14.

Lovelock, J. (1979) *Gaia.* Oxford University Press, Oxford.

Lovelock, J. (1995) *The Ages of Gaia.* 2nd ed. Oxford University Press, Oxford.

Lynch, F. (1990) Mynydd Rhiw axe factory. In Addison, K., Edge, M.J. & Watkins, R. (eds) *North Wales Field Guide.* Quaternary Research Association, Coventry, 48–51.

MacArthur, R.H. & Wilson, E.O. (1967) *The Theory of Island Biogeography.* Princeton University Press, Princeton.

MacEwen, A. & MacEwen, M. (1987) *Greenprints for the Countryside: The Story of Britain's National Parks.* Allen & Unwin, London.

MacFadyen, C. (1999) *Fossil Collecting in Scotland.* Information & Advisory Note 110, Scottish Natural Heritage, Edinburgh.

MacFadyen, C. (2001a) Getting to grips with asset strippers. *Earth Heritage*, **15**, 10.

MacFadyen, C. (2001b) Plain-speaking blueprints. *Earth Heritage*, **15**, 11.

MacFadyen, C. & McMillan, A. (2002) The links between our natural and built heritage: how we can use urban landscape and geological resources to aid understanding and appreciation of the wider landscape. *Abstract for Natural And Cultural Landscapes Conference*. Royal Irish Academy, Dublin, 64.

Mackenzie, A.F.D. (1998). *Land, Ecology and Resistance in Kenya, 1880-1952*. Routledge, London.

Madsen, B.L. (1995) *Danish Watercourses: Ten Years with the New Watercourse Act: Collected Examples of Maintenance and Restoration*. Danish Environmental Protection Agency, Denmark.

MAFF, (1993) *Code of Good Agricultural Practice for the Protection of Soils*. Ministry of Agriculture, Fisheries & Food, London.

MAFF, (1995) *Shoreline Management Plans: A Guide for Coastal Defence Authorities*. Ministry of Agriculture, Fisheries & Food, London.

Magome, H. & Murombedzi, J. (2003) Sharing South Africa's national parks: community land and conservation in a democratic South Africa. In Adams, W.M. & Mulligan, M. (eds) *Decolonizing Nature*. Earthscan, London, 108–134.

Maltby, E. & Proctor, M.C.F. (1996). Peatlands: their nature and role in the biosphere. In Lappalainen, E. (ed) *Global Peat Resources*. International Peat Society, Finland, 11–19.

Maltman, A. (1998) *Geological Maps*. 2nd ed. Wiley, Chichester.

Maltman, A. (2003) Wine, beer and whisky: the role of geology. *Geology Today*, **19**(1), 22–29.

Manning, A. (2001) Towards a new synthesis of the biological and earth sciences. In Gordon, J.E. & Leys, K.F. (eds) *Earth Science and the Natural Heritage*. Stationery Office, Edinburgh, 19–25.

Markovics, G. (1994) National parks and geological heritage interpretation: examples from North America and applications to Australia. In O'Halloran, D., Green, C., Harley, M., Stanley, M. & Knill, J. (eds) *Geological and Landscape Conservation*. Geological Society, London, 413–416.

Marsh, W.M. (1997) *Landscape Planning: Environmental Applications*. 3rd ed. Wiley, New York.

Marshak, S. (2001) *Earth: Portrait of a Planet*. W.W. Norton & Co, New York.

Martini, G. (1991) Bilan general de la protection du patrimoine geologique en France (abstract). *Terra Nova*, **3**, 2.

Martini, G. (1994) Actes du Premier Symposium International sur la Protection du Patrimoine Géologique. *Mémoires de la Société Géologique de France*, **165**.

Martini, I.P., Brookfield, M.E. & Sadura, S. (2001) *Principles of Glacial Geomorphology and Geology*. Prentice Hall, New Jersey.

Masalu, D.C.P. (2002) Coastal erosion and its social and environmental aspects in Tanzania: a case study in illegal sand mining. *Coastal Management*, **30**, 347–359.

Mason, V. & Stanley, M. (2001) *RIGS Handbook*. Royal Society for Nature Conservation, London.

Masselink, G. & Hughes, M. (2003) *Introduction to Coastal Processes and Geomorphology*. Arnold, London.

Mather, A.S. & Chapman, K. (1995) *Environmental Resources*. Longman, Harlow.

McCoy, E.D. & Bell, S.S. (1991) Habitat structure: the evolution and diversification of a complex topic. In Bell, S.S., McCoy, E.D. & Mushinsky, H.R. (eds) *Habitat Structure: The Physical Arrangement of Objects in Space*. Chapman & Hall, London.

McGuire, W.J., Kilburn, C.R. & Mason, I.M. (2002) *Natural Hazards and Environmental Change*. Hodder Arnold, London.

McHarg, I. (1995) *Design with Nature*. Natural History Press, New York.

McIntosh, A. (2001) *Soil and Soul: People Versus Corporate Power*. Aurum Press, London.

McKee, E.D. (ed) (1979) A study of global sand seas. *US Geological Survey Professional Paper* No. 1052.

McKeever, P.J. & Gallagher, E. (2001) Landscapes from stone. In Gordon, J.E. & Leys, K.F. (eds) *Earth Science and the Natural Heritage*. Stationery Office, Edinburgh, 262–270.

McKeever, P.J. & Morris, J. (2002) *European Geoparks Network Abstract for Natural And Cultural Landscapes Conference*. Royal Irish Academy, Dublin, 73.

McKenzie, D.I. (1994) Earth science assessments for heritage waterways: the French River and others in Canada. In O'Halloran, D., Green, C., Harley, M., Stanley, M. & Knill, J. (eds) *Geological and Landscape Conservation*. Geological Society, London, 127–131.

McKinney, M.L. & Schoch, R.M. (1998) *Environmental Science: Systems and Solutions*. Jones & Bartlett, Sudbury, Massachusetts.

McKirdy, A. (1990) Resolving the Naze conflict: the crenulate bay model. *Earth Science Conservation*, **27**, 16–17.

McKirdy, A. (1993) Coastal superquarries: a blot on the Scottish landscape. *Earth Science Conservation*, **33**, 18–19.

McKirdy, A.P. (1996) Recycling waste and secondary material. *Earth Heritage*, **5**, 18–19.

McKirdy, A.P., Threadgold, R. & Finlay, J. (2001) Geotourism: an emerging rural development opportunity. In Gordon, J.E. & Leys, K.F. (eds) *Earth Science and the Natural Heritage*. Stationery Office, Edinburgh, 255–261.

McMillan, A.A., Gillanders, R.J. & Fairhurst, J.A. (1999) *The Building Stones of Edinburgh*. 2nd ed. Edinburgh Geological Society, Edinburgh.

McNeely, J. (1988) Protected areas. In Pitt, D.C. (ed) *The Future of the Environment: The Social Dimensions of Conservation and Ecological Alternatives*. Routledge, London, 126–144.

McNeely, J. (1989) Protected areas and human ecology: how national parks can contribute to sustaining societies of the twenty-first century. In Western, D. & Pearl, M. (eds) *Conservation for the Twenty-first Century*. Oxford University Press, Oxford, 150–157.

Melendez, G. & Soria, M. (1994) The legal framework and scientific procedure for the protection of palaeontological sites in Spain: recovery of some special sites affected by human activity in Aragón. In O'Halloran, D., Green, C., Harley, M., Stanley, M. & Knill, J. (eds) *Geological and Landscape Conservation*. Geological Society, London, 329–334.

Mellor, D. (2001) *American Rock*. Countryman Press, Woodstock.

Metsähallitus (2000) *The Principles of Protected are Management in Finland*. Metsähallitus, Vantaa.

Midgley, M. (2001) Does the earth concern us? *Gaia Circular*, Winter 2001/Spring 2002, 4–9.

Miller, A.D. (2003) Walking and talking. *Earth Heritage*, **19**, 17.

Miller, G.T. (1998) *Living in the Environment*. 10th ed. Wadsworth, Belmont, California.

Milne Home, D. (1872a) Scheme for the conservation of remarkable boulders in Scotland, and for the indication of their position on maps. *Proceedings of the Royal Society of Edinburgh*, **7**, 475–488.

Milne Home, D. (1872b) First report by the committee on boulders appointed by the society. *Proceedings of the Royal Society of Edinburgh*, **7**, 703–775.

Milross, J. & Lipkewich, M. (2001) Millenium connections: fostering healthy earth science education and industry links. In Gordon, J.E. & Leys, K.F. (eds) *Earth Science and the Natural Heritage*. Stationery Office, Edinburgh, 250–254.

Milton, K. (2002) *Loving Nature: Towards an Ecology of Emotion*. Routledge, London.

Mitchell, C. (2001) Natural heritage zones: planning the sustainable use of Scotland's natural diversity. In Gordon, J.E. & Leys, K.F. (eds) *Earth Science and the Natural Heritage: Interactions and Integrated Management*. Stationery Office, Edinburgh, 234–238.

MMSD, (2002) *Breaking New Ground: The Report of the Mining, Minerals and Sustainable Development Project*. Earthscan, London.

Moles, N.R.& Moles, R.T. (2002) Influence of geology, glacial processes and land use on soil composition and quaternary landscape evolution in The Burren National Park, Ireland. *Catena*, **47**, 291–321.

Monhemius, J. (2002) New processing technologies for sustainability, Paper presented to the *"Minerals 2002" Conference of the Institute of Quarrying*. Macclesfield, England, October 2002.

Monro, S.K. & Davison, D. (2001) Development of integrated educational resources: the dynamic earth experience. In Gordon, J.E. & Leys, K.F. (eds) *Earth Science and the Natural Heritage*. Stationery Office, Edinburgh, 242–249.

Muir, J. (1901) *Our National Parks*. Houghton Mifflin, Boston.

Murck, B., Skinner, B.J. & Porter, S.J. (1997) *Dangerous Earth: An Introduction to Geological Hazards*. Wiley, Chichester.

Murphy, D.D., Freas, K.E. & Weiss, S.B. (1988) An environment-metapopulation approach to population viability analysis for a threatened invertebrate. *Conservation Biology*, **4**, 41–51.

Murphy, M. (2001) Minerals in the hands of the collectors. *Earth Heritage*, **15**, 14–15.

Murphy, M. (2002) Face lift reclaims geological SSSIs. *Earth Heritage*, **18**, 8–9.

Mutka, K. (1996) Environmental use of peat. In Lappalainen, E. (ed) *Global Peat Resources*. International Peat Society, Finland, 335–338.

Myers, N. (2002) Biodiversity and biodepletion: a paradigm shift. In O'Riordan, T. & Stoll- Kleemann, S. (eds) *Biodiversity, Sustainability and Human Communities: Protecting Beyond the Protected*. Cambridge University Press, Cambridge, 46–60.

Nash, R.F. (1990) *The Rights of Nature: A History of Environmental Ethics*. Primavera Press, New South Wales.

Nathanail, P. (1997) Contaminated land management: a UK perspective. In Marinos, P.G., Koukis, G.C., Tsiambaos, G.C. & Stournaras, G.C. (eds) *Engineering Geology and the Environment*. A.A. Balkema, Rotterdam, 2039–2045.

National Research Council, (2001) *Disposition of High-level Waste and spent Nuclear Fuel: The Continuing Societal and Technical Challenges*. National Academy Press, Washington DC.

Nature Conservancy Council, (1984) *Nature Conservation in Great Britain*. Nature Conservancy Council, Peterborough.

Nature Conservancy Council, (1990) *Earth Science Conservation in Great Britain: A Strategy.* Nature Conservancy Council, Peterborough.

Nichols, R.F. & Leatherman, S.P. (1995) Sea-level rise and coastal management. In McGregor, D.F.M. & Thompson, D.A. (eds) *Geomorphology and Land Management in a Changing Environment.* Wiley, Chichester, 229–244.

Nichols, W.F., Killingbeck, K.T. & August, P.V. (1998) The influence of geomorphological heterogeneity on biodiversity. II. A landscape perspective. *Conservation Biology,* **12,** 371–379.

Nielsen, M.B. (1996) Lowland stream restoration in Denmark. In Brookes, A. & Shields, F.D. (eds) (1996a) *River Channel Restoration.* Wiley, Chichester, 269–289.

Nordic Council of Ministers (2003) *Diversity in Nature.* Nordic Council of Ministers, Copenhagen.

Norman, D.B. (1994) Fossil collecting: international issues, perspectives and a prospectus. In O'Halloran, D., Green, C., Harley, M., Stanley, M. & Knill, J. (eds) *Geological and Landscape Conservation* Geological Society, London, 63–68.

Norton, B. (1988) Commodity, amenity, and morality: the limits of quantification in valuing biodiversity. In Wilson, E.O. (ed) *Biodiversity.* National Academy Press, Washington DC, 200–205.

Nusipov, E., Fishman, I.L. & Kazakowa, Y.I. (2001) *Geosites of Kazakhstan.* Kazakh Scientific Institute of Mineral Resources, Almaty.

Nyborg, T. & Santucci, V. (2000) Track site conservation through track replication and *in situ* management of late miocene vertebrate tracks in Death Valley national park, California. *SBCMA Quarterly,* **47**(2), 83–84.

O'Connor, N.A. (1991) The effect of habitat complexity on the macrovertebrates colonizing wood substrates in a lowland stream. *Oecologia,* **85,** 504–512.

O'Donoghue, M. (1988) *Gemstones.* Chapman & Hall, London.

O'Halloran, D. (1990) Caves and agriculture: an impact study. *Earth Science Conservation,* **28,** 21–23.

O'Halloran, D., Green, C., Harley, M., Stanley, M. & Knill, J. (1994) *Geological and Landscape Conservation.* Geological Society, London.

O'Riordan, T. (ed) (1981) *Environmentalism.* 2nd ed. Pion, London.

Okruszko, H. (1996) Agricultural use of peatlands. In Lappalainen, E. (ed) *Global Peat Resources.* International Peat Society, Finland, 303–309.

Oliver, P. & Allbutt, M. (2000) RIGS groups join forces for a face lift. *Earth Heritage,* **14,** 22–23.

Outhet, D., Brooks, A., Hader, W., Corlis, P., Raine, A., Broderick, T., Avery, E. & Babakaiff, S. (2001a) Gravel bed river riffle restoration in New South Wales. In Rutherford, I., Sheldon, F., Brierley, G. & Kenyon, C. (eds) *Third Australian Stream Management Conference.* Co-operative Research Centre for Catchment Hydrology, Monash, 483–488.

Outhet, D., Friys, K., Massey, C. & Brierley, G. (2001b) Application of the river styles framework to river management programs in New South Wales. In Rutherford, I., Sheldon, F., Brierley, G. & Kenyon, C. (eds) *Third Australian Stream Management Conference.* Co-operative Research Centre for Catchment Hydrology, Monash, 489–492.

Owens, S. & Cowell, R. (1996) *Rocks and Hard Places: Mineral Resources Planning and Sustainability.* Council for the Protection of Rural England, London.

Page, K.N. (1998) England's earth heritage resources: an asset for everyone. In Hooke, J. (ed) *Coastal Defence and Earth Science Conservation*. Geological Society, London, 196–209.

Päivänen, J. (1996) Forestry use of peatlands. In Lappalainen, E. (ed) *Global Peat Resources*. International Peat Society, Finland, 311–314.

Parkes, M.A. (2001) Valencia island tetrapod trackway – a case study. In Bassett, M.G., King, A.H., Larwood, J.G. & Parkinson, N.A. (eds) *A Future for Fossils*. National Museums & Galleries of Wales, Cardiff, 71–73.

Parkes, M.A. (2002) Graves and grave stones: geological influence on culture. *Abstract for Natural and Cultural Landscapes Conference*. Royal Irish Academy, Dublin.

Parkes, M.A. & Morris, J.H. (2001) Earth science conservation in Ireland: the Irish geological heritage programme. *Irish Journal of Earth Sciences*, **19**, 79–90.

Parsons, A.J. (1988) *Hillslope Form*. Routledge, London.

Parsons, D.J. (2002) Editorial: understanding and managing impacts of recreation use in mountain environments. *Arctic & Alpine Research*, **34**, 363–364.

Passmore, J. (1980) *Man's Responsibility for Nature*. 2nd ed. Duckworth, London.

Patzak, M. & Eder, W. (1998) "UNESCO Geopark". A new programme – a new UNESCO label. *Geologica Balcanica*, **28**, 33–35.

Pemberton, M. (2001a) Conserving Geodiversity, the Importance of Valuing our Geological Heritage. Paper presented to the *Geological Society of Australia National Conference*, 2001.

Pemberton, M. (2001b) *Macquarie island – geodiversity and geoconservation values*. Unpublished paper.

Pepper, D. (1984) *The Roots of Modern Environmentalism*. Routledge, London.

Perkins, D. (2002) *Mineralogy*. 2nd ed. Pearson Education, New Jersey.

Perry, D.J. (1962) General report on lands of the Alice springs area 1956-7. *Land Research Series*, No. 6 CSIRO, Australia.

Peterborough Environment City Trust, (1999) *The Peterborough Geology Audit*. Peterborough Environment City Trust, Peterborough.

Pethick, J. (1994) Managing managed retreat. In Parker, A. (ed) *Environmental Sedimentology: Interaction of Water, Sediment and Waste*. University of Reading.

Pfeffer, K. (2003) Integrating spatio-temporal environmental models for planning ski runs. *Netherlands Geographical Studies*, No. 311 University of Utrecht, Utrecht.

Phillips, M. & Mighall, T. (2000) *Society and Exploitation through Nature*. Pearson Education, Harlow.

Pigram, J.J. (2000) Options for rehabilitation of Australia's Snowy River: an economic perspective. *Regulated Rivers: Research & Management*, **16**, 363–373.

Pimental, D. (1993) *World Soil Erosion and Conservation*. Cambridge University Press, Cambridge.

Pizzuto, J. (2002) Effects of dam removal on river form and process. *Bioscience*, **52**, 683–692.

Powell, K. (2002) Open the floodgates. *Nature*, **420**, 356–358.

Prasad, K.N. (1994) Geological and landscape conservation in India. In O'Halloran, D., Green, C., Harley, M., Stanley, M. & Knill, J. (eds) *Geological and Landscape Conservation*. Geological Society, London, 255–258.

Prentice, J.E. (1990) *Geology of Construction Materials*. Chapman & Hall, London.

Press, F. & Siever, R. (2000) *Understanding Earth*. 3rd ed. W.H. Freeman, New York.

Price, M. (2002) Sacred peak. *Geographical*, **74**(5), 9.

Price, R.J. (1989) *Scotland's Golf Courses*. Aberdeen University Press, Aberdeen.

Price, R.J. (2002) *Scotland's Golf Courses*. 2nd ed. Mercat Press, Edinburgh.

Pritchard, J.A. (1999) *Preserving Yellowstone's Natural Conditions: Science and the Perception of Nature*. University of Nebraska Press, Lincoln.

Prosser, C. (2001a) Spectacular coastline saved. *Earth Heritage*, **16**, 4–5.

Prosser, C. (2001b) *Progress Review of English Nature's Geological Conservation Strategy*. Report to Council of English Nature, 4 December 2001.

Prosser, C. (2001c) Geological conservation and the minerals industry – breaking new ground. In Addison, K. & Reynolds, J.R. (eds) *Proceedings of the Fourth UK RIGS Conference*. UKRIGS, Wirksworth, 1–12.

Prosser, C. (2002a) Terms of endearment. *Earth Heritage*, **17**, 12–13.

Prosser, C. (2002b) Your chance to strike grant gold. *Earth Heritage*, **18**, 11.

Prosser, C. (2002c) Terminology: speaking the same language. *Earth Heritage*, **18**, 24–25.

Prosser, C. (2003) Going, going, gone .... but not forgotten: Webster's Clay Pit SSSI. *Earth Heritage*, **19**, 12.

Prosser, C. & Hughes, I. (2001) CROW flies for geological conservation. *Earth Heritage*, **16**, 19.

Prosser, C. & King, A.H. (1999) The conservation of historically important geological and geomorphological sites in England. *The Geological Curator*, **7**, 27–33.

Prosser, C.D. & Larwood, J.G. (1994) Urban site conservation – an area to build on? In O'Halloran, D., Green, C., Harley, M., Stanley, M. & Knill, J. (eds) *Geological and Landscape Conservation*. Geological Society, London, 347–355.

Prosser, I.P., Hughes, A., Rustomji, P., Young, B. & Moran, C. (2001) Predictions of the sediment regime of Australian rivers. In Rutherford, I., Sheldon, F., Brierley, G. & Kenyon, C. (eds) *Third Australian Stream Management Conference*. Co-operative Research Centre for Catchment Hydrology, Monash, 529–536.

Prothero, D.R. & Schwab, F. (1996) *Sedimentary Geology*. W.H. Freeman, New York.

Purbeck District Council (1997) *Purbeck Local Plan, Deposit Draft*. Purbeck District Council, Wareham.

Puri, G., Willson, T. & Woolgar, P. (2001) The sustainable use of soil. In Gordon, J.E. & Leys, K.F. (eds) *Earth Science and the Natural Heritage*. Stationery Office, Edinburgh, 190–196.

Putnis, A. (1992) *Introduction to Mineral Sciences*. Cambridge University Press, Cambridge.

Raudsep, R. (1994) Geological protected areas and features in Estonia. In O'Halloran, D., Green, C., Harley, M., Stanley, M. & Knill, J. (eds) *Geological and Landscape Conservation*. Geological Society, London, 237–241.

Raven, P.J., Fox, P., Everard, M., Holmes, N.T.H. & Dawson, F.H. (1997) River habitat survey: a new system for classifying rivers according to their habitat quality. In Boon, P.J. & Howell, D.L. (eds) *Freshwater Quality: Defining the Indefinable?* Stationery Office, Edinburgh, 215–234.

Read, P.G. (1999) *Gemmology*. 2nd ed. Blackwell, Oxord.

Reading, H.G. (1996) *Sedimentary Environments*. 3rd ed. Blackwell Science, Oxford.

Reaka-Kudla, L., Wilson, D.E. & Wilson E.O. (eds) (1997) *Biodiversity II: Understanding and Protecting our Biological Resources*. Joseph Henry Press, Washington DC.

Reed, M.A. (1998) Making good use of re-use. *Planning*, **1253**, 12–13.

Regnauld, H., Lemasson, L. & Dubreuil, V. (1998) The mobility of coastal landforms under climatic changes: issues for geomorphological and archaeological conservation. In Hooke, J. (ed) *Coastal Defence and Earth Science Conservation*. Geological Society, London, 103–114.

Reid, W.V. (ed) (1993) *Biodiversity Prospecting: Using Genetic Resources for Sustainable Development*. World Resources Institute, Washington DC.

Rhoades, B.L. & Herricks, E.E. (1996) Naturalization of headwater streams in Illinois: challenges and possibilities. In Brookes, A. & Shields, F.D. (eds) *River Channel Restoration*. Wiley, Chichester, 331–367.

Rice, C.M., Trewin, N.H. & Anderson, L.I. (2002) Geological setting of the early Devonian rhynie cherts, Aberdeenshire, Scotland: an early terrestrial hot spring system. *Journal of the Geological of London*, **159**, 203–214.

Richards, K.L., Brasington, J. & Highes, F. (2002) Geomorphic dynamics of floodplains: ecological implications and a potential modelling strategy. *Freshwater Biology*, **47**, 559–579.

Richards, L. (1996) Reclamation – making the best of derelict land. *Earth Heritage*, **6**, 16–18.

Rifkin, J. (2002) *The Hydrogen Economy*. Polity Press, Cambridge.

Ritchie, D. & Gates, A.E. (2001) *Encyclopedia of Earthquakes and Volcanoes*. 2nd ed. Checkmark Books, New York.

Rittmann, A. (1962) *Volcanoes and their Activity*. Wiley, Chichester.

Rivas, V., Rix, K., Francés, E., Cendrero, A. & Collison, A. (1995a) Geomorphology and environmental impact assessment in Spain and Great Britain: a brief review of legislation and practice. In Marchetti, M., Panizza, M., Soldati, M. & Barani, D. (eds) *Geomorphology and Environmental Impact Assessment*. CNR–CS Geodinamica Alpina e Quaternaria, Milan, 83–97.

Rivas, V., Rix, K., Francés, E., Cendrero, A. & Brunsden, D. (1995b) The use of indicators for the assessment of environmental impacts on geomorphological features. In Marchetti, M., Panizza, M., Soldati, M. & Barani, D. (eds) *Geomorphology and Environmental Impact Assessment*. CNR–CS Geodinamica Alpina e Quaternaria, Milan, 157–180.

Robert, A. (2003) *River Processes: An Introduction to Fluvial Dynamics*. Arnold, London.

Roberts, D. (2002) The west coast fossil park: serving science and the community. *Geoclips*, **2**, 1–2.

Roberts, I. (2003) Car wars. *The Guardian*, 18 January 2003, 20.

Robertson, R.M. (1995) *Selected Highland Folk Tales*. House of Lochar, Colonsay.

Robinson, E. (1984) *London – Illustrated Geological Walks*. Scottish Academic Press, Edinburgh.

Robinson, E. (1985) *London – Illustrated Geological Walks*. Book 2. Scottish Academic Press, Edinburgh.

Robinson, E. (1987) A geology of the Albert memorial and vicinity. *Proceedings of the Geologists' Association*, **98**, 19–37.

Robinson, E. (1996) 'The paths of glory . . . .' In Bennett, M.R., Doyle, P., Larwood, C.D. & Prosser, C.D. (eds) *Geology on your Doorstep*. Geological Society, London, 39–58.

Rochefort, L. & Campeau, S. (1997) Rehabilitation work on post-harvested bogs in south eastern Canada. In Parkyn, L., Stoneman, R.E. & Ingram, H.A.P. (eds) *Conserving Peatlands*. CAB International, Oxford, 287–294.

Rodwell, J.S. (1991) *British Plant Communities, Vol .1: Woodlands and Scrub*. Cambridge University Press, Cambridge.

Rogers, T. (2000) Report on the activities of the Association of Welsh RIGS Groups. In Oliver, P. (ed) *Proceedings of the second UK RIGS Conference*. Hereford & Worcester RIGS Group, Worcester, 133–135.

Rolfe, I. (1990) Expert controls for valuable fossils: the trials and tribulations of "Lizzie". *Earth Science Conservation*, **27**, 20–21.

Rose, E.P.F. & Nathanail, C.P. (2000) *Geology and Warfare: Examples of the Influence of Terrain and Geologists on Military Operations*. Geological Society, London.

Rose, J. & Letzer, J.M. (1977) Superimposed drumlins. *Journal of Glaciology*, **18**, 471–480.

Rosengren, N.J. (1994) The newer volcanic province of Victoria, Australia: the use of an inventory of scientific significance in the management of scoria and tuff quarrying. In O'Halloran, D., Green, C., Harley, M., Stanley, M. & Knill, J. (eds) *Geological and Landscape Conservation*. Geological Society, London, 105–110.

Rosgen, D. (1996) *Applied River Morphology*. Wildland Hydrology, Pagosa Springs.

Royal Commission on Environmental Pollution, (1996) *Sustainable Use of Soil*. Royal Commission on Environmental Pollution, 19th Report, HMSO, London.

Rushton, B.S. (ed) (2000) *Biodiversity: The Irish Dimension*. Royal Irish Academy, Dublin.

Rutherfurd, I., Sheldon, F., Brierley, G. & Kenyon, C. (eds) (2001) *Third Australian Stream Management Conference*. Co-operatice Research Centre for Catchment Hydrology, Monash.

Sacks, J. (2002) *The Dignity of Difference*. Continuum Books, London.

Sadler, J.P., Bell, D. & Fowles, A. (2004) The status and habitat of beetles (Coleoptera) of exposed riverine sediments in England and Wales. In press.

Saiu, E.M. & McManus, J. (1998) Impacts of coal mining on coastal stability in Fife. In Hooke, J. (ed) *Coastal Defence and Earth Science Conservation*. Geological Society, London, 58–66.

Salo, J., Kalliola, R., Häkkinen, I., Mäkinen, Y., Wiemelä, P., Puhakka, M. & Coley, P.D. (1986) River dynamics and diversity of Amazon lowland forest. *Nature*, **322**, 254–258.

Santos, J.J. (2000) La Palma: history, landscapes and customs. J.J. Santos, La Palma.

Santucci, V.L. (1999) *Palaeontological Resource Protection Survey Report*. National Park Service, Ranger Activities Division and Geologic Resources Division.

Santucci, V.L. (2000a) Battling vandalism: geologic resource protection training increases park vigilance. *Natural Resources Year in Review 1999*.

Santucci, V.L. (2000b) A survey of the palaeontological resources from the national parks and monuments in Utah. In Sprinkel, D.A., Chidsey, T.C. & Anderson, P.B. (eds) *Geology of Utah's Parks and Monuments*. Utah Geological Association Publication No. 28, Salt Lake City, 535–556.

Santucci, V.L. & Hughes, M. (1998) *Fossil Cycad National Monument: A Case of Palaeontological Resource Mismanagement*. Technical Report NPS/NRGRD/GRDTR-98/1.

Santucci, V.L., Hunt, A.P. & Norby, L. (2001a) Oil and gas management planning and the protection of palaeontological resources. *Park Science*, **21**, 36–38.

Santucci, V.L., Kenworthy, J. & Kerbo, R. (2001b) *An Inventory of Palaeontological Resources Associated with National Park Service Caves*. Technical Report NPS/NRGRD/GRDTR-01/02.

Sax, J.L. (2001) Implementing the public trust in palaeontological resources. In Santucci, V.L. & McClelland, L. (eds) *Proceedings of the 6th Fossil Resource Conference*. Technical Report NPS/NRGRD/GRDTR-01/01, 173–177.

Schlüchter, C. (1994) A model of consensus in aggregate mining and landscape restoration: science-industry-conservation. In O'Halloran, D., Green, C., Harley, M., Stanley, M. & Knill, J. (eds) *Geological and Landscape Conservation*. Geological Society, London, 39–42.

Schmilewski, G.K. (1996) Horticultural use of peat. In Lappalainen, E. (ed) *Global Peat Resources*. International Peat Society, Finland, 327–334.

Schofield, N.J., Collier, K.J., Quinn, J., Sheldon, F. & Thoms, M.C. (2000) River conservation in Australia and New Zealand. In Boon, P.J., Davies, B.R. & Petts, G.E. (eds) *Global Perspectives on River Conservation*. Wiley, Chichester, 311–333.

Schullery, P. (1999) *Searching for Yellowstone: Ecology and Wonder in the Last Wilderness*. Mariner Books, Boston.

Schulze, E.D. & Mooney, H.A. (eds) (1994) *Biodiversity and Ecosystem Function*. Springer-Verlag, Berlin.

Schumm, S.A. (1979) Geomorphic thresholds: the concept and its applications. *Transactions of the Institute of British Geographers*, **NS4**, 485–515.

Schumm, S.A., Dumont, J.F. & Holbrook, J.M. (2000) *Active Tectonics and Alluvial Rivers*. Cambridge University Press, Cambridge.

Scott, R.C. (1991) *Essentials of Physical Geography*. West Publishing Co, St Paul.

Scott, R.C. (1992) *Physical Geography*. West Publishing Co, St Paul.

Scottish Natural Heritage, (2000a) *Minerals and the Natural Heritage in Scotland's Midland Valley*. Scottish Natural Heritage, Edinburgh.

Scottish Natural Heritage, (2000b) *Car Parks in the Countryside: A Practical Guide to Planning, Design and Construction*. Scottish Natural Heritage, Edinburgh.

Scottish Natural Heritage, (2000c) *A Guide to Managing Coastal Erosion in Beach/Dune Systems*. Scottish Natural Heritage, Edinburgh.

Scottish Natural Heritage, (2002) *Natural Heritage Futures* (21 Prospectuses, 6 Themes & CD ROM), Scottish Natural Heritage, Edinburgh.

Scottish Natural Heritage & Scottish Golf Course Wildlife Group, (2000) *Nature Conservation and Golf Course Development: Best Practice Advice*. Scottish Golf Course Wildlife Group, Dalkeith.

Scottish Wildlife Trust, (2002) *Geodiversity: Policy Summary*. Scottish Wildlife Trust, Edinburgh.

Sear, D.J. (1994) River restoration and geomorphology. *Aquatic Conservation: Marine and Freshwater Ecosystems*, **4**, 169–177.

Selby, M.J. (1993) *Hillslope Materials and Processes*. 2nd ed. Oxford University Press, Oxford.

Sellars, R.W. (1997) *Preserving Nature in the National Parks: A History*. Yale University Press, New Haven.

Selley, R.C. (1996) *Ancient Sedimentary Environments*. 4th ed. Chapman & Hall, London.

Selley, R.C. (2002) Geological and climatic controls on english wines. *GA*, **1**(2), 6–7.

Semeniuk, V. (1997) The linkage between biodiversity and geodiversity. In Eberhard, R. (ed) *Pattern & Process: Towards a Regional Approach to National Estate Assessment of Geodiversity*. Australian Heritage Commission, Canberra, 51–58.

Sharples, C. (1993) *A Methodology for the Identification of Significant Landforms and Geological Sites for Geoconservation Purposes*. Forestry Commission, Tasmania.

Sharples, C. (1995) Geoconservation in forest management: principles and procedures. *Tasforest*, **7**, 37–50.

Sharples, C. (1997) *A Reconnaissance of Landforms and Geological Sites of Geoconservation Significance in the Western Derwent Forest District*. Forestry Tasmania.

Sharples, C. (2002a) *Concepts and Principles of Geoconservation*. PDF Document, Tasmanian Parks & Wildlife Service website.

Sharples, C. (2002b) *Some Basic Principles of Geoconservation*. Unpublished paper.

Shell (2001) *Energy Needs, Choices and Possibilities: Scenarios to 2050*. Shell International, London.

Shelley, D. (1993) *Igneous and Metamorphic Rocks under the Microscope*. Chapman & Hall, London.

Siegesmund, S., Vollbrecht, A. & Weiss, T. (2001) *Natural Stone, Weathering Phenomena, Conservation Strategies and Case Studies*. Special Publication 205, Geological Society, London.

Silvestru, E. (1994) Karst and environment: a Romanian approach. In O'Halloran, D., Green, C., Harley, M., Stanley, M. & Knill, J. (eds) *Geological and Landscape Conservation*. Geological Society, London, 221–225.

Simonson, W. & Thomas, R. (1999) *Biodiversity: Making the Links*. English Nature, Peterborough.

Skinner, B.J. & Porter, S.C. (1997) *The Dynamic Earth*. Wiley, New York.

Smith, B. (2002) Eating soil. *Planet Earth*. Autumn 2002, Natural Environment Research Council, Swindon, 21–22.

Smith, K. (2000) *Environmental Hazards: Assessing Risk and Reducing Disaster*. Routledge, London.

Smith, R.B. (2000) Windows into Yellowstone. *Yellowstone Science*, **8**(4), 2–13.

Smith, R.B. & Siegel, L.J. (2000) *Windows into the Earth: The Geologic Story of Yellowstone and Grand Teton National Parks*. Oxford University Press, Oxford.

Smith-Meyer, S. (1994) Protection of Norwegian river systems. *Mémoires de la Société Géologique de France*, **165**, 217–219.

Sorensen, J. (1993) The international proliferation of integrated coastal management efforts. *Ocean and Coastal Management*, **21**, 45–80.

Sorensen, J. (2000) *Building a Global Database of ICM Efforts*. University of Massachusetts Press, Boston.

Soulsby, C. & Boon, P.J. (2001) Freshwater environments: an earth science perspective on the natural heritage of Scotland's rivers. In Gordon, J.E. & Leys, K.F. (eds) *Earth Science and the Natural Heritage*. Stationery Office, Edinburgh, 82–104.

South Norfolk Council (2003) *South Norfolk Local Plan*. South Norfolk Council, Long Stratton.

Spate, O.H.K. & Learmonth, A.T.A. (1967) *India and Pakistan*. Methuen, London.

Speight, M.C.D. (1973) *Outdoor Recreation and its Ecological Effects*. Discussion Papers in Conservation 4, University College London.

Sprinkel, D.A., Chidsey, T.C. & Anderson, P.B. (2000) *Geology of Utah's Parks and Monuments*. Utah Geological Association, Utah.

Stanley, M. (2000) Geodiversity *Earth Heritage*, **14**, 15–18.

Stanley, M. (2001) Editorial. *Geodiversity Update*, **1**, 1.

Stanley, M. (2002) Geodiversity – linking people, landscapes and their culture. *Abstract for Natural And Cultural Landscapes Conference*. Royal Irish Academy, Dublin, 14.

Steers, J.A. (1946) Coastal preservation and planning. *Geographical Journal*, **107**, 57–60.

Stephenson, M. & Penn, I. (2003) Rebuilding a survey. *Geoscientist*, **13**(3), 14–16.

Stevens, C., Gordon, J.E., Green, C.P. & Macklin, M.G. (eds) (1994) *Conserving our Landscape*. English Nature, Peterborough.

Stephens, S. (1998) Case study: slate quarries. *Landscape Design*, **269**, 15.

Stewart, G.A. & Perry, R.A. (1953) *Survey of Townsville-Bowen region (1950)*. Land Research Series 2 (CSIRO 1120, Australia).

Stock, E. (1997) Geo-processes as heritage. In Eberhard, R. (ed) *Pattern & Process: Towards a Regional Approach to National Estate Assessment of Geodiversity*. Australian Heritage Commission, Canberra, 41–50.

Stocking, M. (2000) Soil erosion and land degradation. In O'Riordan, T. (ed) *Environmental Science for Environmental Management*. 2nd ed. Pearson Education, Harlow, 288–321.

Stout, G. (2002) *Newgrange and the Bend of the Boyne*. Cork University Press, Cork.

Strahler, A.H. & Strahler, A.N. (2002) *Introducing Physical Geography*. 3rd ed. Wiley, Chichester.

Stürm, B. (1994) The geotope concept: geological nature conservation by town and country planning. In O'Halloran, D., Green, C., Harley, M., Stanley, M. & Knill, J. (eds) *Geological and Landscape Conservation*. Geological Society, London, 27–31.

Sugden, D.E. (1982) *Arctic and Antarctic: A Modern Geographical Synthesis*. Blackwell, Oxford.

Sugden, D.E. & John, B.S. (1976) *Glaciers and Landscape: A Geomorphological Approach*. Edward Arnold, London.

Summerfield, M.A. (1991) *Global Geomorphology*. Longman, Harlow.

Summerfield, M.A. (2000). *Geomorphology and Global Tectonics*. Wiley, Chichester.

Sutherland, D.G. (1984) Geomorphology and mineral exploration: some examples from exploration of diamondiferous placer deposits. *Zeitschrift fur Geomorphologie, Suppleentary Band*, **51**, 95–108.

Sutherland, M. & Scarsbrick, B. (2001) Land care in Australia. In Stott, D.E., Mohtar, R.H. & Steinhardt, G.C. (eds) *Sustaining the Global Farm*. Selected papers from the 10th International Soil Conservation Organization Meeting, Purdue University and USDA-ARS Soil Erosion Research Laboratory.

Swanwick, C. & Land Use Consultants (1999) *Interim Landscape Character Assessment Guidance*. Countryside Agency and Scottish Natural Heritage.

Swanwick, C. & Land Use Consultants (2002) *Landscape Character Assessment: Guidance for Scotland and England*. Countryside Agency and Scottish Natural Heritage.

Swart, R. (1994) Conservation and management of geological monuments in south Australia. In O'Halloran, D., Green, C., Harley, M., Stanley, M. & Knill, J. (eds) *Geological and Landscape Conservation*. Geological Society, London, 319–322.

Tarbuck, E.J. & Lutgens, F.K. (1999) *Earth*. 6th ed. Prentice Hall, New Jersey.

Taylor, A.G. (1995) Environmental problems associated with soil in Britain: a review. *Scottish Natural Heritage Review*, No. 55.

Taylor, A.G. (1997) *Soils and Land Use Planning*. Information & Advisory Note 85. Scottish Natural Heritage, Edinburgh.

Taylor, A.G., Gordon, J.E. & Usher, M.B. (eds) (1996) *Soils, Sustainability and the Natural Heritage*. Scottish Natural Heritage, Edinburgh.

Taylor, G. & Eggleton, R.A. (2001) *Regolith Geology and Geomorphology*. Wiley, Chichester.

Tayside Biodiversity Partnership (2002) *Tayside Biodiversity Action Plan*. Tayside Biodiversity Partnership, Perth.

Teodorascu, D. (1994) Landscape conservation and the national parks of Romania. In O'Halloran, D., Green, C., Harley, M., Stanley, M. & Knill, J. (eds) *Geological and Landscape Conservation*. Geological Society, London, 157–160.

Therivel, R. & Thompson, S. (1996) *Strategic Environmental Assessment and Nature Conservation*. English Nature, Peterborough.

Thiéry, J.M., D'Herbès, J.M. & Valentin, C. (1995) A model simulating the genesis of banded vegetation in Niger. *Journal of Ecology*, **83**, 497–507.

Thomas, K. (1983) *Man and the Natural World: Changing Attitudes in England 1500–1800*. Oxford University Press, Oxford.

Thomas, L. (2002) *Coal Geology*. Wiley, Chichester.

Thompson, A., Hine, P.D., Greig, J.R. & Poole, J.S. (1998) *Environmental Geology in Land Use Planning: Emerging Issues*. Department of Environment, Transport & the Regions, London.

Thompson, D.B.A., Gordon, J.E. & Horsfield, D. (2001) Montane landscapes in Scotland: are these natural artefacts or complex relics? In Gordon, J.E. & Leys, K.F. (eds) *Earth Science and the Natural Heritage*. Stationery Office, Edinburgh, 105–119.

Thompson, T.R.E. & Bullock, P. (1994) The justification for a soil conservation policy for the UK. In O'Halloran, D., Green, C., Harley, M., Stanley, M. & Knill, J. (eds) *Geological and Landscape Conservation*. Geological Society, London, 53–55.

Thomson, J.R., Taylor, M.P., Fryirs, K.A. & Brierley, G.J. (2001) A geomorphological framework for river characterization and habitat assessment. *Aquatic Conservation: Marine and Freshwater Ecosystems*, **11**, 373–389.

Thorne, C.R. (1998) *Stream Reconnaissance Handbook: Geomorphological Investigation and Analysis of River Channels*. Wiley, Chichester.

Thorvardardottir, G. & Thoroddsson, T.F. (1994) Protected volcanoes in Iceland: conservation and threats. In O'Halloran, D., Green, C., Harley, M., Stanley, M. & Knill, J. (eds) *Geological and Landscape Conservation*. Geological Society, London, 227–230.

Todorov, T.A. (1994) Earth science conservation in Bulgaria. In O'Halloran, D., Green, C., Harley, M., Stanley, M. & Knill, J. (eds) *Geological and Landscape Conservation*. Geological Society, London, 247–248.

Toghill, P. (1972) Geological conservation in Shropshire. *Journal of the Geological Society*, **128**, 513–515.

Toth, L.A. (1996) Restoring the hydrogeomorphology of the channelized Kissimmee river. In Brookes, A. & Shields, F.D. (eds) (1996a) *River Channel Restoration*. Wiley, Chichester, 369–383.

Trewin, N.H. (2001) Scotland's foundations: our geological inheritance. In Gordon, J.E. & Leys, K.F. (eds) *Earth Science and the Natural Heritage*. Stationery Office, Edinburgh, 59–67.

Trimmel, H. (1994) Sixty-five years of legislative cave conservation in Austria: experience and results. In O'Halloran, D., Green, C., Harley, M., Stanley, M. & Knill, J. (eds) *Geological and Landscape Conservation* Geological Society, London, 213–214.

Trudgill, S. (2001) *The Terrestrial Biosphere: Environmental Change, Ecosystem Science, Attitudes and Value.* Pearson Education, Harlow.

Tucker, M.E. (1996) *Sedimentary Rocks in the Field.* 2nd ed. Wiley, Chichester.

Tucker, M.E. (2001) *Sedimentary Petrology.* 3rd ed. Blackwell Science, Oxford.

Turner, B.L., Clark, W.C., Kates, R.W., Richards, J.F., Mathews, J.T. & Meyer, W.B. (eds) (1990) *The Earth as Transformed by Human Action.* Cambridge University Press, Cambridge.

Turner, T. (1998) *Landscape Planning and Environmental Impact Design.* 2nd ed. UCL Press, London.

UNEP-WCMC (2002) *Mountain Watch.* United Nations Environment Programme/World Conservation Monitoring Centre, Cambridge.

UNESCO, (2002) Progress report on the analysis of the world heritage list and tentative lists. *26th Session of the UNESCO Convention Concerning the Protection of the World Cultural and Natural Heritage.* UNESCO, Paris.

US Department of the Interior, (1971) *Impacts of Strip Mining.* US Department of the Interior, Washington, DC.

Usher, M.B. (2001) Earth science and the natural heritage: a synthesis. In Gordon, J.E. & Leys, K.F. (eds) *Earth Science and the Natural Heritage.* Stationery Office, Edinburgh, 315–324.

Valiūnas, J. (1994) Environmental geology maps for national parks and geomorphological reserves in Lithuania. In O'Halloran, D., Green, C., Harley, M., Stanley, M. & Knill, J. (eds) *Geological and Landscape Conservation.* Geological Society, London, 273–277.

Varnes, D.J. (1978) Slope movement and types and processes. In Schuster, R.L. & Krizek (eds) *Landslides: Analysis and Control.* Transportation Research Board, Special Report 176, National Academy of Sciences, Washington DC, 11–33.

Vasileva, D. (1997) Devil's town: truth and legend. *Earth Heritage,* **8**, 17.

Vernon, R.H. (2000) *Beneath our Feet: The Rocks of Planet Earth.* Cambridge University Press, Cambridge.

Viney, C. (2001) *Macquarie Island.* Department of Primary Industries, Water & Environment, Tasmania.

de Waal, L.C., Large, A.R.G. & Wade, P.M. (eds) (1998) *Rehabilitation of Rivers: Principles and Implementation.* Wiley, Chichester.

Waley, P. (2000) Following the flow of Japan's river culture. *Japan Forum,* **12**, 199–217.

Walker, S. (2000) *Cairngorms Rivers Project: Phase 1 Report.* University of Aberdeen, Aberdeen.

Wang, S. Lee, K.C. & King, A. (1999) Eastern promise. *Earth Heritage,* **12**, 12–13.

Wang, S., Sheu, L.Y. & Tang, H.Y. (1994) Conservation of geomorphological landscapes in Taiwan. In O'Halloran, D., Green, C., Harley, M., Stanley, M. & Knill, J. (eds) *Geological and Landscape Conservation.* Geological Society, London, 113–115.

Ward, S.D. & Evans, D.F. (1976) Conservation assessment of British limestone pavements based on floristic criteria. *Biological Conservation,* **9**, 217–233.

Ward, S.N. & Day, S. (2001) Cumbre Vieja volcano – potential collapse and tsunami at La Palma, Canary Islands. *Geophysical Research Letters,* **28**, 3397–3400.

Warnock, S. & Brown, N. (1998) A vision for the countryside. *Landscape Design*, **269**, 22–26.

Warren, A. (2001) Valley-side slopes. In Warren, A. & French, J.R. (eds) *Habitat Conservation: Managing the Physical Environment*. Wiley, Chichester, 39–66.

Warren, A. & French, J.R. (2001) Relations between nature conservation and the physical environment. In Warren, A. & French, J.R. (eds) *Habitat Conservation: Managing the Physical Environment*. Wiley, Chichester, 1–5.

Warren, C. (2002) Of superquarries and mountain railways: recurring themes in Scottish environmental conflict. *Scottish Geographical Journal*, **118**, 101–127.

Washburn, A.L. (1979) *Geocryology: A Survey of Periglacial Processes and Environments*. Edward Arnold, London.

Watson, A. (1984) A survey of vehicular tracks in north-east Scotland for land use planning. *Journal of Environmental Management*, **18**, 345–353.

Webber, M. (2001) The sustainability of a threatened fossil resource:lower Jurassic Caloceras Beds of Doniford Bay, Somerset. In Bassett, M.G., King, A.H. Larwood, J.G. & Parkinson, N.A. (eds) *A Future for Fossils*. National Museums & Galleries of Wales, Cardiff, 108–112.

Webb, S. (2001) Limestone pavements – on stony ground? *Earth Heritage*, **16**, 8–9.

Webb, S. & Glading, C. (1998) The ecology and conservation of limestone pavements in Britain. *British Wildlife*, **10**, 103–113.

Werritty, A. & Brazier, V. (1991) *The Geomorphology, Conservation and Management of the River Feshie SSSI*. Nature Conservancy Council Report.

Werritty, A. & Brazier, V. (1994) Geomorphic sensitivity and the conservation of fluvial geomorphology SSSIs. In Stevens, C., Gordon, J.E., Green, C.P. & Macklin, M.G. (eds) *Conserving our Landscape*. English Nature, Peterborough, 100–109.

Werritty, A., Duck, R.W. & Kirkbride, M.P. (1998) *Development of a Conceptual and Methodological Framework for Monitoring Site Condition in Geomorphological Systems*. Research, Survey and Monitoring Report 105, Scottish Natural Heritage, Edinburgh.

Werritty, A. & Leys, K.F. (2001) The sensitivity of scottish rivers and upland valley floors to recent environmental change. *Catena*, **42**, 251–274.

Westing, A. & Pfeiffer, E.W. (1972) The cratering of Indo China. *Scientific American*, **226**(5), 21–29.

Westmacott, S. (2002) Where should the focus be in tropical integrated coastal management? *Coastal Management*, **30**, 67–84.

Weston, J. (ed) (1997) *Planning and Environmental Impact Assessment in Practice*. Longman, Harlow.

White, B. (1997) *Principles and Practice of Soil Science*. 3rd ed. Blackwell, Oxford.

White, L. (1967). The historical roots of our ecologic crisis. *Science*, **155**, 1203–1207.

Whitney, G.G. (1994) *From Coastal Wilderness to Fruited Plain: A History of Environmental Change in Temperate North America from 1500 to the Present*. Cambridge University Press, Cambridge.

Whittow, J. (1992) *Geology and Scenery in Britain*. Chapman & Hall, London.

Wiedenbien, F.W. (1994). Origin and use of the term 'geotope' in German-speaking countries. In O'Halloran, D., Green, C., Harley, M., Stanley, M. & Knill, J. (eds) *Geological and Landscape Conservation*. Geological Society, London, 117–120.

Wilkinson, S.N., Keller, R.J. & Rutherford, I.D. (2001) Rehabilitating pool and riffle sequences: the role of alternate bar dynamics and flow constriction. In Rutherford, I.,

Sheldon, F., Brierley, G. & Kenyon, C. (eds) *Third Australian Stream Management Conference*. Co-operative Research Centre for Catchment Hydrology, Monash, 681–688.

Williams, D. (2001) Things you can only do with replica fossils. In Bassett, M.G., King, A.H., Larwood, J.G. & Parkinson, N.A. (eds) *A Future for Fossils*. National Museums & Galleries of Wales, Cardiff, 105–108.

Wilson, C. (ed) (1994) *Earth Heritage Conservation*. Geological Society London & Open University, Milton Keynes.

Wilson, E.O. (ed) (1988) *Biodiversity*. National Academy Press, Washington DC.

Wilson, E.O. (1997) Introduction. In Reaka-Kudla, L., Wilson, D.E. & Wilson E.O. (eds) *Biodiversity II: Understanding and Protecting our Biological Resources*. Joseph Henry Press, Washington DC, 1–3.

Winchester, S. (2002) *The Map that Changed the World*. Penguin, London.

Windley, B.F. (1995) *The Evolving Continents*. 3rd ed. Wiley, Chichester.

Wojciechowski, K. (1994) Landscape parks and other protected areas: the polish experience of landscape conservation. In O'Halloran, D., Green, C., Harley, M., Stanley, M. & Knill, J. (eds) *Geological and Landscape Conservation*. Geological Society, London, 181–184.

Wood, C. (1995) *Environmental Impact Assessment: A Comparative Review*. Longman, Harlow.

Woodcock, N. (1994) *Geology and the Environment in Britain and Ireland*. UCL Press, London.

Wordsworth, W. (1952) *A Guide through the District of the Lakes*. Indiana University Press, Bloomington.

Workman, D.R. (1994) Geological and landscape conservation in Hong Kong. In O'Halloran, D., Green, C., Harley, M., Stanley, M. & Knill, J. (eds) *Geological and Landscape Conservation*. Geological Society, London, 291–296.

World Conservation Monitoring Centre, (1992) *Global Biodiversity: Status of the Earth's Living Resources*. Chapman & Hall, London.

Wright, J.R. & Price, G. (1994) Cave conservation plans: an integrated approach to cave conservation. In Stevens, C., Gordon, J.E., Green, C.P. & Macklin, M.G. (eds) *Conserving our Landscape*. English Nature, Peterborough, 195–197.

Xu, X., Zhou, Z., Wang, X., Kuang, X., Zhang, F. & Du, X. (2003) Four-winged dinosaurs from China. *Nature*, **421**, 335–340.

Xun, Z. & Milly, W. (2002) National geoparks initiated in China: putting geoscience in the service of society. *Episodes*, **25**, 33–37.

Zinko, Y. (1994) Development and management of geological and geomorphological conservation features within the urban areas of the western Ukraine. In O'Halloran, D., Green, C., Harley, M., Stanley, M. & Knill, J. (eds) *Geological and Landscape Conservation*. Geological Society, London, 231–235.

Index

Geodiversity: valuing and conserving abiotic nature   Murray Gray
© 2004 John Wiley & Sons, Ltd   ISBNs: 0-470-84895-2 (HB); 0-470-84896-0 (PB)